THE SAGA OF LEON TROTSKY

His Clandestine Operations and His Assassination

Harry Thayer Mahoney

Marjorie Locke Mahoney

THE SAGA OF LEON TROTSKY

His Clandestine Operations and His Assassination

Harry Thayer Mahoney
Marjorie Locke Mahoney

Austin & Winfield, Publishers
San Francisco - London - Bethesda
1998

Library of Congress Cataloging-in-Publication Data

Mahoney, Harry Thayer, 1922-
The saga of Leon Trotsky : his clandestine operations and his assassination / Harry Thayer Mahoney, Marjorie Locke Mahoney.
p. cm.
Includes bibliographical references and index.
ISBN 1-57292-124-2 (alk. paper)
1. Trotsky, Leon, 1879-1940. 2. Trotsky, Leon, 1879-1940—Assassination. 3. Revolutionaries—Soviet Union—Biography. 4. Soviet Union—Politics and government—1917-1936. 5. Military intelligence—Soviet Union. I. Mahoney, Marjorie Locke.
II. Title.
DK254.T6M34 1998
947.084'092—dc21 --dc21
[b]
[947.084'092] 98-8299
CIP

Editorial Inquiries:
Austin & Winfield, Publishers
7831 Woodmont Avenue, #345
Bethesda, MD 20814
Website: www.interscholars.com
To order: (800) 55-PUBLISH

DEDICATION

This book is gratefully dedicated to our children,
their spouses and their children:

Elizabeth, Rafael, Rosemary, Francisco and Alexander Zuniga.
Paul, Janelle, Daniel, Rachel and Kevin Mahoney
Matthew, Sara, Katherine and Sean Mahoney
Sarah, Craig, Rebecca, Colleen, Maura and Leah Antas

Table of Contents

APPRECIATION

This book could never have been written without the help of so many friends, fellow historians and Mexican Government officials in Mexico City and its suburbs, especially those in Coyoacan, Desierto de los Leones and Huixquiliucan — also, the professional intelligence sources from Langely and Falls Church, Virginia. Most of those people who helped us wish to remain anonymous and to name only those who do not care about their reputations would not be fair. It must suffice to note that we express our most sincere thanks to all of them. They all know who they are and how true it is that this book could never have been written without them. It was not only their substantive assistance that we appreciate but their good-natured critisms were invaluable (at least we took them to be good-natured).

The creation of a book can never belong solely to the authors. There are those unsung (and too often unknown) and publicly unappreciated people who put the raw manuscript into a comprehensible publishible form. Therefore, we accord some of this well deservewd public acclaim to our senior editor and publisher (the real "jefe maximo") Dr. Robert West. There is a word in Spanish "simpatico" which roughly translates into caring, understanding, friendly and to helpfully nurture. Bob West is "muy simpatico".

Similarly, Ms. Susan Peabody, a paragon of perfectionism, patience and knowledge about how to change what authors submit into a finished product ready for publication. She has the knack of doing this without bruising the authors' pride. Susan has our most sincere appreciation and she is also "muy simpatico".

With all of these kudos the authors reserve none for themselves but assume the complete responsibility for any errors that may have crept into the book. We have checked and double checked but if any errors remain we apologize. Any differences of interpretation of events, or contrary opinions about what Trotsky did, the motives ascribed to him or his intentions rest solely with the authors.

Our fondest wish is that each reader enjoy our book.

Harry Thayer Mahoney
Marjorie Locke Mahoney
Bethany, Delaware and Colonia Roma, Mexico City

INTRODUCTION

> With the passage of time historical figures either shrink or grow in stature. In the case of Leon Trotsky, time, after a brief eclipse, has increased his image so that he appears today, for good or evil, as one of the giants of the first half of the twentieth century.
>
> Irving H. Smith in *Trotsky*

In researching the intricacies and causative factors of Trotsky's assassination, some explanation of our choice of material is pertinent.[1] Our choices were essentially the most notable events, with an admitted penchant for the seldom reported and of little renown but, in retrospect, interesting vignettes of history as well as those seldom heralded crises along the way of Trotsky's life that led up to, and almost preordained, his assassination.[2]

It has been said that a crises is a situation in which a previously tolerable set of circumstances is suddenly, by the addition of an unforeseen factor, rendered wholly intolerable. Trotsky had long been perceived by Stalin to be a personal threat to his continued existence. When Stalin learned that Trotsky's scathing biography of him was nearing completion, this might have become the catalytic factor which Stalin found wholly intolerable.

Another axiom claims that all crises are essentially the same; they all have their beginnings long before the actual onset, they are made either by man or acts of God. If made by man they cannot help but reflect the preconceptions, prejudices, propensities and strivings of the man. Stalin was obsessed with a personal hatred for Trotsky almost from their first meeting before the Bolshevik revolution.

Finally, one sage observer noted there is a disquieting similarity to all crises, and that is their utter and complete predictability in retrospect. This is also true in the Trotsky case.

Along the way to Trotsky's denouement in Coyoacan there were many fateful and historically noteworthy encounters. Some were orchestrated by Trotsky but most were directed by external forces over which he had no control. The progression of these events contained many subplots and value changes. For example, in Russia, Trotsky was perceived as a savior to some and the devil incarnate to others. To the rest of the world he was either a benign politician or bloody megalomaniac.

Russia has historically been divided between its Asiatic and its European impulses; between its emotional Slavophiles and the countering, more rational, Westernized peoples. Consistently, after every period of Westernization, the Slavophiles have again dominated and turned Russian inward, purging itself of the foreign influences in an almost ritualistic blood letting.

Reduced to its common denominator, the struggle between Trotsky and Stalin was another version of this old conflict and the outcome was true to the ancient pattern.[3] Trotsky, the intellectual steeped in Western European ideology, pitted against Stalin, the provincially minded Asiatic for whom the West and its ways were a mystery to be shunned and detested. The slogans were new, however, with Trotsky's "world revolution" versus Stalin's "socialism in one country."

There are many occurrences that shook the world and personages that dominated the international stage that seemed relevant for inclusion in this story. Many of these great events helped shape Trotsky's life and they therefore have to be presented with a sense of perspective. However, the major factors that directly pertain to Trotsky's assassination, and the paladin who committed the execution, must remain as the controlling elements of this story.

Stalin murdered Trotsky. The actual killing was done by a surrogate, but it was Stalin who conceived and nurtured the plot. And although it was a Spanish communist who wielded the Alpinstock that pierced Trotsky's skull it was Stalin's hand that guided the blow. Stalin hated Trotsky but, more importantly, Stalin feared Trotsky and that is what fed Stalin's paranoia.

Trotsky was imbued with the spirit of a true revolutionary — the fanatical zeal of the dedicated terrorist of the new order with the reasoning intellect and charisma to lead and persuade the masses. Trotsky was described at the time as steel-hard, dedicated and determined. He had no moral conscience, very little sympathy or understanding of people, and he was incapable of playing a minor role; he belonged, as though by right of birth, to a revolutionary elite.[4] Stalin had none of these qualities and, probably in his rational moments, realized his own inadequacies.

There was no question that Trotsky was the acknowledged revolutionary and the designated heir apparent to Lenin's mantle of leadership of the Bolshevik government and the Russian revolution. Stalin feared this legacy and was apprehensive of what support the Russian masses, and communists throughout the world, might pledge to Trotsky's claim for Lenin's bequest.

Stalin irrationally feared that his position as dictator of the Soviet Union, and as the leader of the world communist movement, was threatened by a nebulous Trotskyite movement.[5] Mostly though, in his paranoia, Stalin felt personally threatened by the mere fact of Trotsky's continued existence.

Trotsky's opposition to Stalin's power also frightened Stalin. He firmly believed that behind Trotsky's facade of the courageous, determined, charismatic Jew, there lurked a nucleus of a large cohesive fighting force of dedicated Trotskyites.

According to Stalin's perception, Trotsky might also have enough of a popular following inside Russia itself to effectively lead a revolt against his tenuous base of power. Stalin was obsessed with these fears. He not only feared the political threat posed by this scrappy, nettlesome exile but he deeply resented that what he considered to be an inferior human being, a Jew, could pose such a threat to a superior gentile like him.

The story of Trotsky in Mexico is relatively simple and straightforward. There is, however, a labyrinth of intricate plots and counterplots leading up to these events. The getting there and the discovery of how it all finally came about is where the real fascination lies. This is pure history, and without resorting to that old saw about the truth being stranger than fiction, the events that ultimately untangle and culminate in Trotsky's murder do make for compelling historical drama. (Frankly, I have been involved in some of the peripheral aspects of the aftermath of the Trotsky assassination. Thereafter I will always have a keen interest in what led up to this bizarre murder.)

The planning and execution of a clandestine operation such as an assasination is an arduous task and requires an inordinate amount of time, manpower, money and operational support assets. There are literally thousands of intricate pieces that have to be conceived, nurtured, pulled together and fitted into place, usually under the guise (or cover) of pretending to be doing something completely different. Agents have to be spotted; and this means they must already have access to the target, or a reasonable chance of achieving access. They must also have the capabilities, once they achieved access to the target, of accomplishing their objective. Very few agents who are initially spotted are ultimately recruited because they lack these essential qualities.

To carry out an assassination a newly-spotted agent has to be recruited, trained, motivated and supported over a long period of time by extensive backup mechanisms. Funding must be set up through covert channels and supply and communications networks have to be established, tested and maintained in constant readiness.

Furthermore, cover stories ("legends" as the Brits would say) must be backstopped with genuine or spurious documents. These documents, in turn, should be backstopped in the official archives. For example, an agent's spurious cover birth certificate must be backstopped in the birth registry usually with a forged doctor's certificate attesting to the birth. Better cover documentation would be a genuine birth certificate of someone about the same age and description as the agent. Then the original doctor's genuine birth certificate in the registry provides legitimate backstopping for the agent's cover story.

Alternate emergency plans must also be prepared for any contingency that can be anticipated, and even for those contingencies that cannot be expected. These alternate plans must be field tested and honed and held in readiness for instant activation should anything go wrong with the secure and efficient functioning of the basic plan. And fall back positions must be prepared for the redirection of the objective should a glitch occur in the operation of the principal plan. Propaganda camouflage must also be ready for publication to support plausible denial releases by the sponsoring government if the agent is arrested or turned.

In addition, funding channels must be instituted, tested by deposits and withdrawals, backstopped and made secure against compromise. Communications networks must be technically perfect in order not to fail at the critical moment and alternate channels of communication must be prepared, tested and held in readiness in the event of a breakdown of the primary message channel.

Couriers, if necessary, must go over their routes until no obstacle to their swift completion exists; the movement of operational supplies to forward bases near the area of operations may have to be started years ahead of the target date; escape and evasion routes for compromised agents must be set up with safe houses manned by trustworthy sleeper housekeeping personnel and must be held in readiness; resettlement, or keeping blown agents on ice until the heat blows over must also be considered; and a myriad of other equally detailed requirements have to go into in the planning of an operation as intricate as an assassination.

* * *

In order to dispel any contrary notion that some readers may have, Stalin did not suddenly decide that Trotsky had to be liquidated. He did not suddenly turn his thumbs down in the old Roman gesture as a signal to his GPU gladiator agents to proceed with the execution.

> The frenzy with which he [Stalin] pursued the feud [with Trotsky], making it the paramount preoccupation of international communism as well as of the Soviet Union and subordinating to it all political, tactical, intellectual and other interests, beggars description: there is in the whole of history hardly another case in which such immense resources of power and propaganda were employed against a single individual."[6]

Stalin had to have trusted subordinates and specialized staffs planning the Trotsky execution for years before it actually took place. As Trotsky moved from country to country in exile, time and again, he unwittingly frustrated the GPU plans prepared for that area. Imagine, if you will, that as soon as the GPU planners had a viable assassination plan ready to go in country X, with indigenous agents trained and in place, perhaps the police bribed to look the other way and the hit man on his way to infiltrate Trotsky's entourage, Trotsky would be expelled from country X, or decide to move of his own volition. There had to follow a scurried re-planning by the GPU to re-create the assassination team with new indigenous agents and a compatible ambient of bribed government officials for the murder operation to begin again in the new country. The GPU would have to start over with spotting, recruiting, funding, training, greasing of the palms of the local police and so forth.

There were, to be sure, certain moveable assets, mostly from the communist party cadres that could be moved from one country to another if they would fit into the operational environment of the new country. The support assets, however, such as safe house keepers, surveillance teams, propaganda journalists in the local media, etc. would, in all probability, have to be indigenous to the new country.

In 1920 Lenin stated an axiom, the precept that Stalin would rely on to marshal a cadre of GPU agents to assassinate Trotsky:

> The scientific concept of dictatorship means nothing more, nor less than unlimited power resting directly on force, not limited by anything, nor restrained by any bans or any laws or any absolute rules.[7]

The catalyst for Stalin's decision to murder Trotsky was the articles in the mid-thirties that called for the overthrow of Stalin. When mass repression began in the Soviet Union, Trotsky came to the conclusion that it was necessary to destroy Stalin because he was a tyrant and destroying the Soviet Union. It was then that Stalin ordered the NKVD to arrange Trotsky's assassination.[8]

The question remains why should Stalin risk the rupture of diplomatic relations with Mexico, one of the few countries in Latin America still friendly to the Soviets? Possibly more important, why should Stalin risk the more dire official and diplomatic condemnation, and possible retaliation, from other countries around the world? With World War Two beginning, the position of the Soviet Union, from any point of view, was precarious and Stalin needed all the allies he could get!

Stalin, and it must be personalized as Stalin the individual and not the generic term to denote the figurehead of the Soviet state at that time, was oblivious to all such risks of censure and diplomatic slaps-on-the-wrist. Stalin, personally and viscerally, wanted Trotsky dead and was prepared to pay any price for himself and the Soviet Union to accomplish that purpose.

Claire Sterling, the noted authority on such matters, states there has been a change in Soviet policy since Stalin ordered Trotsky's murder. In their quest for respectability, Soviet leaders shifted to a less openly incriminating methods. John Barron, the author of several authoritative works on Soviet assassination techniques reports,

> The Soviet Union decided in late 1962 or early 1963 to entrust future assassinations not to Soviet personnel, but to hired foreign criminals and illegal agents of other nationalities, who could not be easily linked to the Soviet Union.
>
> The one known exception to that rule was the murder of the late president of Afghanistan, Hafizullah Amin, on 27 December 1979 in his Palace in Kabul. The assassination team also mowed down everybody else in sight. The shocking carnage was a vicious sign of how badly the KGB team had botched the job.[9]

Trying to find the answers to these questions has required an intensive study of Russian history. Trotsky's early life did not provide answers to these questions. As Trotsky became more and more the pro-Bolshevik revolutionary leader, the activist he would become slowly began to emerge. Trotsky's preeminent leadership of the Bolshevik revolution, while Stalin was relegated to a comparatively insignificant role (despite the later facetious claims of re-written Soviet history that Stalin won the Bolshevik revolution almost single handedly), reasons came to the fore that partially explained some of Stalin's hatred of Trotsky.

As this story unfolds, the reader will note how Stalin's penchant for stealth, cheating, lying and deceit were dramatically revealed by how quickly he, and the entire Bolshevik apparat, became adept at clandestine operations, conspiratorial tradecraft and black propaganda. The very nature of the Bolshevik inception in an underground and strongly oppositionist environment required that they be proficient in self-protective tradecraft. The sophistication practiced in their day-to-day operations revealed a thorough knowledge of advanced operational security.

Stalin and Trotsky were both deeply imbued with the idea that clandestinity was normal and the only rational way of revolutionary life; certainly it was the only logical reality that insured continued political existence in the environment of the Czar's very efficient counter-revolutionary anti-terrorist police. However, unlike all other Russian conspiratorial organizations, neither Trotsky's immediate entourage nor his followers in the Trotskyite Fourth International, ever created a counterintelligence department to watch over the security of the movement and to prevent hostile penetrations.[10]

There is a common thread that predominates in the backgrounds of almost all the participants in the Trotsky assassination. They all had been involved, directly or indirectly, with the Spanish Civil War. Certainly Ramon Mercader, the assassin, and his mother and co-conspirator, Caridad Mercader, were both natives of Spain (Caridad had been born in Cuba but was considered to be a Spaniard) and they had both been soldiers in the Republican army during the Civil War. Another example, David Alfaro Siqueiros, the leader of the 24 May 1940 raid on Trotsky's house in Coyoacan near Mexico City, had been a brevet colonel in the International Brigade in Spain. Leonid Eitingon, the on-site GPU field commander in Mexico of the assassination operation, had been a undercover GPU officer in Spain as deputy to the legendary Soviet intelligence officer, Alexander Orlov.

Many of the other conspirators of the assassination team had been recruited and trained for their assignments to assist in the murder of Trotsky while they were fighting fascism in Spain with the communist Republicans. The roster of this murderous clique of GPU assassins reads almost like a reunion of the Soviet element of the Spanish Republican army.

The saga of the many Americans who went to Spain with the Abraham Lincoln Battalion of the communist International Brigade provokes a feeling of empathy and compassion. Some were imbued with the righteousness of their cause for freedom for the Spanish people and genuinely concerned that fascism not prevail. Unfortunately they were duped by the communist cadres. Some of these naive idealistic heroes of the Abraham Lincoln Battalion were murdered by the same communist leaders whom they had regarded as their brothers-in-arms. Some of these murderers, such as George Mink and Carlos Contreras, became part of the Trotsky assassination team.

While the Trotsky assassin and his henchmen were being readied in Spain, the GPU agents in the United States were preparing a support structure for the operation, wherever it might eventually take place. The GPU were searching for a plausible entre' for the assassin to use in order to get close to Trotsky. They were convinced that an American agent might make Trotsky lower his defenses.

The four causative phases of preparation for the assassination of Trotsky culminated in Mexico. The first phase began in Russia when Trotsky and Stalin became enemies, in-fighting over which faction would control the post-Bolshevik government. The second phase is esoteric and problematical but poses the possibility that the Allies themselves may have created a permissive ambient of condoning Bolshevik "dirty tricks." The Allies may have caused Stalin to subconsciously assume, in his peasant ignorance, that even after the urgent necessity of revolutionary pressures had ceased to require illegal covert action and scheming, he could continue to use these same illegal tactics in the world arena to dispose of Trotsky. Stalin could be complacent that other nations would not condemn his lethal actions against Trotsky since they had not condemned the Allies use of illegal operations in 1917 and 18 in Russia.

The third causative factor was the training of the hard core GPU cadre in the Spanish Civil War. Also, much of the administrative support for the assassination team came out of the Spanish Republican army. One example of this would be the manner in which the GPU confiscated the legitimate passports from the members of the International Brigade as they arrived in Spain. The GPU technicians altered these legitimate passports to fit the description of future GPU deep cover agents so they could send them into the field on an espionage assignment with a bona fide new identity. Ramon Mercader, the assassin, used one of these altered passports obtained in Spain to enter Mexico to kill Trotsky.

The last causative factor was the extensive use of American agent assets of the GPU for the support of the operation in Mexico. Professional intelligence officers will agree that Stalin did not personally order the exposure and compromise of these assets. The charge that Stalin did compromise these assets still applies, however, indirectly. In his rage against Trotsky and his obsession to have him eliminated, he caused the GPU to proceed with the operation to murder Trotsky without regard to the cost in money or agent assets.

Further, given the pre-war tenor of the times, and the reckless ambient the GPU had created for itself with the flagrant murders of Trotsky's supporters in Europe, they had become inured to the threat of exposure. The GPU had murdered, among others, Reiss, Klement, Sniveilent, and Sedov without any genuine attempts to conceal its work but no zealous attempt had been made by police authorities to place the blame for these atrocities on the GPU. All of these Trotsky supporters and comrades were murdered in haphazard, non-professional ways, which indicated the GPU had no real concern about exposing a few more agents in their determination to murder Trotsky.

Conversely, and in mitigation of the GPU planners, they could not have foreseen that Mercader would lose his nerve and panic at the crucial moment that caused the Trotsky assassination to end in a fiasco. If, on the other hand, had the GPU planners been competent, they would have planned for just such a contingency and how best to circumvent the disastrous consequences.

It is estimated that the Trotsky assassination operation cost the Soviet government upwards of $600,000 in nineteen-forty United States dollars.[11] This figure in 1998 would exceed $20,000,000. Louis Fisher, a journalist, possibly said it best,

> Tens of thousands of men and women have been shot and imprisoned or sent to rot in foreign wastes because of this war between Trotsky and Stalin. Soviet domestic and foreign policies were perverted by it. History has been rewritten and changed because of it. And, the whole labor movement of the world was rent and weakened by the Trotsky-Stalin enmity.[12]

It is very easy to love Mexico and everything Mexican. Next to the United States it is the most personally compatible country in the hemisphere with the friendliest people anywhere. The Mexican sources on the Trotsky assassination are many and some of those who were personally involved are still alive and vividly recall the events.

There is no question, for example according to the sources, about the facts in the case. Siqueiros led the 24 May raid on Trotsky's house in Mexico because Siqueiros himself freely admitted it. There was no question that Trotsky was murdered in his home in Coyoacan in Mexico because the killer, Mercader, admitted it. Also, there was no question that Jaime Ramon Mercader del Rio Hernandez spent the next nineteen years and six months in a Mexican jail for the murder.

There are other questions about who helped Mercader work out the confession he carried in his pocket when he was arrested at the scene of the murder? Who tried to arrange Mercader's escape from the prison in Mexico — the Black Palace of Lecumberri? How thoroughly was

Trotsky's household infiltrated by GPU agents, in addition to Mercader? Who was the so-called "French Jew" who was with Siqueiros and directed the 24 May raid on Trotsky's house? If Joe Hanson, Trotsky's secretary, was an FBI informant, could he have saved Trotsky if he told his FBI case officer what he suspected about Mercader's lack of bona fides? Could Hanson have warned Trotsky more forcefully about Mercader and, more importantly, would Trotsky have heeded the warning or would he have deprecated it as he had all similar warnings concerning his own safety?Should Alfred Rosmer, a close friend and house guest at Trotsky's house just before the assassination, have warned Trotsky about his doubts concerning Mercader's language ability to speak the native language he claimed?

Of course, it is impossible to answer all of these questions due to the passage of time and the clandestine nature of the GPU's way of operating. Recent "tell all" books written by former GPU officers have helped but most have only raised more questions than they have answered. It is a fervent hope that this book does not do the same!

PART I

THE FORMATIVE YEARS

> At Karlsruhe in 1903, Medem asked his debating opponent, Trotsky, "When it comes to classifying yourself, you certainly cannot ignore the fact of national allegiance. You consider yourself, I take it, to be either a Russian or a Jew. "No," cried Trotsky, "you are mistaken. I am a Social Democrat and that is all!"
>
> The Biography of Valdimir Medem

Trotsky's fate is rooted in the legacy of oppression against the Jews that was firmly imbedded in traditional Russian culture. From the reign of Ivan the Terrible in the sixteenth century, Russian rulers sought to reaffirm Russian ethnic purity by blaming the Jews for the nation's misfortunes.

In 1762 Catherine the Great opened Russia to immigration for all foreigners but specifically prohibited Jews. Those Jews, who were already in Russia were forced to live under the ever present threat of pogroms. Mobs congregated with the slightest provocation to attack individual Jews or invade and destroy Jewish villages and ghetto districts in the larger cities.

During these urban incursions the mobs pillaged and burned the hovels that the Jews called home, the younger women were raped and the systematic murder of Jewish men and older women was common place. This genocide was carried out with impunity while the police and municipal authorities conveniently ignored the rioters and their heinous crimes.[13]

The only relief the Jews could hope for from these seemingly ceaseless brutalities was a revolt. They thought that armed resistance would change the tyranny of persecution from the nobility, and the indifference of the middle classes, against the pervasive bigotry they encountered in their everyday life. The focus of their hatred and their resentment against the indifference was directed against the Czar. Who else but the father figure of the government could have permitted such senseless bigotry and prejudice?

The Minister of Justice, as a member of the Czar's government, had reported that the threat of a Jewish revolt in Russia against the Czar was a possibility. Actually there existed a much more serious omen, the Czar's secret police, the Okhrana, reported that Jews made up over ninety percent of all radicals known to be fomenting strife and revolution in Russia.[14]

Popular discontent, and particularly the very vocal cries for tolerance of the Jewish minority, increased dramatically in those years that proceeded the Russian revolution. But Jewish radical rhetoric only brought forth harsher measures of repression and persecution by the Okhrana against the Jews, their enclaves and their few gentile supporters.

The free exchange of ideas, speech and assembly were stifled at the institutions of higher learning by heresy hunting pedagogues. The military was purged of those junior officers who were suspected by the Czar's more senior officers of conspiring against the government. Spurious charges of insubordination and treasonous activities were routinely brought by senior Russian officers against those junior officers who were suspected of adherence to any revolutionary philosophy.

Alexander II, in one of his rare benign moments, in an attempt to alleviate the suffering of the Russian peasants, freed some twenty million serfs from feudal bondage. Alexander II related his motive which was not one of charitable compassion for the serfs:

> The present position cannot last, and it is better to abolish serfdom from above, then to wait till it begins to be abolished from below.[15]

This gesture unfortunately was a catastrophic failure and too little too late. Moreover, relatively no provision was made to provide rural infrastructure nor financing to assist the serfs to meet the initial expenses necessary to begin productive farming. Many of the serfs, when confronted with these seemingly insurmountable obstacles, wished to return to the paternal feudal master-servant relationship instead of starving as free men.

In many instances the serfs actually received less land to cultivate as free men than they had been allotted for their own crops under the former feudal system. The disappointment of the serfs was apparent and assumed crisis proportions throughout Russia. They considered the token gesture as another example of Czarist deceit. Rural disturbances, tantamount to riots, further inflamed the smoldering urban unrest. Stringent government counter-terrorist measures temporarily suppressed the rioting but also increased the resentment seething beneath the restored calm.

The fact that any discontent at all had occurred effectively stopped any further attempts by the Czar to institute additional reforms. Okhrana repressions to control the rioting only engendered more hatred against the government. Discontent quickly developed into resentment in which nothing the government could do to attempt to ameliorate the situation could have changed the temper of the people.

In the cities the situation was much the same. In 1878, for example, a young Jewish woman, Vera Zasulich, shot the chief of police of St. Petersburg, General Trepov, who had been notorious for his barbaric treatment of revolutionary prisoners. Police abuses, instead of the issue of the woman's guilt or innocence, became the crucial element in the verdict of the court.

The police abuses were considered more offensive than her crime and she was acquitted. The spectators in the courtroom greeted the verdict by rioting in support of her freedom. When the police attempted to re-arrest the young Jewess outside the courtroom, the elated crowd repulsed the police and helped the woman to escape in the resulting melee. Such was the tenor of the sympathies, and popular will, of the average citizen in pre-revolutionary Russia.

Around this time, many secret anti-government organizations were formed in Russia by the discontented peasants and urban workers. One such group was the highly secret "Land and Liberty Society" ("Zemlya i Volya") founded in 1876 as a student revolutionary organization.[16] Internal factional fighting in 1879 among the members of this group resulted in a split. One faction, headed by Georgi Pelkhanov, left Russia for Switzerland where it became the nucleus for the Russian Marxists in exile. The other faction called, "People's Liberty" ("Narodnaya Volya") decided to remain in Russia to fight against the Czar's government as an underground terrorist movement.[17]

Tumultuous times, this year of 1879, and these revolutionary events may have been a way of foretelling the birth of one of the world's greatest revolutionaries and a molder of even more tumultuous times to come. In that year of 1879 there seemed to be increased revolutionary riots in Russia. There were two attempts to assassinate the Czar. The head of the political police was almost assassinated by revolutionaries and Prince Kroporkin, the governor of Karkov, was shot by radical leftist dissidents. Amidst this revolutionary fervor, on 25 October 1879, Leon Davidovich Bronstein, who was later to become internationally famous as Leon Trotsky, was born.[18] They called him Lev and Lyova, the Russian word for lion.[19]

The father, David Leontievich Bronstein, was a Jewish patriarch and dirt farmer who came from the Jewish ghetto of Paltova in the Ukraine.[20] David had moved to Yankova where his son, Leon, was born. He worked about 250 acres of his own land and about 400 additional acres that he rented from a former wealthy, ex-army Colonel by the name of Yankovsky, who was the patron of the village. Yankovsky was known locally as a decadent drunkard who, although formerly respected as a wealthy citizen, was now drinking away his estate.[21]

The Bronsteins lived in a mud hut of five rooms with low ceilings made of straw that leaked when it rained and an earthen floor that became a quagmire when the roof leaked. Certainly these were not palatial lodgings but this humble abode and bare subsistence living were Trotsky's earliest memories of life under the Czar. After Trotsky left home, his father built a house of brick with a roof of tin. After the Bolshevik Revolution the house was taken over by the government and used as a school.[22]

Trotsky's mother, Anna, was a city bred woman whose parents had disapproved of her marriage to a dirt farmer. Anna soon discovered that being a farmer's wife was strenuous and, except for the joys of a family, unrewarding for the most part. However, she was a wonderfully resilient woman who adapted solidly and lovingly to her duties as a wife and mother. Trotsky was the fifth child of eight that Anna bore David, but only two boys and two girls survived the rigors of this arduous peasant existence.[23]

Trotsky's parents' marriage was described by him as a "strong relationship of labor" with little display of tenderness or frivolity. Unfortunately, this was a marital relationship that did not provide the young maturing Trotsky with the nourishment he needed for the development of his natural affection and outgoing spirit. As a consequence, Trotsky developed a formidable hostility toward those whom he believed permitted such conditions to prevail. To his young mind this could only be the nebulous father-figure of the Czar and the government. His parents were forced, he believed, to live a way of life that so preoccupied them with subsistence living that they could not find time to show him the love and affection he obviously needed, and so desperately believed he deserved.

While Trotsky's father had been a fugitive from the Czar's anti-Jewish campaign, as the family prospered he gradually dropped out of the local synagogue as well as the observance of the Jewish Sabbath. Eventually he out came openly in favor of atheism.[24]

In the summer of 1888, when Trotsky was nine years old, he went to the Black Sea resort of Odessa to live with his mother's nephew, Moisey Filippovich Shpentser. During this time Shpentser became Trotsky's mentor, teaching him city manners and the niceties of social demeanor and graces. Trotsky thus acquired a veneer of sophistication and took a step on the way to becoming a man of the world. They had nightly talks around the fire in the small kitchen after the evening meal. Shpentser was a worldly scholar and socialist. He converted Trotsky to a strongly focused repugnance of the harsh treatment of the farm laborers, city workers and Jewish workers in all fields of endeavor.

There were almost inviolate prohibitions against Jews entering any private school in Russia. However, Trotsky was able to conceal his Jewishness and enroll in St. Paul's Realschule.[25] At this private school the curriculum was mostly science, mathematics and modern languages. This early training explains Trotsky's later facility with languages, his bent for scientific inquiries and his penchant for precise expository writing.

There was also a highly developed artistic and intellectual atmosphere in the Shpentser household provided by Shpentser's wife who was the principal of the State School for Jewish Girls. She was much more sophisticated and cultured than her husband. Trotsky became enraptured with classical literature and developed a passion for the theater and journalism. The results of these early interests emerge clearly in Trotsky's later life, particularly during those years of exile in Prinkipo and Mexico when he fought boredom with his writing and recourse to the classics of music and literature.

In 1894, Alexander III, the Czar of all Russia, died. Contrary to the aspirations of the Russian liberals that there might be a relaxation of the government's oppression against the farmers, urban workers and Jews, the new ruler, Nicholas II, not only reinforced the same despotic edicts of his predecessor but promulgated additional laws restricting the peoples' liberties. Like everyone else Trotsky had fervently hoped, during the change of rulers and before Nicholas II's position had become known, that Nicholas II would act decisively to relieve the suffering of the Russian people. Trotsky wanted Nicholas II to bring about, as he later wrote, some "realization of my nonsensical dreams."

In 1895 when Trotsky was sixteen years-of-age, Frederick Engels died. Engels had been the co-founder, together with Karl Marx, of the doctrines of communism. Trotsky later recalled how he had avidly listened to the reports read in Engels's honor at secret meetings, although at the time he knew nothing about the philosophy of communism.

In order to graduate in 1896 Trotsky spent his final year of school in the nearby town of Nikolayev. He lived there with a family whose two sons were about Trotsky's age but who were years ahead of him in political maturity. These boys, Trotsky later wrote, were obsessed with the new revolutionary ideas of social utopias. Through them Trotsky was drawn into a group of Narodniks, so-called because of their membership in the terrorist organization of Peoples' Liberty. Trotsky was entranced with their populist philosophy and he greatly admired their vows to change the future.

Another reason that made this association pleasant for Trotsky was the attendance at the conspiratorial meetings of an attractive, Marxist-oriented young woman by the name of Alexandra Lyovna Sokolovskaya. Although she was six years older than Trotsky, he fell deeply and passionately in love with her. They argued political philosophy and fought over the relative merits of various social systems — he asserting populist liberalism and she espousing rigid communist Marxism.

The slavish, utterly degrading, and brutally oppressive, working conditions of the dock and factory workers in Nikolayev were observed by Trotsky on a daily basis. He was appalled that such conditions were permitted to exist. As a result, he formed a secret group called "The South Russian Workers Union" to agitate for better working conditions for these workers.[26]

This organization is remembered chiefly for a leaflet it distributed called *Our Cause*.[27] It was almost two years after the first issue of *Our Cause* had been published that the police tracked down the humble origins of this flamingly treasonous pamphlet. In January 1898 Trotsky and Alexandra were arrested in a police raid with four other hardcore Narodniks and several hundred hangers-on.[28]

In 1897 Trotsky graduated with honors from his academic endeavors while he continued to print seditious leaflets, working with the Narodniks and writing material for *Our Cause*. He continued his studies after his arrest. He was often in solitary confinement for infractions of prison rules for helping other prisoners. He read everything that was available inside the prison and whatever he could arrange to have smuggled in to him. He read Darwin and steeped himself in the Marxist literature that was smuggled into the Kherson prison for Alexandra by her communist comrades.

While awaiting trial, Trotsky was transferred to a prison at Odessa. There was a guard at this prison named Trotsky. A few years later, Trotsky needed a name to use on a false passport and chose the name of this guard in Odessa. (Prior to this time, the person we know as Trotsky had been known as Leon Davidovich Bronstein.)

In early 1900 Alexandra and Trotsky were each sentenced, along with the other arrested Narodniks, to four years in exile in eastern Siberia. While waiting for his trial, Trotsky studied Marxism and read some of Lenin's works on Russian capitalism. This was a most important intellectual period for Trotsky; it molded his thinking away from abstract socialism into a deeper, more action oriented, Marxist communist doctrine. Nineteen hundred was also important in another way, because he and Alexandra were married in prison.[29]

The newlyweds settled into exile at Ust-Kutas as well as could be expected. Important events occurred during their married life in eastern Siberia: a daughter, Zena, was born; Trotsky intensified his study of Marx's *Das Kapital*; he organized a union to protest conditions in his forced labor camp, and he contributed vast amounts of literary criticism and vivid colloquial sketches of the local countryside to a newspaper the *Eastern Review* in Irkutsk.

In 1902, after two years in exile, Nina, a second daughter was born. Trotsky was studying Lenin's latest books, *What is to be Done?* and *The Development of Capitalism in Russia*. He admired Lenin's concept of the creation of a centralized cadre of professional revolutionaries who would be bound together by an iron rule of action. As a result, he felt impelled to escape from Siberia because he desperately wanted to join

Lenin's dedicated group. However, he was constrained from putting his desire into action because he realized the awful consequences his escape could have for Alexandra and his two daughters.

Trotsky had not considered Alexandra's iron determination. She was a true revolutionary, unconcerned for her own fate or that of her daughters. She felt that Trotsky was striving to reach the common goal of the revolutionary socialized millennium which was an ideal she strongly shared with Trotsky. She therefore exhorted him to leave and assured him that she would take care of herself and their daughters. He was to go and do his duty for the revolution!

Trotsky placed a dummy in his bed to deceive the guards when they made the routine evening count of prisoners. He then escaped from the prison compound with a fellow woman prisoner hiding under a load of straw in the cart of a peasant collaborator. Friends at Irkutsk provided him with a false passport in which he used, for the first time, the name Leon Trotsky. He separated from his female companion at the railhead.

Trotsky went by way of the Trans-Siberian Railway to Samora. There he learned that his literary critiques and colorful vignettes of the countryside had provoked favorable comment in the local newspapers. The readers had been so impressed by his writings that they had bestowed upon him the nickname of "The Pen."

Fatefully, almost as soon as he arrived in Samora a message came from Lenin for him, simply addressed to the local newspaper for "The Pen." The letter instructed Trotsky that, as a true revolutionary, he was not to question orders but to report immediately to Vienna. Lenin wanted Trotsky to work with him on the embryonic party newspaper called *Iskra* (*The Spark*).

Trotsky, realizing that his desire to work with Lenin had miraculously occurred, quickly left Samora. He could not know then that leaving his much beloved Russia was a condition that would reoccur throughout his life.

In Vienna, Trotsky was given enough money by Paul Axelrod, a local comrade to continue his journey to Zurich, Switzerland where some of the editors of Iskra were working. They advanced him a little more money to get to Paris with barely enough money remaining to eventually arrive in London where Lenin was waiting for him.[30]

One historically memorable morning in October of 1902 Trotsky presented himself at 30 Halford Square at King's Crossing in London where Lenin lived with his wife, Nadezhda Konstantinovna Krupskaya. Lenin and his wife lived in a one room combination living, dining, bedroom with a tiny kitchen off to one side. They lived in this apartment under the alias of Mr. and Mrs. Richter, They both eagerly welcomed Trotsky into their circle of comrades and found him lodging nearby.

Trotsky began to write revolutionary articles for *Iskra* and his logic and eloquence brought him offers to lecture at Whitechapel. This was an outdoor arena for radical speakers who exhorted, among other things, themes of all political persuasions, faddish health regimes, free love and quirky religions. Trotsky lectured to the curious at Whitechapel on Marxism, historical materialism and the urgent need for social reforms and economic revolution.

Trotsky's success as a speaker grew with his ability to convince the skeptical, and often jeering, hooting audiences, of the panacea of communism. He stressed the inevitability of social revolution which would ultimately overthrow the capitalist myth of economic prosperity. Trotsky was so successful in convincing this ultra-critical audience that Lenin suggested he should go on a tour of the European capitols to preach the gospel of revolutionary Marxism.

Very early Trotsky saw clearly where the Bolshevik system of power would lead — to the ultimate tyranny of one man. Everything that he subsequently claimed to have believed in was sacrificed to the one aim of this monopoly of power by the Communist Party. He played a full part in the Bolshevik deception of its followers. One could compile a long list of statements Trotsky made in the early years when he was one of the top leaders of the party about how their blatant hypocrisy and dishonesty are matched only by Lenin's demagoguery.[31]

Lenin wanted to appoint Trotsky as one of the three editors of *Iskra*. However, soon after assuming the lofty position of a minor assistant editor, Trotsky accused Lenin of putting in too many "I"s in his writings. Lenin replied that so long as he had the ultimate authority he would continue to employ as many "I"s in his writing as he pleased.[32]

In addition to writing and editing the newspaper, Trotsky was assigned to help care for the increasing flow of refugees from Russia who were continually arriving in London. The secretary of the editorial triumvirate, Krupskaya, Lenin's wife, was really the conspiratorial ringmaster handling the refugees. She received the refugee comrades, and when they departed she gave them directions for their next rendezvous on the underground railway.

Krupskaya established rat lines of escape for the refugees and safe houses along the secret routes. She provided accommodation addresses where they could write for help if they became lost or captured, supplied codes and invisible inks, passwords, safety signals, and disposed of compromising documents by burning them in an old pot-belly stove. It was an office joke that when they could smell burning paper, they knew that Krupskaya was getting rid of sensitive messages.

There were continuing attempts, however amateurish, to communicate clandestinely with the local committees in Russia. Fortunately, the detection methods and procedures used by the Okhrana were more amateurish. The letters spoke of "handkerchiefs" (passports), "barrels of beer" (shipments of Lenin's newspaper *Iskra*), "warm fur" (illegal literature); codes names were used beginning with the same initial of the true name to designate places (Ossip for Odessa, Petya for Paltava, Terenty for Tver); women's names were substituted for men's and vice versa. The letters also contained an amazing amount of nature nonsense about animals like "the snail" for Zinaida Krzhizhanovsky; "the bear" for Lenin; "the little bear" for Lenin's sister; and "the horse" for Krassin, etc. It might seem almost plausible that "Papasha" (Papa) was used for Meyer Wallack, better known to history as Maxim Litvinov.[33]

At the same time that Trotsky was helping to edit *Iskra*, a very young William Fisher was helping his father deliver copies of *Iskra* around the neighborhoods of London.[34] Many years later William Fisher became notorious as Colonel Rudolph Ivanovich Abel, the illegal Soviet spy in the United States.[35] Fisher's father was an avowed and dedicated communist who was on the fringes of Lenin's group in London. He did some work of an administrative nature (paying office expenses, keeping the ledgers, sweeping up and similar menial duties) in the Iskra offices. The son, William Fisher, who was to become Abel, also ran errands and did odd jobs around the Iskra office.[36]

In 1902, at Lenin's request, Trotsky went to Paris. There he met the third woman who was to have a major impact on his life.[37] In succession these women were; his mother, Anna, his first wife, Alexandra, and lastly, Natalia Ivanovna Sedova, an attractive brown-haired young lady with a svelte figure and an impish face, who became his mistress for life.

Natalia was born in the Ukraine in 1882, the daughter of a Russian middle class family. She was orphaned at the age of eight. Some of her relatives had been associated with the revolutionary Narodniks and she had accepted their advanced socialist philosophies. She was, therefore, already predisposed to Trotsky's philosophy of revolution throughout the world when they first became conspiratorial confidants and lovers.

Natalia was a feisty girl who had a rebellious background similar to Trotsky's. She had been expelled from a private school for leading a revolt against compulsory religious attendance, and at nineteen she had helped smuggle illegal Marxist propaganda into Russia. She then joined the Russian emigres in Paris who were working on *Iskra*.

Natalia arranged a hideout in Paris for Trotsky in an attic room in the same house where she lived.[38] The two, to the surprise of no one, were soon living together as man and wife. Natalia would remain Trotsky's common law wife and loyal devoted companion for the rest of his life, until his murder in Mexico thirty-eight years later. She was to bear Trotsky two sons, share his few triumphs and suffer with him through his many defeats and final exile. Natalia, only briefly, would regret her relationship with Trotsky every time his womanizing would make her doubt that he really loved her and was truly her's alone.[39]

In the vote on the editorial board of *Iskra* Trotsky was a menshevik (men of the minority) and his rigid minority position on matters of doctrine and political philosophy soon crystallized into what would be known by political activists as Trotskyism. In essence, at this time (1904), Trotsky was passionately dedicated to Marxism and its dogma of permanent revolution, but he was also steadfast in his opposition to both the soft position of the other revolutionaries and to the rigid centralism of Lenin's Bolsheviks (men of the majority).

In a pamphlet entitled, *Our Political Tasks*, published in 1904, Trotsky accused Lenin of trying to impose a "barracks-room regime" on the Russian radicals. In language startlingly similar to that which he was later to use in his attacks on Stalin, the young Trotsky denounced Lenin as "the leader of the reactionary wing of our party."[40]

In this same year Trotsky became friendly with Alexander Helphand, also known by the aliases of Helfand, Melatov and Parvus. Helphand was a Russian Jew living in Germany. He wrote articles for *Iskra* and had a considerable reputation as an economist and a fervid Marxist. Helphand wrote that the Russian proletariat may be the advance guard of a victorious uprising against capitalism in the industrialized countries. Trotsky was captivated by the idea that events in Russia might be the precursor to world revolution.[41]

In 1905 Natalia, Trotsky and Helphand spent some part of the summer hiking and camping in Switzerland. They lived on milk and mountain air, they claimed. When they needed money Helphand and Trotsky wrote articles for one of the Socialist Democratic newspapers.[42]

Helphand (aka Alexander I. Gelfand) who was ten years older than Trotsky, enjoyed an enormous reputation as a Marxist writer and political thinker. He first brought to Trotrsky's attention the fact that "revolution may place a workers' government in power."[43] In Germany Helphand founded a newspaper, *Aus der Weltpolitik,* and became wealthy. In 1917 he returned to Russia where he joined Trotsky in creating the party newspaper *Nachado*. He became more wealthy through speculation and in the war of 1914 became even more wealthy through intrigue.

Helphand also attempted to broker an alliance between Germany and Russia. He was involved in supplying Lenin with the "sealed train" that brought Lenin across Germany back to Russia. It was because of his closeness with Trotsky and Lenin that the charges were made that they were sent back to Russia as recruited German agents. Helphand died from bursting blood vessels in the brain in 1924, the same year as Lenin.[44]

Although a Russian by birth Helphand had always considered himself to be a German. He became deeply involved in smuggling separatist propaganda into the Ukraine. Dr. Diego Bergan, the head of Germany's "political warfare" (intelligence) operations, provided Helphand with a German passport and an initial fund of $250,000 to draw on. Helphand was a German agent in the true sense of the word and he subtly subsidized Lenin and Trotsky to further German strategic objectives, but the question remains, did they realize it? The answer, almost certainly, is a resounding affirmative.

The money to Lenin and Trotsky came from Helphand through Karl Radek and Jacob Furstenburg who may not have told Lenin and Trotsky the source of the funds. Lenin and Trotsky never acknowledged any return indebtedness to Helphand for the money. In sum, Lenin and Trotsky were probably not controlled German agents but only milking a money source for all they could get without accepting any responsibility for reciprocation.[45]

In 1905, the prestige of the Russian autocracy had been undermined by the ignominious defeat of the Russian military forces in the war with Japan and by the calamitous domestic peasant and worker unrest. Trotsky decided to return to Russia to exploit this unrest. A disguise was needed to deceive the Okhrana, who by this time was on the alert for the "revolutionary, Trotsky." He shaved off his mustache and beard and set out to revolutionize Russia for Marxism and communism.

Before Trotsky returned to Russia, there was labor unrest and demonstrations in St. Petersburg led by a priest, Father Georg Capon. These disturbances culminated in a march on the Czar's Winter Palace on Sunday, 9 January 1905, to present a petition to Nicholas II. Prince Svyatopolk-Mirsky had been left in charge of security in St. Petersburg while Czar Nicolas II and his family went to their estate about fifteen miles away. He ordered the Cossacks to ride through the crowd estimated at 200,000 on a saber charge. They then turned to charge back through the crowd, indiscriminately slashing with their sabers. The infantry then opened fire with rifles and machine guns. The official toll was 130 killed and 300 wounded. The unofficial toll was 4600 casualties.[46] (This date would be 22 January 1905 in the Western calendar. It become known as "Bloody Sunday.")

The effect of this massacre was tremendous. Former apathetic laborers, farm workers and the downtrodden from all walks of life rallied and were galvanized to the cause of the demonstrators. With the aid of Social Revolutionarics Father Gapon fled abroad. He sent a note to the Czar, "Let all the blood which has been shed hangman, fall upon thee and they kindred."[47]

Shortly after Bloody Sunday, the Governor General of Moscow, the Grand Duke Sergie, was assassinated. The resulting turmoil caused by these and similar catastrophic events seemed to have made the chaos ripe for revolution. Trotsky heeded this summons and hurried his plans to return home to Russia.

Natalia also returned to Russia on a passport forged in the name of the wife of a retired army corporal, named Arbuzov. Trotsky joined her in February in Kiev but soon left Kiev because he felt he could do more in St. Petersburg to make the revolution happen. In St. Petersburg he immediately became notorious as the writer of stirring revolutionary propaganda and as an activist who was not afraid to personally lead the workers' militant demonstrations against the Czar's forces. For all of this tireless writing, preaching and activism Trotsky was acclaimed by his fellow revolutionaries with the nom de plume of "Peter Petrovich" which is something akin to "John Doe—Man on the Street."[48]

Natalia was arrested during an illegal demonstration at a May Day rally and Trotsky was forced to go into hiding because of his known association with her. Trotsky was hounded everywhere by the Okhrana and his imminent capture was a distinct possibility. Rather than risk the curtailment of his revolutionary work he decided to again go into exile. This time he would continue the fight for the principles of the Marxist revolution from Finland.

In June of that year a workers' strike in Odessa, which had been called in sympathy with the mutinous sailors on the battleship Potemkin, was promptly followed by a printers' strike in Moscow. Similar sympathy strikes spread throughout Russia and the revolutionary tempo increased. Russian workers demanded constitutional reforms, higher wages, better working conditions, shorter working hours, relief from government oppression and the right to have a grievance procedure against employers.

On 14 October the first "Soviet of Workers" was formed at the Technological Institute in St., Petersburg.[49] Trotsky, using false documents under the name Yankovsky returned to Russia to help this fledgling soviet grow into a viable political force. Trotsky made a fiery speech to this new group which was received with tumultuous acclaim. Trotsky was thereupon elected chairman of the new soviet and he was soon recognized as the dynamic force behind its increased revolutionary activity. Trotsky at the time was twenty-four years-of-age.

Nicholas II realizing that he must make some concessions to appease the rebellious workers promised a new constitution, civil liberties and universal suffrage. His words, however, were obvious to the people as nothing more than empty promises. In response, Trotsky led a mob of frustrated citizens from the Technological Institute to the University where he denounced the Czar and his paper promises.

Natalia, meanwhile, had been released from prison and joined Trotsky in St. Petersburg. They lived as Mr. and Mrs. Vikentiev for the fifty days that the first soviet lasted. On 2 December Trotsky issued a proclamation through the St. Petersburg soviet that if they were victorious in the revolution overthrowing the Czar they would not be responsible for the debts of the Czarist government. He called upon all Russian people to stop paying taxes to the Czar's government and ordered them to withdraw their savings from Russian banks.

The day after he issued this proclamation Trotsky was arrested with the other leading members of the soviet. He was then put into solitary confinement at the Peter and Paul Fortress and then transferred to the House of Preliminary Detention to await trial.

In the House of Preliminary Detention the prisoners were allowed considerable freedom. They could play games in the garden, take walks together since the cells were not locked during the day. Natalia was permitted to visit Trotsky twice a week. Although there was no provision for the exercise of connubial interplay inside the prison, knowing Trotsky's penchant for such things, he probably found some way to exercise this function.[50]

At his trial Trotsky was sentenced to the loss of his civic rights and to enforced resettlement in Siberia, which meant virtual deportation for life. (Helphand had begun to plot an escape for Trotsky but Trotsky refused to join. He wanted to go on trial and use it as a forum for revolutionary ideas.)

On 10 January 1907, Trotsky, dressed in a gray prison uniform, started the arduous journey to the penal colony of Obdorsk located on the Polar Circle which was over a thousand miles from the nearest railroad. When the convicts changed into the gray prison uniforms, they were allowed to keep their own underwear and shoes. This was particularly fortunate for Trotsky since he had a passport concealed between the soles of his shoes and gold coins in the heel.[51]

Thirty-three days into the journey, Trotsky and his fellow prisoners stopped at Beresov, about 300 miles from Odborsk near the Artic Circle.[52] Realizing that this was his last chance to escape before the oblivion of Siberia, he decided that despite the many risks involved he would have to seize the opportunity. He knew he would never again have the chance to escape from the Siberian isolation without the massive assistance of a large support apparatus and there was little chance of that happening.

On 12 February 1907, Trotsky wrote his last letter to Natalia from Beresov and planned to escape as soon as the circumstances seemed propitious. In a group of prisoners at Beresov there was an old Social Revolutionary, a friend of Trotsky's, Dr. Feit, who taught him how to fake the symptoms of sciatica. Trotsky, with his usual flair for the dramatic, malingered the symptoms of the disease so well that he convinced the guards and hospital staff that he must be left behind the prisoner party bound for Obdorsk in order to recuperate in the hospital at Beresov.

Trotsky speculated on which route he might take to escape from Beresov and safely reach freedom. His choices were few and all circumscribed by even more dangerous alternatives. If he returned the way the prisoner party had just come, which was the best and had the easiest terrain, the chance of recapture was the greatest since this was the route traveled by the police patrols. He could, alternatively, travel directly westward to the Port of Archangel, through the Ural mountains and hope to board a ship there for passage to freedom. One the other hand, he reasoned, he knew nothing of the conditions in Archangel. He might be literally walking into a more dangerous situation and summary execution as punishment for his attempted escape.

The route recommended by the local land surveyor, Roshkovsky, whom Trotsky had taken into his confidence, was to go southwesterly toward a mining settlement in the Ural mountains where there was a railroad terminus. There, Roshkovsky speculated, Trotsky might get on a train and eventually reach St. Petersburg. This southwesterly route was also seen as the most dangerous.

On the journey Trotsky endured eleven days of physically exhausting travel. The mental anguish, the possibility of recapture and the fear of being shot on the spot as an escapee tormented him. With his indomitable spirit, however, he kept the fears under control and completed his escape. Then, in a state of elation, he telegraphed Natalia, using a code known to them both, to meet him at a railway station where he expected their trains could intersect. Natalia was staying with their new born son in a small Finnish village near St. Petersburg. When she received the telegram she set out immediately. In her excitement she had not noticed that Trotsky, in his zeal, had inadvertently omitted the name of the railway station where they were supposed to meet. Natalia, ever deductive, reasoned that the only town where the two railways crossed had to be Samino. As if directed by fate, she met Trotsky exactly as planned at Samino!

Trotsky returned to St. Petersburg on 2 March 1907. Friends welcomed him and urged him to stay and help them in their organizational work. However, he knew he was still in danger of recapture and decided to leave. Trotsky, Natalia and their new son evaded the travel barriers at the railway station and the passport check points when they boarded the train to Finland. They also managed to avoid the roving security police who traveled aboard the train, checking each passenger's credentials several times. At the border they simply walked through the final security barrier without even having their baggage or passports checked beyond a cursory glance.

Trotsky met Lenin in Finland and through him was able to rent a small house in a village near Helsinki.[53] Trotsky then happily settled into domestic tranquility with Natalia and their son. (In this house, under those pleasant circumstances, Trotsky wrote the full account of his adventures in Siberia and the harrowing tale of his perilous escape. He received money from these literary efforts; enough to permit him to leave Finland for Stockholm, Sweden, where he continued to write. Natalia and their son returned to Russia.)

This was the start of what Trotsky thought would be a temporary exile but it was to be ten years before he returned to Russia in May of 1917.[54] Meanwhile, after working a short time in Stockholm, Trotsky decided that he must return to the center of revolutionary thought and whatever action there was. In April of 1907 he arrived back in London. There he attended the Fifth, and last, Party Congress where he renewed friendships with some of the persons destined to be leaders in the communist revolutionary movement.

TRAINING AS A REVOLUTIONARY

> History is a dangerous counselor if you have lost track of the passage of time and fail to acknowledge that history is a process.... that then is not now, nor the future counterpart of the past. Policies conceived in terms of yesterday's failures merely open the way to failure again.
>
> William Phaff

The delegates arguing Marxian philosophy at the Fifth Party Congress in London were keenly observed by the representative of the Bolshevik workers of the Borchalo district. This mysterious observer with the gimlet eye was originally from Teflis and was known at this meeting as "Koba." His real name was Josef Djugashvili who was later to become known as Josef Stalin.

The reason Stalin's presence as a delegate from the Borchalo district caused some suspicion was because there was no party branch in Borchalo. The question the legitimate delegates were pondering was, if there was no party branch in Borchalo how can there be a delegate? Who was this stranger? Could he be an agent of the Okhrana sent to spy on the congress?

Life magazine published an article by the eminent Russian scholar, Isaac Don Levine, which purported to prove that Stalin had been an agent of the Okhrana reporting on his Bolshevik comrades at the Fifth Party Congress.[55] The magazine illustrated the story with copies of a typewritten document alleged to have been dictated and signed on 12 July 1913 by the chief of the Czar's secret police in St. Petersburg, Colonel Yeremin.[56]

It was subsequently proven that on the date of 12 July 1913 when the colonel was supposedly signing the paper in St. Petersburg authorizing Stalin to be an Okhrana agent, the colonel was in Helsinki. He could not, therefore, it was speculated have signed the paper appointing Stalin as an Okhrana agent. However, no less an expert than the preeminent authority on Soviet affairs, the former United States Ambassador to the Soviet

Union, George Kennan, in his book, *Russia and the West,* had reservations.

Kennan states that he had devoted much study to these documents and he remains ambivalent but apparently leans toward accepting the documents as genuine. Kennan concludes that if the authenticity of the documents has not been demonstrated beyond challenge, "nobody has yet offered a plausible theory as by whom, when and why they should have been concocted as a forgery."

Czarist spy or not, Stalin did little for Bolshevik Party harmony in 1907. When he returned to Russia from the Fifth Party Congress he wrote that the true Bolsheviks at the Congress were the native Russians. He wrote that the Mensheviks, the ones in opposition to Lenin's position, were the spoilers. The Mensheviks, Stalin contended, were the faction of the "dirty Jews."

Stalin elaborated, in writing, that it would be of considerable benefit to the Bolsheviks if they could rid the party of the dirty Jews by initiating a pogrom against them. His enmity toward Trotsky as a Jew, was even then glaringly apparent. Stalin commented further that Trotsky's contribution to the real work done at the Fifth Party Congress had been "beautifully useless."

After the Congress, Trotsky and Natalia moved to Vienna where their second son, Sergei, was born. Trotsky would have preferred going to Berlin where he believed he could do more to further the Bolshevik cause. He correctly reasoned that the police in Germany were relentlessly prosecuting communists and he was more valuable to the cause remaining free.

When World War One broke out in 1914 Trotsky and his growing family moved from Vienna.[57] First they went to Zurich and later to Paris. Trotsky was not a front line war correspondent but he covered the war from the relatively safe base of Paris as a reporter for the periodical *Our Thought.*[58]

In September of 1916 the French police, at the urging of the Embassy of Imperial Russia, banned the newspaper Trotsky was working for. He was then concerned that he might be extradited from France to Russia. A return to Russia would mean certain death for being a revolutionary dedicated to the overthrow of the Czar. Trotsky therefore chose to flee and tried to reach Spain. However, he was arrested by the French police before he could get across the border. Fortunately, he was saved from extradition to Russia when the French government decided to deport him to Spain, where he wanted to go in the first place. Unfortunately, as soon as he was pushed across the border into Spain, the Spanish police arrested him on the charge of being a dangerous anarchist.

The Spanish government was anxious to be rid of the embarrassment of Trotsky's presence on Spanish soil. They put him, Natalia and their two sons aboard the S.S. Monteserrat. This was an old steamer of the Spanish Transatlantic Line that was bound for New York.

The Trotskys landed in New York on 13 January 1917, in the bitter cold of a snow laden day, little suspecting that they would remain in the United States only for two and one-half months. They took up residence in a cold water flat on East 164th street at the corner of Stebbins avenue.[59] The rent was $18 per week, an expensive rent in those days, and Natalia paid in advance for three months. According to Trotsky, his only income during his New York exile were small payments for a few newspaper articles.

Trotsky wrote some commentaries for *Novy Mir*, a Russian language newspaper published by Russian emigres in New York.[60] He also wrote a few short pieces for *Nashe Slovo*, another Russian language newspaper that was published in Paris. Other income, admitted by Trotsky to be meager, came into the family coffers from whatever Trotsky could get from his occasional lectures.

By any accounting, Trotsky's comfortable expenditures while he and his family were in New York considerably exceeded his precarious income. It is rumored that he worked as an extra in a motion picture for a short time at the Fox Film Studios in New York. There are some vintage scenes from the Fox Film archives that do show an extra ambling slowly across a scene who looks remarkably like Trotsky, but it isn't Trotsky.

There is another rumor that Trotsky worked at these studios as an electrician but this is equally false. Trotsky himself discounted these stories, and, although they make for interesting anecdotes, they simply are not true. There is no evidence, and Trotsky himself proclaimed that he had no income from any so-called regular employment while he was in New York.

Trotsky claimed that he studied American political and economic life during his free time while he was in New York. However, this period from 13 January to 27 March 1917, or a total of 73 days, was the only time Trotsky was ever in the United States. So despite his lavish claims, it is doubted that he really had the intimate knowledge of the peculiarly America brand of mercantile competition in our capitalistic society. Nor would be he able to completely grasp the manner in which the average American lived or the diverse, fire-eating, rough and tumble political spectrum then in existence. There is little doubt that Trotsky, despite his claims to the contrary, never really understood the American scene with any profundity.

It would be erroneous to presume that Trotsky gained any correct appreciation of the morally rich, sociologically honest and economically equitable nature of the American traditions and civilization from his brief encounter with it on the eastern fringes of the continent. The nuances of the American way of life, many times incomprehensible to natives steeped in their own lore, with all of the spiritual wealth of the philanthropic and religious institutions, remained essentially, an unfathomable mystery to Trotsky.

It is known that Trotsky and his family partook of many aspects of the "good life" stemming from the capitalist style while they were living in New York. The expenditures for this high-style living had to be considerably more than Trotsky admitted was his total income (less than a total of $300). Without belaboring the point, allegations were made that Trotsky was being subsidized during this period by the government of Germany through the back door by the German Intelligence Service.

This subsidization could have amounted to the difference between Trotsky's ostensible income of $300 and the amount of his obvious expenditures of close to several thousands of dollars, especially when the purchase of the family's outbound steamship tickets is added to the amount. It may also have amounted to a great deal more which allowed Trotsky to build up a surplus of over $10,000 which he had with him on the ship when he was arrested at Halifax, Nova Scotia.

On balance it seems unlikely that the Germans would subsidize Trotsky as a subservient field agent or that Trotsky would agree to work for them in this menial capacity. Trotsky's detractors and adherents make the same point on this issue and maintain that, at this early stage of his political life, he probably did not have sufficient stature to be an attractive candidate for recruitment by the Germans as an agent of influence.

If he was an agent, both sides agree, he probably would have been a lowly paid agent most likely being paid on a piece work basis. On the other hand, the Germans might have subsidized him if they anticipated he would achieve a position of power within the Bolshevik hierarchy. The Germans needed to hedge all bets and Trotsky was a small wager in their total gamble. They might have reasoned that if he ever did achieve any degree of eminence in Bolshevik politics he would then be beholden to them as a matter of conscience.

On the other hand, it is difficult to believe that Trotsky would ever have agreed to work for any foreign power against the Bolshevik cause and his conspiratorial comrades. It is more probable that he thought it was he, and not the Germans, who were really in charge of the relationship. If the Germans offered Trotsky money in New York, for whatever reason, Trotsky probably honestly believed that he could repudiate any agreement if he wanted to when the time came for him to reciprocate. As long as the fundamental objectives of the Germans and Trotsky were synonymous, Trotsky would work with them!

There are some indications that Trotsky may have become an agent of influence for the Germans at a later time but this was only because their goals and his goals coincided. The crux of this joint goal, if Trotsky and the Germans really shared a common purpose at some time, was for Trotsky to use whatever influence he might have to urge the Russian military to withdraw from the war. Failing to promote a full Russian withdrawal from the war, the German government hoped that he could, at least, stabilize the Russian military position as a non-combatant force. This meant that, as a military power or fighting threat to Germany, Russia would effectively be out of the war.

When the time came, Trotsky wholeheartedly agreed with this course of action but for an entirely different reason, Trotsky wanted Russia out of the war so that the badly weakened and ravaged Russian motherland could heal itself. The country needed to lick its internal wounds, start to rebuild its shambled economy and regain a measure of political stability so that the development of Bolshevik communism could flourish.

If the German government needed to be assured that the Russians were no longer threatening their eastern front, Trotsky could give them that assurance. This would permit the German military to turn the full force of their war effort against the Allies on the western front. The whole prospect of an "unleashed Germany" was one which the Allies anticipated with horror and which the Germans conversely viewed as their ultimate hope of survival. Both sides recognized that Trotsky could be pivotal in making the final decision in the Bolshevik camp.

Back in New York, before he returned to Russia, Trotsky took an active part in the political manipulation and the recruiting of the Russian emigres for the Bolshevik cause. On 5 March 1917 newspapers in New York reported the increasing prospect that the United States might be forced to enter the war against Germany. The same evening, Trotsky spoke at a meeting of the New York County Socialist Party. He took this opportunity to respond to the possibility of the United States entering the war on the side of the Allies. In a fiery speech he urged the assembled socialists to encourage strikes in the American war industries and to take every measure possible to resist military induction in the event the United States entered the war against Germany.

The obvious bias of this speech, and the seditious nature of the actions proposed, might be interpreted as supporting evidence that Trotsky was a German agent or, at the very least, a German sympathizer. Regardless, if he was not a paid agent of the German government when he delivered that speech, a bona fide German agent could not have done a better pro-German job!

Working with Trotsky on that pro-German exhortation at the New York Socialist Party meeting was Louis C. Fraina, one of the founders of the future Communist Party of the United States of America (CPUSA). Many of the leaders of the future CPUSA were originally closely allied with Trotsky during his stay in New York. He was considered by them to have the potential to become the American Lenin and leader of the future CPUSA. Subsequently, these same CPUSA leaders became, on Stalin's orders, a major element in actively supporting Trotsky's assassination.[61]

It should be noted that even though the revolutionary parties were finding a growing response in Russia, in January 1917, Nikolai Chkheidze, a member of the Duma and a Menshevik leader argued,

> At this time there can be no hope of a successful revolution. I know that the police are trying to organize some simulated revolutionary outbursts in order to draw the workers into the streets [so they can] attack them."[62]

On 15 March 1917 when Czar Nicholas II of Russia abdicated, Trotsky, still in New York, was convinced the Marxist triumph was close at hand.[63] He therefore jubilantly prepared to leave New York to return to Russia to help his Bolshevik comrades wrest power from the provisional government of Alexander Kerensky. The provisional government had been created to fill the governing void left by the Czar's abdication until a more stable and permanent government could be formed.

In New York, Trotsky had been essentially a stateless person. At least he was persona non grata as far as obtaining a passport or a citizen's identification from his native Russia. At the time it is alleged that Trotsky mysteriously received a passport (or some document that permitted his travel) ostensibly from the American government.

If this is true, and there is considerable doubt as to the story's authenticity, President Wilson's purpose in facilitating Trotsky's travel is unknown. Some historians suggest that the American government might have relented to the pressures of American businessmen who supposedly wanted Trotsky back in Russia to hasten the overthrow of the Kerensky provisional government which had been vehemently against American business ventures in Russia.

On 26 March 1917 Trotsky and his family left New York aboard the S.S. Kristianiafjord bound ultimately for Petrograd. When the ship stopped at Halifax, Nova Scotia, on 3 April, Trotsky and his family were removed from the ship by Canadian authorities. They charged Trotsky with being a subversive and interned him at Amherst prison with German prisoners of war. This internment was demeaning and under abdominally filthy conditions. Trotsky's subsequent intense hatred of Britain, and all things British, stem from the indignities of this incarceration.[64]

The British Admiralty in London had informed their representative at Halifax to expect Trotsky on the ship. He was alleged to be accompanied by five other men, described as Russian socialists. All were supposedly on their way to Russia to start a revolution against the established Russian government. The British Admiralty also informed the Canadian government to expect to find $10,000 in United States dollars on Trotsky's person. They said that the money was presumably held by Trotsky for the purpose of financing the socialist revolution against the existing Kerensky government. The British government told the Canadian authorities that this $10,000 had been given to Trotsky by the New York Socialists and the German government.

On 21 April, after many protests by Trotsky, the order for his release came from London. It was alleged that several prominent Americans, international socialists and anti-Czarist Russian officials in the United States, had taken an active interest in securing Trotsky's release.

Traveling with Trotsky aboard the S.S. Kristianiafjord, before the ship reached Halifax, was Charles Richard Crane, a former chairman of the American Democratic Party's Finance Committee. Crane had considerable experience in matters of Russian-American commerce and had organized, before the Czar's abdication, the Westinghouse Company of Russia. He represented the essence of the intrepid American businessman trying to create a commercial success from the Russian market regardless of an inconsequential impediment like a revolution.

Some historians attempt to make much of unplanned meetings and coincidental encounters between Trotsky and Crane. They imply sinister motives and contrive wild tales of the result of such occurrences. Others, possibly more knowledgeable that these things occasionally do happen without pre-arrangement, say let the facts tell the story and leave the speculation to those who will believe the worst regardless of the truth.

There is no evidence that Crane and Trotsky being on board the same ship bound for Russia at that particular time, in a period when clandestinity and subterfuge were epidemic, was anything but a coincidence. It is known that Trotsky desperately wanted the support of the United States government for the Bolshevik cause. During the voyage he may have requested Crane's support but there is no record.

Still, it is known that Crane was subsequently a member of the United States government's Root Commission to Russia. It is a fact that some members of the Root Commission in Russia were engaged in covert political action, black propaganda, intelligence collection and counter-espionage operations. Therefore, Crane would have been a logical choice of the United States government, and especially of President Wilson, for a clandestine meeting with one of the leading figures of the incipient Bolshevik power in Russia under the guise of an innocent shipboard acquaintanceship.

Lincoln Steffans, a noted American journalist and ardent socialist, was also a passenger aboard the S.S. Kristianiafjord. Steffans wrote that Trotsky and Crame did meet socially on board the ship. They did agree that the Russian revolution was in its first phase only and that it would develop into at least one more stage. Crane and Trotsky, reported

Steffans, thought that they would all be in Petrograd for the re-revolution, that is, the deposing of the Kerensky provisional government by the Bolsheviks. There is no record, however, of any more conspiratorial subject matter of their meeting.[65]

Adding to this confusion concerning Trotsky's status at this time, we have the report of William Dodd, who later was the United States Ambassador to Germany during the Hitlerian period in the 1930s. Dodd, an eminent scholar in his own right, wrote that Charles Crane had an active clandestine role in the deposing of the Kerensky provisional government. According to Dodd, Crane was a member of a quasi-official intelligence organization, the Root Commission and was an active participant in supporting the Bolshevik revolution.[66]

There was a definite empathy for the overthrow of the Kerensky provisional government within the power structure of the United States government. The quasi-intelligence organization, the Root Commission, and most official United States representatives in Russia, all were desirous of replacing the Kerensky provisional government with something else. They reasoned that if the do-nothing Kerensky government was only replaced and a more commercially ambitious Bolshevik government was in power, business between Russia and the United States would flourish. Further, where business is good, government relations are bound to become more harmonious. A Bolshevik government could therefore be convinced to remain in the war against Germany and not jeopardize the Allied position on the Western front.

During this period while American businessmen and the United States government were helping depose the Kerensky provisional government, Trotsky was apparently playing a double role. He would lend or withhold his support and ultimately cooperate with the side that offered him and the Bolsheviks the most advantageous opportunity.

A sketch of Trotsky's personality at the time is helpful in understanding the man:

> Flashing out of that remarkable gathering [Democratic Congress of 14 September 1917] was the striking personality of Leon Trotsky. Like a Marat, vehement, serpent-like, he swayed the assembly as a strong wind stirs the long grass. No other man created such an uproar, such a hatred at the

> slightest utterance, used such strong words and yet underneath it all carries such a cool head.[67]

Trotsky scrupulously rejected any government to government alignment with the United States. He preferred to seek non-attributable assistance from the United States agencies in Russia before and after he became a powerful official in the Bolshevik government.[68]

The charges, current during the initial phase of the Bolshevik victory, that Trotsky cooperated with the German intelligence service during World War One, were revived by Stalin in the so-called Purge Trials of the 1930s. When Stalin levied these charges the Dewey commission supposedly "proved them to be ludicrously false." (The investigative-judicial body known as the Dewey Commission was summoned at Trotsky's insistence to answer Stalin's charges.) The Commission accepted all of Trotsky's rebuttals and allegedly rendered an unbiased and impartial decision that Stalin's charges against Trotsky were false and that Trotsky's rebuttal that he had not worked for German intelligence during World War One was accepted as the truth. This vindication, however, believed by skeptics to be contrived, has not laid the matter to rest!

Another charge that Stalin laid against Trotsky was that he was a paid agent of international capitalism, specifically conspiring with international Jewish financiers to subvert the Bolshevik revolution and the Soviet government. These charges were also found to be false according to the verdict of the Dewey Commission.

It should be noted, however, that the allegations that Trotsky may have cooperated with the German intelligence service in 1917-18 without actually being a paid agent and that he connived with the United States commercial and government interests during the same period were not specifically investigated nor refuted by the Dewey Commission. It can correctly be surmised that these allegations will probably remain in historical limbo.

The Dewey Commission wanted to do everything possible to clear Trotsky of any such taint but skirted some issues they felt better left uninvestigated. Chairman Dewey himself, a noted American scholar, was too honest to report falsely that Trotsky had been cleared of this more subtle issue and preferred to simply ignore the matter.

About the same time that Trotsky was traveling toward Russia on the S.S. Krisianiafjord, Lenin and a party of thirty-two Bolshevik comrades left Switzerland aboard a sealed German train destined for Russia. The group with Lenin traveled through Germany with the permission and active assistance of the German General Staff abetted by German intelligence.

There is no ambiguity here since the sole purpose of the German connivance with Lenin was to help the Bolsheviks achieve control of the Russian government. This assistance to Lenin by Germany opened him and his associates to the same charges of being paid agents in the service of German intelligence as were levied against Trotsky.

The fact that Lenin did receive financial aid and material assistance from the German government is acknowledged by the Soviet government. It is dismissed, however, with the apologia that on Lenin's part, at least, it was only a war-time expediency, unlike the same charges against Trotsky brought by Stalin in the Purge Trials. This pandering and possibly treasonous aspect of Lenin's past conduct have never been condemned in the official history of the Bolshevik revolution.

The Germans' ostensible sole reason for aiding Lenin and Trotsky was that they wanted to remove Russia from the war. There was, however, another and possibly more pragmatic reason. The Germans, the same as the Americans, wanted to be assured of a preferred trading position in the post-war exploitation of the lucrative commercial markets of Russia.

After Lenin accepted the offer from Germany of the sealed train, the allegations that he was a German agent was generally accepted as true by most knowledgeable people on both sides of World War One. There were, however, some skeptics on both sides who thought the charges ethereal and lacking in solid proof — that is lacking in documentary evidence until the appearance of the Sisson Documents.

A few months after Trotsky and Lenin returned to Russia, a scandal erupted concerning both of these Bolshevik leaders. Several hundred official German documents had been purchased by Edgar Sisson, a United States government official working in Russia. These documents came from a clandestine source who supposedly had excellent contacts inside the German government.

Edgar Sisson, the Petrograd representative of the United States Committee on Public Information (CPI), had purchased these documents in apparent good faith. The documents purported to prove that Lenin, Trotsky and many other high ranking Bolsheviks were the paid agents of the German intelligence service. Most of these documents were quickly relegated to a category the eminent document examiners and research historians categorized as "being unable to make any confident declaration as to their genuineness." Fifty-three of these documents, however, were characterized as "absolutely genuine" by the American document examiners. There being no questions in the minds of these professional historians that these fifty-three documents were exactly what they purported to be, that is, official reports of the German government. They unquestionably contained absolute proof that Lenin and Trotsky were paid agents of the German intelligence service before, during and after the Bolshevik revolution.

Eventually, all of these Sisson documents were proved, by collateral and unimpeachable evidence to be either clever forgeries or blatant frauds. The Bolsheviks and Party cadres were of course relieved that their leaders had been cleared and vindicated. Subsequently, suspicions were directed at the Americans who had surfaced these documents and, as a consequence, all American intentions and declarations were thereafter suspect in Russia.

Those who had originally been elated at the discovery of these damning documents, and anti-Bolsheviks everywhere, agreed that the Sisson documents could not be used to prove that Lenin and Trotsky had been paid agents of German intelligence. Conversely the fact that these particular documents were false, did not disprove the basic charges themselves. Trotsky and Lenin still could have been paid agents of German intelligence, the skeptics asserted, and the falsity of the Sisson documents in no way lessened the possibility that other evidence would be found that could conclusively prove the truth of these suspicions.

Sisson's place in history, as an unwitting dupe, became assured with the publication of these documents as genuine by the United States government. The Committee on Public Information published a pamphlet on 18 October 1918 before they knew that the documents were spurious, claiming that the documents proved:

> Lenin and Trotsky are German agents. They [the Sisson documents] show the Bolshevik revolution was arranged for by the German General Staff and financed by the German

> Imperial Bank. They show the treaty of Brest-Litovsk was a betrayal of the Russian people by the German agents Lenin and Trotsky.... They show, in short, that the present Bolshevik government [is] acting solely in the interests of Germany and betraying the Russian people... for the benefit of the Imperial German government alone."

There is no doubt that all of the Sisson documents were forgeries. Sir Basil Thompson, chief of the renowned Scotland Yard of Great Britain, traced the origin of the documents through the twisted path of concealed attribution. The documents originated with a Russian journalist, Anton M. Ossendowski, who was a professional propagandist with elastic morals who would do anything for money.

Ossendowski passed the documents to Alexander Gumberg, a Russian Jew who was working as a source for both the American missions in Russia and, at the same time, for the Bolshevik secret police, the Cheka.

Gumberg was probably being subtly directed by Trotsky as a penetration agent of the unwitting American missions in Russia. It would not be beyond the wily Trotsky to allow himself and Lenin to initially be smeared as German agents in the Sisson documents. This would give the documents meaningful credencc to entice the Americans to accept them as genuine. Lenin and Trotsky were already being cast as German agents so why not let the Americans fall into this trap? Moreover, once the documents were exposed as forgeries it would take a great deal of the believability out of any lingering accusations that they were German agents!

There were many allegations that the American Red Cross Mission in Russia was squandering large sums of money to finance quack disinformation and other propaganda operations. Gumberg was one of several people who did odd jobs for the naive Americans in the Red Cross Mission and helped advise them on propaganda themes. Gumberg was, in the minds of the Red Cross officers, a loyal informative asset working solely for the good of the American Red Cross Mission. The fact that he was also working for the Bolsheviks was not even suspected according to available sources. Further, and more revealing, no one in the Red Cross Mission was apparently aware of the fact that the Bolsheviks might even try to run a penetration into their Mission.

There is no proof that Ossendowski was paid by any United States agency to create the Sisson documents in an attempt to discredit the Bolsheviks. However, this would not have been the first time that a misconceived, out of control, poorly coordinated and ineptly conducted propaganda operation of the United State's disinformation efforts backfired.

Gumberg was apparently a man of many talents. In addition to everything he did in Russia for the American agencies, he was a confidential advisor to the New York Chase National Bank in Russia and the Atlas Corporation in Russia. Gumberg took an active role in the diplomatic and intelligence fields in Russia by arranging commercial and financial advantages for both sides.

Of course, the Bolsheviks vociferously denounced the Sisson documents. Their plea for the repudiation of these documents was accepted by those who had seen the rebutting evidence. The embarrassment of the United States when the documents were proved to be forgeries was joyously reported by John Reed and other American pro-Bolshevik reporters. Reed, an American journalist and dedicated Bolshevik agent, wrote from Russia for the magazine *Metropolitan* which, oddly enough, was owned by one of New York's leading capitalists, J.P. Morgan.

TROTSKY'S LOSS OF POWER

> He would be the lunatic of one idea in a world of ideas, who would have all of the people live, work, suffer and die for that idea.
>
> Victor Serge, *The Revolution at Dead End*

> Why do they say those things about Papa? How dare they![1]

After Lenin and Trotsky were reunited in Russia, with any theoretical differences that may have existed between them forgotten, they worked together toward the goal of revolutionary victory. Alexander Kerensky headed the provisional government, a coalition of many factions, but he could not unite them to support his leadership that was necessary to govern the country. Kerensky was probably doomed from the beginning because the Bolshevik, and other more radical revolutionary forces, were aligned against him.

As the leader, Kerensky had been notably unsuccessful in prosecuting the war against Germany. The Bolsheviks were vociferous in their denunciations of Kerensky's overall military strategy and especially his wartime tactical procedures. Kerensky tried to personally direct the military operations rather than allowing the professional military to guide the effort. Opposition to Kerensky gained popular support until eventually he provided his own undoing.

Hoping to discredit the Bolsheviks, Kerensky had published documents purporting to prove that Lenin was a German spy working for the downfall of the Bolshevik revolution. Trotsky wrote a scathing, but very cogent, refutation of Kerensky's charges against Lenin, and, as a result, on 23 July 1917, Trotsky was arrested by Kerensky's police.

---

[1] Trotsky's sons after reading a Stalin inspired attack on their father printed in *Pravda*.

On 4 September 1917 Trotsky was released from prison and immediately renewed his fight against the Kerensky provisional government.[69] While he had been in prison, Trotsky had been elected to the Central Committee of the Bolshevik Party. By this time the Bolsheviks had a majority in the soviets of most of the large cities of Russia. Trotsky claimed that each member of a soviet represented a thousand workers.[70]

Trotsky, recognized as a proven and loyal Bolshevik, was elected president of the Petrograd soviet. Lenin sneaked out of hiding long enough to propose, together with Trotsky, the need for armed insurrection against the Kerensky government. He then went back into hiding, leaving Trotsky to take the responsibility for his call to arms.

Trotsky willingly stepped into the vacuum created by Lenin's disappearance and was the chief organizer, motivator, and leader of the Bolshevik insurrection against the Kerensky government. At the beginning of the Bolshevik revolution the part played by former Czarist officers was noteworthy. According to Trotsky, the officers were interested in overthrowing Keresnky but they also seemed to sympathize with the views of the Bolsheviks.[71] Unfortunately, Trotsky is not given the credit he is rightfully due for his leadership and it is Lenin who emerged as the sole leader of the Bolshevik revolution.

In the final analysis it was not the Bolshevik Party of Lenin but the Petrograd Soviet under Trotsky's direction that forged the weapons that made an armed uprising possible.[72]

During those historic times Lenin and Trotsky did not have time to remember their earlier literary polemics in which Lenin had called Trotsky a "little Judas" and Trotsky had labeled Lenin the "exploiter of everything backward in the Russian workers' movement." As long as Lenin had to remain in hiding Trotsky was without doubt the central figure and motivator in preparations for the October uprising.[73]

Lenin and Trotsky were self-deluding demagogues in their different ways. They believed that they alone knew what was good for mankind. They were thereby licensed to take whatever measures seemed necessary, at whatever cost in misery and bloodshed, to win power for themselves.[74]

On 26 October 1917 the Kerensky provisional government was meeting in the Czar's former Winter Palace when it was captured by a Bolshevik commando raid that had been instigated by Trotsky. Lenin was so pleased with Trotsky's success that he proposed Trotsky be appointed the head of the new Bolshevik government as the Chairman of the Soviet of the Peoples' Commissars.

The phenomenon of the Soviet Union is a supreme example of the triumph over reason of the shibboleth, the slogan and the political catchwords. "All power to the people" and others were afterwards discovered either to have no meaning whatsoever or else to be evidence of some arbitrary variable and unforeseen sense. Similarly, words such as "worker," "bourgeois," "proletariat" are endowed by mob action everywhere with arbitrary significance, meaning one thing one day and another the next as the occasion demands.[75]

With all of the double-talk and scheming it is not surprising that such an evil man as Stalin could emerge. It must be kept in mind, however, that the Soviet Union in all of its nastiness was the creation not of Stalin but of Lenin—who, of course, created Stalin, too.[76] So, who is ultimately to blame?

When the verbal duel between Trotsky and Stalin reached a fever pitch, Trotsky cried out, "Stalin, you are the grave digger of the revolution!" This proved to be a fatal error for Trotsky. "What have you done," Trotsky's friend Yuri Pyatakov cried out, "Stalin will never, never forget: he will forgive neither you nor your children, nor your grandchildren for it." Pyatakov's words were prophetic![77]

Louise Bryant wrote in a series of articles for Hearst's King Features Syndicate about Trotsky:

> Lenin's plans, no matter how carefully thought out, could not have materialized without the solid base of Trotsky's bayonets. No one will overshadow his [Trotsky's] eminence in the history of the revolution except Lenin. They will remain the two most distinguished personalities and they are complementary figures.
>
> When in 1918 Lenin was shot and terribly wounded by an assassin, an emotional speech by Trotsky made the nation grieve and rage. He [Trotsky] 'lighted the torch of Red Terror as a backfire against the White Terror, already so far under way that the very life of the revolution was at stake'. Bryant concludes about Trotsky, '.... it is well for Trotsky and for the revolution that Trotsky did not direct that terror:

> he is too passionate and too thorough a soul to have been entrusted with such a conflagration.

Trotsky refused the honor of being Chairman of the Soviet of the Peoples' Commissars but Lenin persisted and begged him to accept a less demanding position as Commissar of Home Affairs. Again Trotsky refused, feeling that his Jewish ancestry would be a handicap to the Bolshevik cause. Trotsky finally accepted Lenin's offer to be the Commissar of Foreign Affairs. This new position immediately required his presence at the town of Brest-Litovsk to meet with the German Peace delegation. Germany dominated the factionalized country of Russia and was insisting that Russia, under the Bolsheviks, sign a unilateral peace agreement with them.

On 13 November 1917 Lenin sent his first official message to the revolutionary proletariat of the world: "Comrades! We call you to arms for the first international social revolution!" Trotsky was assigned the task of spreading the revolution to Germany. The plan worked out by the Soviets required the assassination of the Kaiser by a German. After a year's search Trotsky was forced to admit that it had been proved impossible to find a German, regardless of political persuasion, or lack of such, who was willing to assassinate the Kaiser—not out of fear, but out of an inborn reverence for authority.[78]

At the so-called peace parley at Brest-Litovsk, Germany's demands for the annexation of Poland and the Baltic states as a condition to stop their invasion of Russia were unacceptable to Trotsky. However, he had seen from his railroad car window on the way to the peace conference the empty Russian trenches and the ravages of hunger on the civilian population with no prospect of a harvest. His own observations had convinced Trotsky that the new Bolshevik Soviet government could not withstand another German military advance.

Trotsky confided in R.A. Bruce Lockhart, the British intelligence officer in Russia, that surrendering to terms of either the Anglo-American capitalist or the demands of the German militarists was equally detestable. Trotsky considered Germany the deadlier enemy and asked Lockhart to arrange Allied support and military aid in order that he might reject the German demands. When the aid from the Allies was refused, Trotsky and the Bolsheviks were left with no course other than to make the best peace they could with Germany.[79]

Marx had insisted that the diplomatic rule of noninterference must be ignored in order to incite the proletariat of a country to rise up against their leaders. The Soviet diplomats to the Brest-Litovsk meeting with the Germans hurled revolutionary pamphlets from the windows of their special train and delivered fiery harangues to the German troops to rise up against their German masters.[80]

The Allies were determined to keep Russia in the war, even though they knew Russian arms were not a formidable threat to the Germans. The Allied High Command reasoned that the continued Russian military presence represented a threat that required the German High Command to keep troops stationed at the eastern front and away from the Allied position on the western front.

The Allies refused to recognize any peace treaty the Soviet government might negotiate under duress with the German High Command. The Allies claimed that the Bolshevik Soviets did not represent the Czar's government of Russia (the de jure government of Russia) nor the will of the Russian people. Therefore any peace treaty the Bolsheviks entered into on behalf of Russia was not binding.

Despite the Allied opposition, on 3 March 1918, the Bolsheviks signed a peace treaty with the Germans. Trotsky had resigned his post as Commissar of Foreign Affairs before he had to suffer the humiliation of signing this demeaning document. According to Trotsky, the concessions demanded by the Germans were unrealistic and he refused to be a part of it.

To accept Germany's terms would be a complete surrender of the Bolshevik program. The Bolshevik leaders accepted Trotsky's formula of their surrender in theory, "...the war is to be discontinued; peace is not to be signed, and the army is to be demobilized."[81]

In December of 1917, after the Soviets had signed a separate peace with Germany, the Allies intended to punish them for their unilateral action. Trotsky, the master counter-puncher of international intrigue devised a propaganda campaign designed to reveal the predatory plans of the Allies supporting his denunciations with the publication of secret treaties that were discovered in the Czar's Foreign Office archives. The Allied governments were afraid of the effect of these disclosures might have on the morale of their own people. Therefore, they tempered their original intention to severely punish the Soviets.[82]

Lenin then appointed Trotsky as the Commissar of War with the awesome task of trying to revive the moribund Russian military to fight the White Russian counter-revolutionary forces. On 29 May 1918 martial law was declared and Trotsky, as Commissar of War, issued the following order:

> If any unit retreats voluntarily, first the commissar shall be shot, then the commander. Courageous brave soldiers will be rewarded for their service and put in posts of command. Cowards, skin-savers and traitors will not escape a bullet.[83]

The reconstituted Bolshevik army (Red Army) would have to fight the Russian counter-revolutionaries (White Army). There was also the ever present threat that the Allied governments, angered at the unilateral peace treaty the Bolsheviks had signed with the Germans, might themselves launch an invasion of Russia.

In 1918 the Red Army under Trotsky relinquished the military counterintelligence function to the Cheka. This special military intelligence department (OO) was formed by Felix Dzerzhinsky as part of the Cheka with the strong support of Trotsky. This facilitated Party control of the military which was Trotsky's intention.[84]

Trotsky faced a formidable task in his new position as Commissar of War. The country was in chaos, with no infrastructure remaining. Railways were out of operation and no food could be delivered into the cities. Manufacturing, especially war goods, was at a standstill with no hope of obtaining raw material. Peasants were leaving the farms and there was no trained manpower to operate the factories, even if there had been raw material. There were no trained, or even inexperienced, maintenance technicians to repair the factories that had been scuttled by the Germans.

Meanwhile, Felix Dzerzhinski created a foreign espionage section of the Cheka in 1921. Soon there was a proliferation of intelligence organizations directed against foreign governments. The Foreign Office inaugurated "legal" networks of espionage agents, with diplomatic officials doubling as intelligence collection agents. Trotsky, as Peoples' Commissar of War set up another espionage network to collect military information.[85]

The morale of the Russia people was at its lowest during this time. Their willingness to continue the war was non-existent and impossible to revive. Food was scare and the counter-revolution against the Bolsheviks was beginning to assume the major proportions of a war. There were many people who were adversely affected by the Bolshevik revolution who hoped that the Bolsheviks would be defeated by the White armies. All of Russia's problems were blamed on the Bolsheviks and, like an unstoppable tide, additional calamities seemed to engulf them. Trotsky recognized these monumental obstacles to Bolshevism's ultimate success. He intended to take the battle to the counter-revolutionaries to demonstrate the invincibility of Bolshevism.

At the same time, the counter-revolutionary White Guards exploited Trotsky's Jewishness saying he was unfit to command the Bolshevik army. However, the average man-on-the-street apparently did not care if Trotsky were Jewish or not. The prevailing attitude seemed to be, so what if he is Jewish! The people said, "Trotsky is a Russian and a good fighter, so who cares if he is a Jew or not?" The average Russian helped to create a firm body of support behind Trotsky in his position as Commissar of War.

According to Trotsky, Stalin's attitude toward anti-Semitism was one of formal antagonism. He used the tactic of broadcasting to the Soviet press, a reminder, "Do not forget that the leaders of the opposition [like Trotsky] are Jews!"

At this time Trotsky, Natalia and their two sons were living in an apartment in the Kremlin. Natalia had a job with the Commissar of Education in charge of preserving museums and ancient monuments. In the two and one-half years of the civil war Trotsky was constantly on the move at the front with his troops and seldom had the time to be with his family.

One day, in July of 1918, just before leaving again for the front, Trotsky asked a friend in the Kremlin where the Czar was. He was told that Lenin had ordered the Czar and his family shot because he did not want to have a live symbol of the Romanov dynasty survive for the White forces to rally around. Trotsky was genuinely surprised at this announcement. Contrary to some historians' contention that Trotsky was instrumental in arranging the murders of the Czar and his family, Trotsky had nothing to do with these executions.

This is as good a place as any to relate a fanciful tale that persists, although proved untrue by recent discoveries of bones and other remains of the Romanov family. There was supposedly a secret codicil to the Brest-Litovsk Treaty by which the Germans stipulated a "safe passage" for the Romanovs. The Kaiser was the cousin of the Czar and in correspondence they referred to each other as "Dear Willy" and "Dear Nicky." The fate of the Romanovs was supposedly a bone of contention between Trotsky and Lenin and that is why Trotsky walked out on the first session of the Brest-Litovsk negotiations. Lenin had to send a replacement, Sokolnikov, to sign the treaty that contained the secret codicil that the Romanovs would be allowed safe passage to England. Supposedly they resided in England in complete secrecy until their deaths.[86]

Shortly after the massacre of the Romanovs there was an attempt on Lenin's life. This, plus the slaying of a Cheka officer in Petrograd, was evidence of continuing counter-revolutionary terrorism that had to be stopped. The Cheka unleashed a massive terrorist repression. These harsh counter-measures went far beyond any previous viciousness, or the usual sadistic brutality, of the Cheka.

Only two of Lenin's commissars were accorded unqualified respect from him: Trotsky and Dzerzhinsky. Trotsky because he was endowed with revolutionary talents and Dzerzhinsky because he was totally dedicated. As head of the Cheka, Dzerzhinsky was absolutely merciless. He did not enjoy killing the enemies of the state but he claimed that he loved human beings too much to pardon them for their sins.[87]

Historically, Trotsky has been blamed for these Cheka excesses but, as in similar incidents, it can be proved that he was not personally involved. He did not have anything to do with the chain of command that ordered the Cheka atrocities and was not even aware of the terror campaign against the civilian population. Trotsky was totally occupied at the front, engaged in heavy fighting against the White forces and knew nothing of the Cheka blood bath of the citizens.

Trotsky traveled to the front by a train which was virtually an armored car on wheels. This railroad car had spartan but adequate living quarters for Trotsky's personal use and was his home for two and one-half years. He lived on the train, visiting every fighting front and area of action where the Red army was engaged. He did this without thought for his own personal safety or physical fatigue. His presence at the front inspired the troops to do his will, to keep fighting despite hardship and to win the civil war. Trotsky's personal bravery greatly increased the morale of the Bolshevik troops and helped forge them into a cohesive indomitable force.

Wherever there was a battle Trotsky's armored train would appear as if by magic. The train was equipped with radio, a printing press, agitprop technicians, food and clothing for the soldiers and thousands of blank order forms signed by Lenin which Trotsky used to give added force to his orders. These forms enabled Trotsky to win a war which Lenin very nearly lost.[88]

It is interesting to note that during this time Trotsky sat for a bust by the American sculptress, Clare Sheridan. She reported that she was pleasantly surprised to discover that Trotsky was a courteous and gentle person and not at all like the devil incarnate she had heard about. (Trotsky in all probability either had, or certainly tried to have, an intimate relation with Sheridan).[89]

Back in England, Mrs. Clare Sheridan, cousin of Winston Churchill, reported that she had just returned from Russia where she sculpted a bust of Trotsky. She described Trotsky as having "the head of a Mephistopheles" and a "most wonderful personality."[90] However, Mrs. Sheridan's persistent questioning during the sitting produced the routine response from the Bolshevik hierarchy and she was denounced as "a spy."[91] Her friends, Trotsky and Lenin, managed to get the charges quashed. "Trotsky sat for the final session and then insisted on seeing the Widow Sheridan back to her hotel room."[92]

Lenin and Trotsky disagreed on the position of trade unions in the Bolshevik communist revolution. This dispute gave Stalin an opportunity to take a position with Lenin which permitted him to garner more power at Trotsky's expense.

It was thought by many at the time that Stalin's divisiveness almost caused the loss of the civil war for the Bolsheviks. What was needed to win the civil war was unity, forbearance and party loyalty. In the midst of this incipient Stalin tempest came another irritating disruption for the Bolsheviks. In February of 1921, the sailors at the naval base at Kronstadt mutinied against the Bolsheviks.

The mutiny at Kronstadt was a sad day for Bolshevik aspirations. The mutinous sailors were almost impregnable in their large armored battleship with the island's armament surrounding them for added protection. The Bolsheviks could not allow such a challenge to their authority to exist without retaliation. However, they never imagined what kind of retribution was to be meted out by their local commander, General Tuchachevsky.

Initially, General Tuchachevsky's troops had been repulsed several times by the withering fire from the entrenched navy guns aboard the battleship. Tuchachevsky had tried to take the island and battleship by going across the ice without any cover for his men. Eventually, after several severe set-backs, he dressed his troops in white so they were camouflaged against the snow.

Tuchachevsky was ultimately able to overcome the sailors' resistance and led his men aboard the mutinous battleship. He then proceeded to wantonly slaughter most of the mutinous sailors without mercy or the nicety of a trial, This ruthless repression was to be assigned by the pro-Stalin and anti-Trotsky forces as additional evidence of Trotsky's instability and vicious Jewish nature.

Nearly all the sailors who survived the mutiny were subsequently shot. Lenin wanted the entire naval base at Kronstadt to be destroyed and the ships of the Baltic fleet sunk. He explained that the sailors served no useful purpose and consumed food and wore clothing which the country needed elsewhere.[93]

Years later, world leaders who should have known better, including Roosevelt and Churchill, cited Kronstadt as evidence of Trotsky's bestial nature. Even today, when historical research has completely exonerated Trotsky from any responsibility for what happened at Kronstadt, critics unfairly cite this savage slaughter of the sailors as evidence of Trotsky's real character.

Trotsky always denied any direct participation in ordering the excesses that Tuchachevsky's troops committed against the sailors at Kronstadt. Trotsky also denied any complicity for the subsequent repressive measures taken by the Cheka against the few surviving sailors from Kronstadt who escaped Tuchachevsky's slaughter.

Conversely, to Trotsky's credit, he did admit that the very nature of revolution was "excesses," which in itself was an excess of history. He could not therefore reject revolution and he said, "In that sense I carry full and complete responsibility for the suppression of the Kronstadt rebellion." He was saying that since he had instigated the Bolshevik revolution he was equally responsible for its aftermath.

When the civil war had been partially won, the Bolsheviks under Lenin and Trotsky were firmly in charge of a relatively stable government. Famines, transportation failures, riots, droughts, plagues of locusts and freezing blizzards of unremitting intensity descended upon

Russia. Estimates of 35,000,000 people suffering the effects of starvation and on the verge of death prompted the humanitarian concerns of other nations. The United States in particular sent food and supplies to Russia under the supervision of the American Relief Administration directed by future president Herbert Hoover.

The famine was more widespread than that of 1891. The peasants abandoned their farms and sought sanctuary in the cities which could neither feed them or give them shelter. The horror of the famine exceeded anything that had happened in four years of the World War and three years of civil war in Russia.[94]

Beyond the sphere of his political mandate, Trotsky worked to develop the agriculture, manufacturing and the overall economy of the new nation under the aegis of Lenin's New Economic Policy. His principal preoccupation, however, was the Communist International. He strove to foster the concept of continuous revolution of all the working classes throughout the world. The Comintern, as the international communist organization was called, was a Soviet controlled instrument of subversion that received most of Trotsky's attention. Throughout this post-revolutionary period Trotsky was in all respects Lenin's deputy, principal confidant and heir apparent.

In 1919 Lenin and Trotsky launched the Third International. "There was a make-believe side to it," said Arthur Ransome, a British journalist. There were only thirty-five delegates in the small hall of the Court of Justice of the Third International. Angelica Balabanoff was appointed the first secretary general and Lenin proclaimed, "The victory of the world communist revolution is certain!"[95]

In March of 1922, the bureaucratic faction of the Fourth Party Congress, that was controlled by Stalin, proposed him for General Secretary of the Central Committee of the Communist/Bolshevik Party. At that time this office was essentially a ceremonial post only. Lenin, in an attempt to appease a hard working bureaucrat like Stalin, approved the assignment.

Lenin, in turn, offered Trotsky the position of Chairman of the Council of the Peoples' Commissar. This position would have formalized Trotsky's position as Lenin's deputy and officially would have placed him next in line for succession to Lenin's position as leader of the Communist Party. However, Trotsky turned down Lenin's offer. He believed that his unofficial position as Lenin's deputy and heir apparent was strong enough not to need any formal recognition, especially when Lenin's health seemed to be promising him a long life.

Some historians report that Lenin did like Stalin for his quality of getting things done, but he did offer Trotsky the higher position and one that would have assured Trotsky becoming heir to the throne. Pride and humility were equally balanced in Trotsky's rejection. The majority of members of the Central Committee were Jewish and Trotsky was acutely conscious of his Jewishness. He was also aware of the anomaly of a small group of Jews ruling a country were the Jews had never been completely assimilated. He strongly believed that the leader of the country should not be a Jew; he was equally convinced that the leader should not be a Georgian like Stalin either.[96]

In May of 1922 Lenin had a stroke but quickly recovered and was back at work by October. Both Lenin and Trotsky were concerned about the growing strength of the party's Organizational Bureau of the Central Committee. This department was presided over by Stalin who was creating an unwieldy and suffocating administrative impasse in getting the routine work of the party through the bureaucratic mess.

In December of 1922 Lenin prepared to abolish Stalin's Organizational Bureau which would have meant a severe curtailment of Stalin's power. On 25 December, Lenin dictated a recommendation and unofficial will for delivery to the Central Committee. It read in part:

> Since Comrade Stalin became Secretary-General he has gathered enormous powers into his own hands and I am not sure that he always knows how to use them with the necessary caution and restraint. On the other hand Comrade Trotsky is distinguished not only by his extraordinary capabilities, as shown in his dispute with the Central Committee on the question of the Peoples' Commissariat of Roads and Traffic, and he is certainly the best qualified man in the present Central Committee. 22 December 1922—Lenin

On 4 January 1923 Lenin added another thought to his will:

> Stalin is too course, and this failing, quite tolerable among us Communists, is completely intolerable in the Secretary-

> General's office. I therefore propose to the comrades that they find some means of removing Stalin from the post and appoint another, differing from Stalin. 4 January 1923.[97]

Under the terms previously set down, this testament was tantamount to assuring Trotsky's ascendancy to Lenin's position as leader of the Party and the country when Lenin died.

Lenin died in 1924 and immediately after his death, Trotsky made his bid for power at the Party Congress of May 1924 and was unanimously defeated. There is a diametrically different story regarding Lenin's will that may explain this defeat of Trotsky's. It is alleged that Max Eastman, Trotsky's official translator first made public the "Lenin Testament" or the "Lenin Will." In 1928 Eastman translated Trotsky's work entitled *The Real Situation in Russia*. In a supplement was the story of "Lenin's will" as outlined above.

It is alleged that Trotsky himself wrote an expose of the "will" story to the New York *Daily Worker* on 8 August 1925:

> As far as the "will," Lenin never left one, and the very nature of his relations with the Party as well as the nature of the Party itself, made such a "will" absolutely impossible. In the guise of a "will" the emigre and foreign bourgeois and Menshevík press have all along been quoting one of Lenin's letters (completely mutilated) which contains a number of advises on questions of organization. All talk of a secret or infringed "will" is so much mischievous invention directed against the real will of Lenin and of the interests of the Party created by him."[98]

Before Lenin died, the terms of his Lenin's will were not generally known outside the tight circle of Lenin's advisors, Stalin worked feverishly to consolidate his power and be prepared to circumvent the terms of Lenin's will.

Unfortunately for the future of mankind, Trotsky contracted malaria in October and on 18 January 1924 his doctors insisted that he go to Sukham on the Black Sea for a rest cure. On 21 January 1924 Trotsky was en route by train to Sukham momentarily paused at the railroad station in Tiflis when one of his aides told him of Lenin's death at 6:50

p.m. that day. He immediately made plans to return to Moscow for the funeral. Stalin, feared Trotsky's presence until he could consolidate his base of power as Lenin's successor, so he cabled Trotsky that he would be too late for the state funeral since Lenin's burial would have taken place before he could get back.

There have always been speculations regarding Lenin's death. In 1958, eighteen years after Trotsky's death, Trotsky's diary in the collection of his papers in the possession of Harvard University was opened. In it Trotsky spoke of Stalin giving poison to Lenin, ten days before his death. Also, ten days before Trotsky was assassinated in Mexico on 20 August 1940 there appeared an article in *Liberty* magazine written by Trotsky in which he suggested that Lenin had been poisoned.[99]

Elizabeth Lermolo in her book *Face of a Victim,* written in 1955, reported that while she was in prison in Russia she met Gavril Volkov who was in charge of the kitchen at Gorky — Lenin's retreat at the time of his death. Volkov said he took Lenin his second breakfast at eleven o'clock in the morning of 21 January 1924. There was no one in the room and Lenin rushed to him and slipped a note into his hand. Just then Lenin's personal physician, Dr. Elistratov came into the room, got Lenin back into bed and gave him an injection to calm him. Outside of the room Garvil Volkov read the note from Lenin. It said:

> Gavrilushka [familiar of Gavril] I've been poisoned. Go fetch Nadya [Lenin's wife] at once.... Tell Trotsky... tell everyone you can.[100]

The statement of Stalin's, that Trotsky could not return to Moscow in time for the funeral of Lenin, was a patent lie, but Trotsky had no way of knowing the depths of Stalin's deceit. Trotsky assumed that in this period of deep mourning, that emotions of good faith and trust would predominate. By not returning to Moscow to attend Lenin's funeral, Trotsky unwittingly ceded forever his rightful place as Lenin's heir, and Stalin assumed that position.

In addition, by not attending Lenin's funeral, Trotsky lost considerable prestige with the party leadership and the Russian people. In the cable dissuading Trotsky from returning to Moscow for the funeral, Stalin had added that the Politburo wanted Trotsky to continue his journey and regain his good health. In retrospect, it is clear that Stalin was creating a position where public opinion would demand Trotsky's ouster from power and approve his subsequent banishment from Russia.

In 1924, shortly after Lenin's death, a volume of Trotsky's collected works, *The Lessons of October,* was published and Trotsky wrote an introduction. In this essay, published as a separate booklet, he wrote that Soviet Russia was backsliding from the proletarian revolution in both domestic and foreign policies. The reason was that the two policies were led by the same two men, Zinoviev and Kamenev, who had lost their nerve in October of 1917 during the Bolshevik revolution. At this time the two men were Stalin's favorites.

The two accused men fought back denouncing as lies everything Trotsky had done in the Bolshevik revolution. From that point on, there was the systematic distortion of the history of the Communist Party and Trotsky's contribution — before, during and after the revolution. The smear campaign grew steadily worse and blackened the memory of every leader, especially Trotsky, who fell afoul of Stalin's ambition.[101]

Stalin's henchmen effectively argued to the party cadre who knew the contents of Lenin's will, that Stalin's conduct was no longer rude and therefore he was no longer the oaf that Lenin had objected to. Thus, they urged, that part of Lenin's will no longer applied and did not prohibit Stalin from taking Lenin's place.

In addition, it was specified by Stalin's supporters, that Trotsky had been absent from Moscow at the time of Lenin's funeral. This was accepted by the party rank and file as proof that Trotsky's ties of loyalty to Lenin were not strong enough for him to be considered as the leader of the party. As a matter of fact, many of the party bureaucrats saw their future more closely allied with Stalin's rising power. Lastly, Stalin denounced the so-called Lenin's will to the party cadre as a forgery perpetrated by Trotsky.

When Stalin began his grand conspiracy campaign against Trotsky, Felix Dzerzhinsky, the head of the Cheka, supported Stalin. Previously the two had been friends. But suddenly Dzerzhinsky developed into a theorist and made long speeches against the heresies of Trotskyism. As a member of the Political Bureau and an old-line Bolshevik, Dzerzhinsky brought his full political clout to bear against Trotsky and contributed in no small measure to Stalin's triumph over him.[102]

Stalin also accused Trotsky of major errors, "Trotsky's error lies in the fact that he has elevated himself into a superman standing above the

Central Committee, above its laws, and above its directors, and in this way, he has provided certain groups within the Party with a pretext for undermining confidence in the Central Committee."[103]

Stalin's secretary who defected across the Persian border on 1 January 1928 with a load of authenticating documents, gave this analysis of Stalin which helps explain his hatred of Trotsky:

> He [Stalin] was extremely secretive and vindictive and never forgave or forgot anything. He would even avenge himself after twenty years.... I could see he had no systematic thinking. He simply searched for the opinion of the majority. He would propose something today which didn't fit at all with what he had proposed yesterday. I know he understood state problems poorly.[104]

During the struggle for the position of Lenin's successor, the evidence that Stalin had been a police spy for the Czarist Okhrana against his Bolshevik comrades was offered to Trotsky to use to denigrate Stalin. Trotsky chose not to use this evidence, however, since it lacked, at that time, absolute proof.[105] Subsequently some proof of the validity of this evidence has been produced.

Winston Churchill, one of Great Britain's foremost statesmen, claimed that Lenin might have saved Russia from the extremities and violence that lay before her: "He alone could have found the way back to the causeway.... Their [the Russian people] worst misfortune was his birth...and their worst—his death."[106]

Stalin continued to consolidate his power to Trotsky's detriment. In January of 1925 the Central Committee on Stalin's orders, denounced Trotskyism as a "falsification of communism." The Central Committee then announced Trotsky's resignation as President of the Revolutionary Military Council and as the Peoples' Commissar of War, just as if his resignations had been voluntary.[107]

Trotsky was then appointed to serve on the Supreme Economic Council. However, Stalin's lackeys already sitting on the Council made certain that all of Trotsky's work was rendered ineffective and deprecated. With virtually nothing to do, Trotsky began writing about foreign affairs and the development of communism in other countries. Unfortunately for posterity, most of his articles during this period were never published.

The editors were controlled by Stalin and were ordered to destroy Trotsky's writings.

The few of Trotsky's articles that survived were known to have caused bitter resentment from indigenous communist leaders in those countries that Trotsky wrote about. This was especially true of his comments regarding the Fabian Society of Great Britain. When Trotsky later sought asylum in Britain, the Fabians, led by Beatrice Webb and George Bernard Shaw, had their revenge. They rallied together all socialists and leftist organizations in Great Britain to prevent the British government from permitting Trotsky to acquire sanctuary there.

In 1926, Stalin consolidated his victories over Zinoviev, Kamenov and, later in the year, Trotsky—all the Jewish members of the Politburo. By the end of 1926 the number of Jews on the Central Committee could be counted on one's fingers. Although the purge resulted in part from Stalin's own anti-Semitism, it also represented a generally popular orientation.[108]

On 14 November 1926, just after the Tenth Anniversary of the Bolshevik revolution, Trotsky was expelled from the party along with several other former stalwarts. Trotsky had already moved his family out of their Kremlin apartment in anticipation of such an ouster. If he could have foretold the future at the time, Trotsky would have seen even more diabolical punishment was to follow in the form of harassment by the GPU, the intelligence arm of Stalin's government, until he would finally be hounded out of Russian itself.

In 1924, Averill Harriman, the American capitalist, signed a fifty-two page agreement with the Soviet government to mine manganese in the Chiatura area of Russia. A year later the world price of manganese dropped and Harriman sought to renegotiate some of the conditions of the agreement. In December of 1926 Trotsky, far past his glory days, was assigned to work with Harriman on the amendments. Harriman remembered Trotsky as cold and aloof but very thorough.[109]

In the summer of 1927, Trotsky made his famous Clemenceau statement.[110] He charged Stalin, and most of the ruling Politburo, with lack of foresight, efficiency and determination. He gave notice that, in an emergency, he would strive to achieve a change in government, so that the country might be rallied and its resources efficiently organized for defense.[111]

Trotsky hoped for some popular support to oppose his ouster but he was years too late. His slogans such as "Let's Fulfill the Testament of Lenin" sounded pompous and went unheeded. The only persons to rally in Trotsky's support were a group of Chinese communist students at the Moscow Sun Yat Sen University who lifted papier-mache dragons off their heads just long enough to throw pro-Trotsky literature in the air and shout "Death to Stalin." The OGPU quickly arrested them and that was the last pro-Trotsky demonstration in Moscow.[112]

After Trotsky's downfall many who had been in awe of him, some who had praised him and all who had hated him, hastened to condemn and denigrate him. One referred to Trotsky's "nonchalant, high and mighty way of speaking"; another chided Trotsky's "tremendous imperiousness and inability or unwillingness to be at all amiable and attentive to people"; and "the old Bolsheviks resented Trotsky or at least were jealous whenever Lenin saw fit to defer publicly to Trotsky."[113]

Angelica Balabanov, an unsympathetic critic of Trotsky's commented:

> His arrogance equals his gifts and capacities. His manner of exercising it in personal relations creates a distance between himself and those about him which excludes both personal warmth and any feeling of sympathy and reciprocity."[114]

Unfortunately Trotsky's departure from the political equation changed forever the fundamental philosophy of the Soviet government. Trotsky represented a European radicalism that did not go beyond the bounds of civilization. Stalin, on the other hand, was a concentrated expression of Asiatic brutality.[115]

AMERICAN BACKGROUND

> Russia's allies — Britain, France and the United States — feared the collapse of the Russian army was at hand. At any moment a million German troops might suddenly be released from the Eastern Front and hurled against the tired Allied forces on the Western Front. Equally alarming was the prospect of Ukranian wheat, Donets coal, Caucasian oil and all the other resources of the Russian land falling into the rapacious maw of Imperial Germany.
>
> *The Great Conspiracy*

Before we follow Trotsky into exile there is a dramatic period of the Bolshevik revolution while it was consolidating power in Russia that epitomizes Trotsky's dynamism. This short period of time helps to personify the tremendous force he represented as an adroit political persuader that later was to cause Stalin to continually fear his omniscience, and over-estimate Trotsky.

One of the leaders of the American Red Cross mission in Petrograd, just after the Bolshevik revolution, referred to Trotsky as a "four square son of a bitch and the greatest Jew since Jesus Christ." This accolade gives an insight into the degree of awe, and possible reverence, the Americans held for Trotsky in his powers as a revolutionary leader.

During the chaotic period after the Bolshevik accession, when Trotsky was building the Bolshevik's power, he had to overcome the obstacles of internal party factional fighting. In addition he had to contend with massive interference from foreign powers who sought to protect their selfish interests in Russia's future.

On 7 November 1917, in the Western Calender, only two weeks after the Bolsheviks wrested power from the Kerensky provisional government, Trotsky, as Minister of Foreign Affairs, sent a letter to all diplomatic missions in Petrograd. In it he announced the creation of the

Soviet government and requested that it be formally recognized as the de facto government of all Russia.

Unfortunately from hindsight, David R. Francis, the United States Ambassador to Russia chose to ignore the letter. Francis believed he knew that the response from Washington to Trotsky's letter would be negative. Still, John Reed, the American communist journalist in Russia, became, with the approval of Ambassador Francis, the backdoor liaison between Raymond Robins of the American Red Cross Mission and Lenin and Trotsky.[116]

This disinclination on the part of the United States government to recognize the new Soviet government was shared by the other Allied governments. They all anxiously desired that Russia remain in the war against Germany and they knew the expressed intent of the new Soviet government was to seek a unilateral peace with Germany.

The Allied governments were also frightened by the violent and unorthodox political philosophy of the Bolsheviks as expressed by their fanatical spokesmen, Lenin and Trotsky.

The Czar, although recognized as despotic and woefully ineffective by the Allies, was at least a political constant in prosecuting the war against Germany. When the Czar had been replaced in March of 1917 by the provisional government of Alexander Kerensky, the Allies had hoped that this constancy of Russian foreign policy would continue. They also hoped the Kerensky government would be accompanied by a semblance of constitutional representative freedom that was more in harmony with the Allies' own ideas of democratic government. The Allies were unprepared for the realities of revolutionary life and the frightening aspects of the Bolsheviks' concept of total centralized power with no regard for democratic principles or the rights of individuals.

The United States government thought the Russian people would be revolted when they became aware of the Bolsheviks' basic cruelties, their abuses of the political process and their fundamental indifferences to the will of the people. They hoped that the Russian peasant, and urban dweller alike, would find common ground to spontaneously rise up against the violence of the Bolsheviks and their prostitution of the Russian peoples' traditional humanistic ideals. The other Allies shared this hope since they too were without any dependable intelligence concerning conditions inside Russia. However, they erroneously attributed more will power to the Russian people than they had.

The lack of any firm policy toward each other caused serious confusion to both parties in the United States-Bolshevik confrontation as to the real intentions of the other. Moreover, as a result of the lack of any properly functioning diplomatic channel of communications the two countries, engaged in spontaneous psychological warfare, abetted by spur-of-the-moment black and grey propaganda.[117]

This sub-rosa skirmishing also regularly employed calculated leaks of information damaging to the other side and genuine attempts at collecting intelligence about the intentions of the opposition. There was the constant and pervasive Cheka counterintelligence surveillance of all foreigners in Russia. (The Cheka was the Bolshevik secret police.) Plainly, the overt lack of trust and the disinclination, or inability, of the officials of both sides to come to any realistic cooperative terms, left no other course available to either side except impromptu sub-rosa intelligence warfare.

From the beginning of the relationship between the Soviet Union and the United States, espionage, adversary intelligence activities and disinformation was to be the norm rather than the exception. In a special supplement to the *New Republic* of 4 August 1920, Walter Lippmann and Charles Merz reported the results of a detailed survey of the *New York Times* from March 1917 to March 1920, which came to about one thousand issues. One conclusion illustrates their findings,

> From the point of view of the professional journalist the reporting of the Russian revolution is nothing short of disastrous. On essential issues the net effect was almost always misleading and misleading news is worse than none at all.
>
> For example, on 12 February 1919, the *New York Times* reported that under Bolshevism, women were nationalized. They were decreed to be nothing but property of the state and made available for indiscriminate use by any Russian man.

This erroneous reporting was given wide currency in the United States and the Western world. Many otherwise well informed and intelligent people believed these lies and it helped make objective dealings with the Soviets more difficult.[118]

On 8 November 1917 Trotsky called for peace between the Allies and Germany. The Allies ignored his overtures. Trotsky decided to proceed without the Allies and began to negotiate a separate peace for Russia with the Germans.

Probably to try and force the Allies to lose face since they would not join his peace initiative, Trotsky, the master of the ploy and dissimulation, published documents that had allegedly just been discovered in the Czar's secret archives. These documents purported to show predatory aggrandizement on the part of the Allies in seeking the Czar's cooperation to prolong the war against Germany. These documents, it was alleged, proved the Allies wanted Russia to remain in the war against Germany so that the Russian economy after the war would be too weak to compete commercially against the Allied businessmen.

On 3 March 1918 Trotsky, disregarding Allied objections and representing only Russia, began peace negotiations with the Germans at Brest-Litovsk. Trotsky purposely temporized at Brest-Litovsk issuing inflammatory statements that the Soviets wanted "Neither peace nor war!" The Germans denounced Trotsky's dilatory tactics and proceeded to launch a military offensive. Trotsky discontinued his delays and admitted that Lenin was probably correct in seeking peace with the Germans.[119]

During these chaotic times the American and British governments lacked adequate reporting sources — people who could supply them with trustworthy information from inside Russia. They therefore had to develop covert reporting sources and channels for the transmission of the information thus obtained back to their respective capitals. The degree of covertness of these hastily created operations, in most instances, was palpably thin. Most of the so-called covert sources were cloaked only with a veneer of cover which left their sponsoring governments only a facade of deniability.

There were essentially three United States espionage networks operating in Russia against the Bolsheviks:

1. State Department: Quasi-legal, under official cover as the Root Mission and the Red Cross Mission. Illegal under commercial cover such as Dmitri Kalamatiano.[120]

2. United States Army's Military Intelligence Department, (MID).[121] Operated under the military attaches system. Colonel James A.

Ruggles, Jr. was the Military Attache in the American Embassy in Petrograd. There was a covert commercial cover unit as part of MID which included Marguerite Harrison working under cover as a reporter for the *Baltimore Sun* and former aviator with the Lafayette Escadrille, Captain Merion C. Cooper flying with the Kusciazko Squadron in Poland. There were many other commercial cover agents working in Russia at this time.[122]

3. Office of Naval Intelligence — The Naval Attache, Vice Admiral Newton A. McCully was assigned to the American Embassy.

Most United States intelligence operations during the early days of the Bolshevik assumption of power were probably undeniable if they were specifically challenged by the Soviets. The lack of confrontation might have been due to the fact that the Soviets knew about most of these operations and decided to leave them alone while they continued to monitor them. It could also have been because the Soviets did not have the time, and maybe not even the inclination, to call the bluff of the United States.

From contemporary reports, many of the United States's operations were palpably ill-run and must have been obvious to the Cheka—the Soviets all purpose counterintelligence and intelligence corps. The Soviets viewed Americans' efforts in the arena of intelligence collection as amateurs trying to compete against professionals and not doing too well. The Bolsheviks, on the other hand, had been steeped in clandestinity and had become adroit in practicing secure tradecraft.

Many of the so-called clandestine operations of the United States government in Russia, and the covert funding channels they tried to use, were probably known to the Cheka. Other United States operations had a comic opera aura, with wolf hounds occupying opera seats alongside their master. This kind of shenanigans did much to discredit the entire American effort in Russia. The diplomatic community in Petrograd, and the Russian man on the street, ridiculed and laughed at the American efforts. The Soviets did not have to do a thing to play up the obvious ineptitude of the Americans in trying to be spies.

Conversely, at the beginning of the Russian revolution, the Bolsheviks did not have any clandestine intelligence gathering mechanism in the United States. Trotsky, and the other Soviet leaders, could not understand the United States's decision not to allow them diplomatic accreditation. They wanted to proceed, as all other normal diplomatic missions did, to collect intelligence semi-overtly through their ambassador and attaches operating out of their own embassy in Washington.

With the entry of the United States into the war, President Wilson and his Cabinet determined on a policy of suppression toward those who opposed war. The Espionage Act was fortified by the Trading With the Enemy Act in October of 1917 and the Sedition Act of 16 May 1918. These prudent measures were perceived by some as draconian. One wag put it, "They can now give you ninety days for quoting the Declaration of Independence, six months for quoting the Bible and pretty soon, someone is going to get a life sentence for quoting a speech by Woodrow Wilson."[123]

What intelligence collection the Bolsheviks eventually developed in the United States during the Bolshevik revolutionary period made no pretense at dissimulation by using any type of cover for their operations. Actually these groups, few in number, were initially only sympathetic to the sufferings of the Russian people under the Czar and idealistically against autocracy. None of these groups could, at the beginning, be considered pro-Bolshevik and consequently motivated to collect intelligence against the United States for the new Soviet government. However, inside each of these groups were certain pro-Bolshevik factions who, from time to time, did try to induce their organizations to collect intelligence. The primary objective of this collection effort was to get information on the intentions of the United States toward the new Soviet regime.

One of these groups, "The Friends of Russian Freedom," who boasted Mark Twain among its members, had some adherents who were sympathetic to the general aims of the Social Revolutionaries in Russia. They had little knowledge of the doctrines of Marxism as espoused by the Bolsheviks, or perhaps they would not have been so cooperative. Regardless of their political persuasion these groups could more accurately be characterized as overt propaganda agencies for the Soviet government and not clandestine intelligence collectors.

The Friends of Russian Freedom fostered an empathy in the United States for the Russian people and a feeling of brotherhood with all the revolutionaries. They made no attempt to distinguish between the Social Revolutionaries, led by Alexander Kerensky whom they genuinely admired, and the Bolshevik communists, who were largely unknown to them. As a result, of the all-encompassing pro-revolutionary propaganda of The Friends, the average American began to feel that there may be a common bond among all Russian revolutionaries. It was believed they were simply fighting together to overthrow the Czar. All that was needed, it seemed to the American members of The Friends, was to be against Czarist tyranny and one was, ipso facto, against the same kind of tyranny the British had exerted over the American colonists before the American Revolutionary War.

Another group that was molding American public opinion was made up principally of Jewish emigres from Russia. These emigres had settled on the eastern seaboard of the United States. This group, of less than one hundred members, made up a very effective vocal presence with close political ties to the government power base in Washington. This Jewish group was essentially the same as The Friends in their political philosophy but were more vehement in their denunciation of the Czar's policies of racial discrimination and social persecution of the Jews. Their propaganda to those ends had a subjective pro-Bolshevik bias.

The net effect of this propaganda barrage from these groups was that the Bolsheviks were the ultimate beneficiaries of the general pro-revolutionary empathy in the minds of the average American. There was therefore a contrived sympathy engendered in America for the success of the Bolshevik revolution. (However, any pro-Bolshevik sympathy among American Jews was soon replaced by enmity when the anti-Jewish discrimination policies of the Bolshevik/Soviet government became evident.)

These pro-Russian groups in the United States were unable to exert any influence against President Wilson's decision to recognize the provisional government. David R. Francis had been the United States Ambassador duly accredited to the Czarist government. After the Czar's abdication, Francis remained accredited to the Kerensky provisional government.

Shortly after the United States recognized the Kerensky government, other circumstances compelled the United States to enter the war against Germany. The United States fully expected that Kerensky, as the leader of an Allied power, would keep Russia in the war against Germany. In order to secure Kerensky's commitment the United States gave the Kerensky government $350 million dollars in credits. To keep the United States government informed concerning conditions inside Russia and how their credits were being used, the United States insisted on sending several groups of observers to monitor the situation in Russia.

The first such group was the Root Mission under the direction of a distinguished lawyer and former Secretary of War, Elihu Root. The ostensible purpose of the Root Mission was to observe whether the Kerensky government was adhering to the terms imposed by the credits. Also, and most implicitly, they were to report to Washington everything they heard or observed about political conditions. Stripped of all niceties, or any veneer of obfuscation, the Root Mission was basically an intelligence gathering operation.

To assure expert and properly evaluated intelligence from the Root Mission, President Wilson selected a formidable team of specialists: James Duncan, the vice president of the American Federation of Labor, was to report on Russian socio-economic matters and the reliability of Russian labor to support their war production; Charles Wilson, a veteran newsman, was to report on civilian morale and the degree of Russian popular support for the war against Germany; Samuel Seton, a banker, was to report on the financial stability of the Kerensky government; and General Hugh Scott, the retired Chief of Staff of the United States army, was to report on Russian military capabilities and their potential to continue the war against Germany.

In addition to the Root Mission, President Wilson sent a Commission of Railway Experts under General John Stevens, to report on the conditions of Russian transportation and its ability to carry the military supplies needed to continue the war against Germany. As it turned out, the Kerensky government was not completely devoid of perspicacity about the true intelligence gathering purposes of the railway experts. Kerensky specifically told the United States government that his government did not want any American railway experts investigating the Russian railway system.

Without a moment's hesitation the Untied States government shot back a reply to Keresnky, "No railway experts — no $350 million dollar credits!" The railway experts were on the next ship to Russia.

Private American groups, ostensibly with no association with the United States government, also sent missions to study the "Russian situation." They reported to their respective parent organizations in the States who, in turn, passed pertinent information to the American government. To contend that some of these groups were not cover for the collection of intelligence is not only unrealistic but denies the sagacity of the American government. The American Red Cross was one such group.

The American Red Cross Mission to Russia was, in addition to its philanthropic purpose, a cover mechanism for American intelligence. Major Raymond Robins was officially the deputy chief of the Red Cross Mission but, in reality, was the chief of the unit of the Intelligence Department of the U.S. Army. His secret directive was to keep Russia in the war against Germany.[124] The real chief of the military section of the Red Cross Mission, Colonel William Boyce Thompson, did not want Robins attached to the Mission but was overruled. (Robins and the chief of British intelligence in Russia, Bruce Lockhart, became close personal friends in addition to professional allies on intelligence matters. They routinely had breakfast together and consulted with each other on their respective operational plans.)[125]

Headed by Dr. Frank Billings, a group of doctors and nurses under the auspices of the American Red Cross arrived in Russia shortly after the start of the Kerensky period. Unfortunately, similar to the railway experts, if there was one thing the Russians had an abundance of, it was doctors and nurses. Keresnky politely told the American Red Cross that he did not want any more doctors and nurses. At the urging of the United States government, who badly needed the broadcst reporting base possible inside Russia, the American Red Cross ignored Kerensky's objections and sent the Red Cross Mission to Russia.

Kersenky adamantly objected that Russia did not need nor want any more American semi-private groups wandering loose inside Russia. His protests went unheeded mainly because no one in the United States government bothered to pay any attention to them.

What the Russians were complaining about can best be illustrated by the activities of the Root Mission. From the time of their arrival in Petrograd the Root Mission had done little except pester members of the Kerensky government for information and lavishly entertain the diplomatic corps at expensive formal dinners. In fact, if President Wilson's intention was for the Root Mission to favorably sway the Kerensky government towards the United States position, in addition to providing intelligence, he was to be bitterly disappointed in both respects. Similarly, there was no worthwhile intelligence forthcoming from the railway experts nor the Red Cross.

The United States Embassy was also unable to consistently provide worthwhile intelligence about events inside Russia. The Embassy was caught unprepared and had not been able to coordinate any long range intelligence forecasts concerning the stability of the Kerensky government. Consequently when Kerensky was overthrown by the Bolsheviks, President Wilson and his policy planners in the Department of State, were surprised. They were unprepared to either assimilate the significance of the event or to sympathetically respond with any degree of understanding.[126]

Ordinarily, in obtaining intelligence about Russia and the impact of the Bolshevik revolution, the United States policy planners would have depended upon reports from the American Ambassador in Russia and his Political Section in the Embassy. Unfortunately, Ambassador Francis was not the most adept reporter, and the Political Section, petulantly at odds with the Ambassador, was not doing a good job of supporting him.

In addition, the credibility of Ambassador Francis's reports suffered from the patina of his personal life. Although no one could doubt the Ambassador's integrity, sincerity or honesty, he had acquired the reputation of a cigar-chomping, cuspidor-spitting, whiskey-drinking, midwest American hick among his tea-sipping, cookie-pushing colleagues in the Petrograd diplomatic community.

Ambassador Francis's subordinates in the Political Section in his own Embassy also tended to ostracize him because he was not really "one of them" (he was not a true professional striped-pants diplomat). This atrocious behavior toward Ambassador Francis by his own staff was based, in part, on the old school tie concept, and soon-to-be-discarded tradition of the American Foreign Service.

Although the Embassy staff should have remembered that Ambassador Francis had been appointed by the President of the United States to be his official representative and the honored sole representative of the country of the United States of America in Russia, the Ambassador, to a large degree, contributed to his own demise by having a torrid, back-door, penny-dreadful, illicit romance with a femme fatal improbably named, Matilda de Crum.

The Petrograd diplomatic community, and the Foreign Service officers' loose tongues inside the American Embassy, wagged mightily about the alleged adulterous relationship of the pompous American Ambassador with the spirited Matilda. Of course, the Bolsheviks too heard the gossip and must have been delighted at the public embarrassment of the staid representative of the great and powerful United States of America.

Ambassador Francis, who was sixty-five years-of-age, and should have known better, had apparently succumbed to the charms of the beguiling Matilda, who was a vixenish forty-six. Under the influence of a shipboard romance while they were both journeying to Russia from the United States, to use a cliche, "love blossomed." (There is evidence that Matilda connived to get on the same ship with Ambassador Francis for the sole purpose of seducing him.)

Matilda had two sons who were Russian naval cadets. She was supposedly returning to Russia to visit her sons. Her current husband, a German national, did not accompany her on this trip. He preferred to remain in the United States and probably with good reason. Apparently Ambassador Francis had not been properly briefed, either by his Political Section or the Department of State, regarding the potential dangers of someone like Matilda could represent.[127]

Maybe the Ambassador's Political Section did not have the right kind of investigative reporting sources who would be able to develop information about Matilda's antecedents or her current highly suspicious associations with German intelligence. However, Trotsky and the Bolsheviks knew all about Matilda and her connections with German intelligence.[128] (Although the Embassy Political Section did carry on a back door liaison of sorts with the Bolsheviks, it was not sufficiently close for the Political Section to inveigle the information about Matilda from Trotsky and the Bolsheviks.[129]) They had reliable information that Matilda was a German espionage agent. Matilda's husband, whose name was probably von Kram, had worked in Russia for a German owned insurance company that was suspected of having been a cover for German espionage, for many years before the Russian revolution.

The allegation that Matilda's husband's company was suspected of being a cover for German intelligence should be considered in its proper perspective. Many times Russian companies circulated such rumors on purpose. The Russian companies published such stories about German and other foreign companies doing business in Russia in an effort to discredit them in the eyes of the Russian people. The Russian companies hoped that the foreign firms thus discredited, with the smear of being a cover company for a foreign intelligence service, would make the Russian government revoke their licenses. Foreign competition, thus deprived of permission to do business in Russia, would have to pack up, pull out and leave the market for domestic Russia business only.

Russian companies also hoped that after the war the average Russian customers would switch their shopping allegiance from the discredited foreign companies and patronize only the home grown Russian firms. On balance, however, when Matilda's operation is added to the existing evidence, von Kram's German insurance company probably was a cover for German intelligence.

Matilda had helped verify the suspicion that she was a German espionage agent by her previous travel from Russia to the United States on the same ship with Mirolyubskaya, a Russian lady of dubious repute. The Soviets later prosecuted Mirolyubskaya as a German spy for revealing Soviet state secrets to German intelligence. It was revealed during her trial that Matilda da Crum had arranged to debrief Mirolyubskaya about Soviet military matters during their ocean voyage. Matilda subsequently passed this information to her case officer in German intelligence.

Eventually, the rumors about Ambassador Francis and Matilda took a more sinister turn. The Ambassador's bedroom in the Embassy was next to the top secret code room. All of the Embassy's classified dispatches, codes and the cable chronos were stored in the code room. It was alleged that the Ambassador and Matilda spent considerable time together in his bedroom.

The use of codes and ciphers for secret communications between the Embassy in Petrograd and the Department of State in Washington (and between the Embassy and lateral Consular posts) made them particularly sensitive. The potential irreparable damage to the United States's negotiating position in all matters affecting relations between the two countries could have been at stake if the ciphers and codes were compromised to the Soviets or Matilda's sponsors in German intelligence.

The potential danger of compromise was well appreciated by all governments and intelligence services of the period. For example, this concern was always uppermost in Trotsky's mind and he zealously guarded the Soviet military and Foreign Ministry communications. As a matter of fact, the first clear victory that can be attributed to signal intelligence in modern warfare occurred during this period of World War One. The Germans used signal intelligence to defeat the Imperial Russian armies of the Czar at the battle of Tannenberg. This debacle occurred when the Russian military codes were inadvertently compromised. The Russian General Staff knew they must change the compromised codes but neglected to distribute, by secure means, their new codes to the subordinate military units. Consequently, orders for a grand offensive had to be radioed by the General Staff to all subordinate units in clear text (plain language) messages. The German radio interceptors thus became fully aware of Russia's plans through their radio monitoring of this Russian traffic. The German General Staff was able to devise an effective counter-strategy to thwart the Russian army's offensive. In the ensuing battle of Tannenberg the Russian armies were blocked at every turn and eventually lost over a quarter of a million soldiers. This massive and tragic loss was the start of the Czar's eventual removal, Russia's downward slide into subsequent revolutionary chaos and the ultimate emergence of Bolshevik communism.

The possibility of a serious diplomatic reversal to the United States government should Matilda da Crum steal the American codes was uppermost in the minds of Ambassador Francis's staff. That this matter was of genuine concern to them cannot be doubted. It was also probably a source of watchful and benign amusement to Trotsky and the Cheka. In addition, if Matilda were to succeed in stealing the American codes, it represented a potential windfall for the German intelligence service.

If the Foreign Service Officers in the Embassy had rallied around their Ambassador in his time of need and given him their support and good counsel, this dangerous situation could have been defused. They could have discreetly suggested that the Ambassador move his trysts with Matilda to another room in the Embassy or, preferably, off the premises to a hotel. Unfortunately, no one on the staff was personally close enough to the Ambassador to have that kind of confidential rapport with him. When he first arrived in Petrograd, Ambassador Francis had made friendly overtures to most of the staff but had been rebuffed.

All sections of an Embassy in a chaotic situation, such as the revolutionary period in Russia should work closely together as a team to obtain the comprehensive reliable information for the planners in the White House to act upon. Instead the Embassy staff, and Matilda da Crum, torpedoed Ambassador Francis. They thus greatly reduced the overall effectiveness of the entire United States Mission in providing President Wilson with a trustworthy intelligence product.

The result of all this bedroom intrigue was a deplorable atmosphere of suspicion about everything the Ambassador reported to Washington. Much to their discredit, this suspicion was fostered and abetted by some of the Ambassador's own subordinates by private correspondence embellishing the Matilda caper to their cronies back in Washington at the State Department's Russian Desk.

Taking their stance from the American Embassy's own staff, the members of the diplomatic community in Petrograd viewed the Ambassador's presence and utterances with barely concealed smirks of bemusement and disinterest. More importantly, the Soviets did everything they could to exacerbate the situation with a calculated disinformation campaign of embellished rumors and innuendo. They tried hard to further discredit the Ambassador with vicious black propaganda attributing other gross sexual depravities to him.

During the Kalpashnikov affair, Trotsky berated Ambassador Francis in a speech for not dealing directly with the Bolshevik government. He claimed that the Ambassador had maintained a "golden silence" as far as any direct dealings with the Bolsheviks were concerned since the time of the revolution. During a long tirade in Trotsky's usual nasal baritone voice, he suddenly broke into an elaborate falsetto voice stating that, "Sir Francis will [now] have to break his golden silence" to answer the charges of the Kalpashnikov matter. The crowds applauded Trotsky's charade and heartily approved this denigration of the American Ambassador with its obvious implication that he was a sissy and possibly a homosexual.[130]

There is no doubt that the Ambassador's problems had an effect on the quality of the intelligence product coming out of the American Embassy in Petrograd. There were other factors, such as a multiplicity of reporting agencies, that further debilitated the reporting efforts of this already confusing and murky dilemma in the Embassy.

For example, originally sent to Russia as a member of the Root Mission, Brigadier General William V. Judson, was co-opted away by the United States army to be the Military Attache of the American Embassy in Petrograd. In addition he was appointed by President Wilson as the Chief of the American Military Mission to Russia.

This duality of positions was both a curse and a blessing, depending on whether it was viewed from Ambassador Francis's position or from General Judson's. As the Military Attache on the Embassy staff, General

Judson was technically subservient to the Ambassador who would normally have approval authority over Judson's cables to Washington as part of the Embassy's overall reporting procedure. As Chief of the Military Mission, Judson was his own authority with an independent code to send his reports back to Washington that the Ambassador was not routinely entitled to read. (Technically, as the President's representative in Russia, Ambassador Francis could have insisted on approving authority over any communication with Washington. The resulting hassle that would probably arise if this request was challenged by the military in Washington made it realistically not worthwhile.)

This confusing situation of military reporting excluded Ambassador Francis from knowing what was being reported about the internal Russian situation through his own Embassy when Judson reported wearing his Chief of Military Mission hat. Effectively the Ambassador was excluded from knowing what intelligence and policy statements were being sent to Washington by his own Military Attache in his other capacity as Chief of the Military Mission. Clearly, this was a situation that the Ambassador should have protested and that Secretary of State, Robert Lansing, or even President Wilson himself should have corrected for their own benefit of eliminating confusion in the reporting coming out of Russia.

Previously Judson had served for five months as the United States observer with the Russian army during the Russo-Japanese War. There, he had become close friends with a Russian military officer named Guchkov. Later, when Kerensky became head of the provisional government, Guchkov was appointed Minister of War. In Petrograd Guchkov was a valuable and prolific source of information for Judson about the Russian military capabilities and the conduct of the war with Germany. Guchkov also kept Judson acquainted with happenings inside the Provisional government.

When Kerensky was deposed, Guchkov also had to leave his position as Minister of War. Judson was a pragmatist and when his source was no longer productive, he realized he would have to use whatever influence he could with the Bolsheviks to keep Russia in the war against Germany.

The only instruction Judson received from Washington regarding his future relations with the Bolsheviks was a message expressly forbidding him to have any contact with Trotsky. Regardless of this admonition Judson was an astute reporter of what was going on in the

Bolshevik camp. He wrote lengthy reports about the wily Trotsky which Washington read with considerable interest. Washington was convinced that Trotsky would succeed Lenin. Despite the validity of Judson's reports he knew that President Wilson was not reading, or being briefed on, the information in his reports.

When Judson returned to Washington in February of 1918 he attempted to give President Wilson an objective assessment of the confused Russian puzzle. President Wilson ignored him.

Judson was an honest and brilliant professional military observer. He believed that cooperation with the new Soviet government was not only possible but highly desirable if that cooperation was honestly and prudently offered by both sides. Judson resented the Foreign Service Officers in the Embassy whom he believed to be basically anti-Soviet. They apparently lacked the sense of responsibility and professional perspective to seek a rapproachment with the Soviets. This seemed to Judson to be preventing the United States government from cooperating with the Soviets, for what he believed would be the ultimate benefit of both nations.

Another semi-official United States group in Russia, the Committee on Public Information (CPI) under the direction of George Creel, was attached to the Embassy, This Committee later became the Office of War Information. At times its reporting was at odds with that of Ambassador Francis and with that of the Embassy's Political Section. The addition of another disruptive reporting source that further riled the informational confusion produced a situation back in Washington that was close to catastrophic.

The Committee on Public Information was originally established to conduct wartime propaganda and censorship. It gradually, but persistently, assumed responsibility for counter-espionage and military and political reporting. It encroached on existing reporting jurisdictions and refused to coordinate its reports with the agencies it duplicated.

George Creel was a feisty journalist and a good choice to head up the CPI with all of its diverse duties. He had the ear of President Wilson and was able to win support for his views on Russia to the detriment of some of the reports from Ambassador Francis and the Political Section of the Embassy. Reports that differed in reasoning and projected outcome are welcomed in the analytical process but they must be weighed as to validity and the sources' possible bias; they must be coordinated within the community of reporting agencies as only a central intelligence type agency can do. Otherwise divergent reports, such as those from the CPI, only add to the confusion and prevent prudent and logical response.

Colonel House, President Wilson's unofficial special assistant at the White House, commented to a friend that he believed Creel's reports were prejudicing President Wilson against Secretary of State Lansing's views on Russia. This preference might be understandable in view of the lack of good reporting coming through Lansing from the Embassy; the reports from the Embassy that Lansing was reading could be no better than what the existing staff (preoccupied with their own problems) at the Embassy produced.

Regardless, if the allegation was true, that President Wilson was spurning the advice and counsel of his own Secretary of State, it was an intolerable situation and ripe for mistakes to be made in foreign policy planning. President Wilson should have been relying on the best reasoned assessments available to him in planning foreign policy and, in turn, being responsible for its execution. He could not carry out these functions without reliable timely intelligence. Unfortunately he had to take his intelligence, well sourced or fabricated, evaluated or ad hoc, slanted or objective, from whatever group he could get it. He had no way of knowing what kind of intelligence he was receiving!

Backstopping Creel on the Washington desk of the CPI was Arthur Ballard, before he also went to Russia on a tour of duty. Ballard was a respected journalist and alleged to have been one of the best informed men about Russia in Washington. Ballard had been an official in the anti-Czarist association, American Friends of Russian Freedom.

When Ballard had been in Russia before his assignment with the CPI in Washington, he had the use of considerable funds supplied by the American Friends of Russian Freedom. He overtly used these funds to help the Social Revolutionaries of Alexander Keresnky's provisional government stay in power. There was nothing subrosa, or any taint of United States covert funding, associated with this financial help.

Ballard himself was an honest, hardworking idealist and apparently was only following his own liberal socialist philosophy during his first tour of Russia. Ballard was a close personal friend of Colonel House. When he was in Russia before World War One he sent voluminous, well-reasoned intelligence reports on the Russian situation back to Colonel House.

Much might be conjectured about such an arrangement today. The situation would probably convince skeptics that some of today's operations in third world countries were not the first time that clandestine intelligence operations were being run out of the Chief Executive's Oval office independently and without the knowledge or concurrence of the Department of State, the War Department (Defense Department today) or both Houses of Congress.

Ballard was only another of the many reporting sources available to President Wilson. Many of the reporting sources had separate agendas and were basically conflicting with reports from the Department of State. Most sources submitted only raw and unevaluated reports and some blatantly advocated a selfish course of action. Some reports were the subtle product of Bolshevik or Allied disinformation and others were from paid informants who would report anything the payor demanded. Of course, there were also some reports that were factual and reported a reasoned point of view. These reports were usually lost in the shuffle and melee that surrounded the influx of paper received by President Wilson each day.

The variety of reports forced on the President served only to confuse him regarding the true situation in Russia. This cornucopia of information might be excused for any one of a number of bureaucratic reasons. In retrospect, however, the plethora of confusing and conflicting reports had a detrimental effect on United States foreign policy planning.

Some historians argue that President Wilson's perception of how to plan wise policy vis-a-vis relations between the United States and the Soviet government was unduly anti-Soviet and biased. This was allegedly due to the prejudices of the sources reflected in the unevaluated reports. This may or may not be true, but for every anti-Soviet report there was usually a corresponding pro-Soviet report from well meaning socialist liberals.

Arthur Ballard's unfavorable report on this rudimentary publicity work being done in Russia, plus the Root Mission's recommendation for a massive propaganda campaign, resulted in Edgar Sisson being sent to Russia as the special representative of the Committee on Public Information. Sisson was the former city editor of the *Chicago Tribune* and a former managing editor of *Colliers*[131] and Randolph Hearst's *Cosmopolitan* magazine.[132]

In 1919 the chief war planner for the Chief of Naval Operations, Captain Harry E. Yarnell, recommended the best method of purging the United States of "sinister Bolshevik influences" was to declare war against the Soviets and conduct treason trials for domestic "communist sympathizers." In December of that year ONI produced a report that there was a "massive nationwide plot to take over the entire country." The plot was to be led by anarchists Emma Goldman and Alexander Berkmanm, along with "German and Russian Jews," "Mexican bandits," "IWW subversives" and a "Japanese master spy." This uprising was scheduled to take place in January or February of 1920. In retrospect it is difficult to imagine that this information was published as an official intelligence report by ONI.[133]

No attempt is made to argue that President Wilson's policy planning took an anti-Soviet stance because of biased reporting. This ignores the basic obstacles to detente previously established in fundamental Marxist doctrine. At that time any harmonious peaceful continuum of understanding sought between Russia and the United States that was rooted in such diverse political philosophies was almost impossible.

In June of 1917 Ballard left Washington for Moscow but moved to Petrograd in November when an office of the CPI was opened there. There is some indication that although Ballard was supposed to be a subordinate employee to Creel, he was secretly in charge of the whole operation from the start of his Russian appointment. Later he was officially confirmed as being in charge of the CPI and this may have been the intent of Colonel House and President Wilson from the start. They wanted a man in charge who they could trust to send them reliable information about Russia.

Meanwhile, Ambassador Francis remained in Russia until July of 1918. Periodically he would issue statements calling upon the Russian people to overthrow the Soviet government. Just before he left Russia for the States, Chicherin, the Soviet Commissar of Foreign Affairs, gave him a message extending greetings to the American people. Ambassador Francis said, "The message was made for pacifists. I did not transmit it."[134]

As well as the sources of information already discussed, many of President Wilson's contacts with foreign governments were accomplished through Colonel Edward M. House, his "irregular agent." Colonel House was President Wilson's aide who was firmly ensconced at the White

House as the President's personal emissary. Colonel House had an influence over President Wilson that was out of all proportion to the wisdom of his counsel.

President Wilson's previous reliance on an unofficial emissary, such as Colonel House, for his background information and negotiations in foreign affairs was well known and accepted in international diplomacy. This previous acceptance may have induced him to rely equally on such unorthodox and unreliable systems of untrained, ill-equipped, informal intelligence gathering organizations that he sent to Russia, such as the Root Mission. It does offer one explanation for the inept and confusing, obviously uncoordinated and unevaluated, information President Wilson had to rely on regarding future intentions of the newly created Soviet government of Lenin and Trotsky.

In sum, President Wilson was not receiving the hard intelligence he needed for proper planning from these unorthodox semi-private American groups in Russia. His official intelligence collection apparatus in Ambassador Francis's Embassy in Petrograd was not producing hard intelligence either. President Wilson turned to Colonel House, upon whom he knew he could rely to at least obtain some information he could trust. House, in turn, relied upon the same semi-private reporting groups such as the Root Mission, the American Red Cross, the Railway Mission and other similar sources.

Meanwhile, in the summer of 1919, without any declaration of war, the armed forces of fourteen countries had to send troops to Russia to preserve the peace and protect their citizens.[135] This entire affair raised monumental problems for Trotsky.[136]

THE BRITISH AND TROTSKY

> The tale of history forms a very strong bulwark against the stream of time, and checks in some measure its irresistible flow, so that, of all things done in it, as many as history has taken over it secures and binds them together, and does not allow them to slip away into the abyss of oblivion.
>
> Duna Comnena

There was much confusion in the American government community in Petrograd regarding the intelligence they were getting out of Russia at the time of the Bolshevik revolution. The British effort suffered under a different sort of problem. Their problem was ineptness compounded by confusion in the collection and the evaluation of their intelligence. The British Military Attache in Petrograd, Colonel Alfred Knox, was "not even civil" to General Hanbury-Williams, the head of the British Military Mission to Russia. Major Archibald Campbell, the representative of British intelligence was supported by the British Ambassador, George Buchanan against his feuding enemies the British Naval Attache, Captain Francis Cromie, and Colonel Knox. The situation got so bad that on 23 March 1915, Cromie telegraphed London demanding either his recall or that of Campbell.[137]

These British errors did them a singular, self-inflicted disservice. It contributed to their indecision about whether to accept the Bolshevik/Soviet government at face value and thereby prolong the sub rosa intelligence war.[138]

The British, however, were apparently satisfied with their intelligence collection effort and, with a typically British self-congratulatory air, commented:

> The best method [of collecting intelligence] seems to have been that used by the British, who selected men of a resourceful kind and allowed them to work out for

themselves the methods by which they found they could best achieve their individual tasks.[139]

Lloyd George, the British Prime Minister, was anxious to keep Russia in the war against Germany. The leader of the Provisional government in Russia, Alexander Kerensky, was committed to a similar position. Lloyd George felt secure as long as Kerensky was in power. He funneled cash to Kerensky though clandestine channels, mostly through Britain's intelligence service branch office in New York City.

Sir Samuel Hoare, the chief of the British Foreign Office intelligence section, returned from Russia to report that a military dictatorship was the best answer to the Russian problem. Hoare recommended Russian Admiral Kolchak who, Hoare said, was the nearest thing to an "English gentleman" he had found in Russia.[140] Major Robins, an American intelligence officer in Russia, protested this proposal but was brusquely told by General Alfred Knox, of the British Military Intelligence service, to keep his mouth shut.

The New York office of British intelligence was headed by Sir William Wiseman who was a close friend and confidant of Colonel House, President Wilson's principal aide. One of Wiseman's case officers was a young man, with pretensions toward a literary career, named W. Somerset Maugham. At that time Maugham was also gathering raw material for his later writing successes about espionage, notably his novel, popularly referred to as *Ashenden*.[141] However, Maugham's primary concern at that time was earning his livelihood as an intelligence operative. He showed none of the kinky homosexual behavior that later dominated his life as a famous author.

Maugham, who had just finished two years of duty as an British intelligence agent in Switzerland, agreed to an assignment in Russia that Wiseman had chosen for him. In June of 1917 he prepared to leave New York to be the British intelligence man in Russia. Wiseman solicited funds for Maugham's assignment from Colonel House, the confidential intelligence aide and general factotum to President Wilson. House contributed $75,000 to Maugham's operation which made Maugham a joint Anglo-American agent.[142]

As the state of the espionage art required in that era, Maugham carried an elaborate code with him for sending his messages. This code identified Trotsky as "Cole," Lenin as "Davis," and Kerensky as "Lane," and it required that Maugham sign all official cables as "Somerville."[143]

When Maugham arrived in Petrograd the British Ambassador, Sir George Buchanan, thought him incompetent, mostly because of his speech impediment (he was a stutterer). His relations with Buchanan improved however after he healed a breach between Buchanan and American Ambassador Francis. Buchanan thought Francis was "something of a roughneck" and Francis thought Buchanan was "a stuffed shirt." Maugham helped each realize they were both "regular guys."[144]

In October, Keresnky, who knew Maugham's true identity as a British intelligence agent, gave him a personal message for Lloyd George. In the message Kersensky described his position as leader of the Provisional government as precarious and asked for large amounts of material and money from Britain. Kerensky also demanded that Lloyd George remove the British Ambassador to Russia, Sir George Buchanan, whom he personally disliked.

Kerensky insisted that this message was so secret that it could not be entrusted to coded transmission. He demanded that his message not be written down but delivered personally, and only verbally, by Maugham to Lloyd George in London.

As soon as Maugham arrived in London, he met with Lloyd George at 10 Downing Street, the Prime Minister's official residence. In spite of Kerensky's admonition that the message was too secret to be committed to paper, Maugham, afraid that his stutter might impair the importance of the message should he try to deliver it verbally, had written it down. Maugham handed the written message to Lloyd George and patiently waited for his answer.

After silently reading Kerensky's message and considering it, Lloyd George raised his head and very quietly said that he could not comply with any of Kerensky's requests. Maugham, concerned about resuming his post in Russia, wondered if Kerensky's reaction to such a negative reply might be personally dangerous for him. Maugham's doubts were premature.

Several days before Maugham was to return to his post in Russia, Kerensky was deposed and Lenin and Trotsky had seized power for the Bolsheviks.

Lenin and Trotsky, and the Bolshevik party in general, were regarded by the British, and Allied missions in Russia, as traitors and agents of the German government.[145]

Lenin's declared policy was the immediate initiation of peace talks with Germany, as soon as the Bolshevik's got the chaotic internal Russian situation under control.

The British government immediately sent Mrs. Emmeline Pankhurst, the champion of women's suffrage in England, to Russia. They entertained the vain hope that, as a British "radical" she might help prevail upon the Bolshevik leaders to keep Russia in the war against Germany. "It must stand as one of the era's more preposterous diplomatic moves!"[146]

British intelligence suspected Lenin's reason for advocating the policy of unilateral peace with Germany. They surmised that he owed Germany a favor for having helped his secret return to Petrograd from Switzerland in the so-called "sealed train."

When the Bolsheviks seized power, the British government ordered the evacuation of all diplomatic personnel from Russian except Bruce Lockhart. They left him behind to be in charge of British representation. Lockhart, the former consul-general of the British Embassy, was ordered to keep the Bolsheviks in the war against Germany and to act as the housekeeper of British interests in the new Soviet communist Russia.

Lockhart originally proposed to the British government that they back Trotsky against Lenin to overthrow the Bolshevik government.[147] Lockhart expanded on this:

> Trotsky was a great organizer and a man of immense physical courage. But, morally, he was as incapable of standing against Lenin as a flea against an elephant. In the Council of Commissars there was not a man who did not consider himself the equal of Trotsky. There was not a Commissar who did not regard Lenin as a demi-god whose decisions were to be accepted without question.[148]

Lockhart's primary objective was to keep Russia in the war against Germany. By the time the Soviets made peace with Germany at Brest-Litovsk on 3 March 1918, London was already considering military intervention against the Bolsheviks. Trotsky told Lockhart scornfully,

"Your Lloyd George is like a man playing roulette and scattering chips on every number."[149]

On 3 December 1917, the British War Cabinet held a secret meeting. It was decided that "assistance should be forthcoming.... to support any responsible body in Russia [counterrevolutionaries] willing to oppose actively the Bolsheviks."[150]

The Director of British Royal Naval Intelligence in Russia, Captain Francis Cromie, was to maintain the official British naval presence in Petrograd. Together with Lockhart, Cromie was to prevent, at all costs, the Russian Baltic fleet from falling under German control. Cromie, to this end, recruited a network of saboteurs whose task it would be to cripple the Russian fleet by demolition if the German seizure appeared imminent.

Any subversion of the Soviet military by the British was made virtually impossible by Trotsky's order in 1919 establishing a system of political controls. Side by side with the military officers was a controlling political officer of the CPUSSR. Though Trotsky limited their authority to "political supervision" their actual duties had a much wider application.[151]

The British were in a critical situation with their diplomatic mission all but gone from Russia and no administrative base in which the remaining representatives could operate. The normal channels of communications for reporting on Russian internal affairs and Soviet intentions were non-existent; the British lion had virtually blindfolded himself in Russia!

In an attempt to correct this alarming deficiency, the British initiated a crash program to recruit Russian speaking British agents who could operate clandestinely in Russia. They used the training syllabus developed by the legendary chief of British intelligence, Menzies.[152]

As part of this recruiting campaign, British intelligence obtained the services of a British officer, Captain George Hill, to be their SIS chief in Petrograd. Hill had perfected his Russian language while serving as a British intelligence agent during the war at the Alexandra Palace Internment camp. Hill was disguised as a Bulgarian prisoner in order to wheedle information from Russian prisoners of war. He was later sent to Petrograd as the official British liaison contact with the Russian armed forces.[153]

In late 1917, soon after he arrived in Petrograd, Hill went on a mission on his own initiative — much to the embarrassment of the Russians. Hill intended to rescue the foreign cash reserves of the Rumanian government which were then stored in Moscow, and he did manage to get the money and return it to Bucharest. The grateful Rumanian government pledged their undying gratitude to Hill while the same action earned for him the undying enmity of the Soviets.

This unorthodox venture was only one of Hill's contributions. He never did become the voluminous reporting source on Soviet capabilities and intentions that British intelligence expected. Further, his Rumanian sojourn made him practically unacceptable to the Soviets.

Trotsky appointed Hill "Inspector of [Soviet] Aviation" and he was given extensive powers in that department. This brought him into close touch with the Soviet aviation personnel and access to all airports. He was also in touch with the Evacuation Committee. According to Hill, he gave Trotsky advice on the formation of a new Soviet air force and other important matters.[154]

At Trotsky's insistence Hill helped the Bolshevik military headquarters organize an intelligence section for the purpose of identifying German units on the Russian front and for keeping troop movements under close observation.[155]

> Within a few weeks, we had a complete net of agents working in all the Eastern territories occupied by the German-Austrian army. Identifications came to me [Hill] every day, and a copy of them was telephoned to the war Office in London. Time and again I was able to warn London that a German division had left the Russian front for the Western front.
>
> I arranged a Bolshevik counter-espionage service to spy on German secret service [agents] and missions in Petrograd and Moscow. Our interception organization worked well. We deciphered German codes, opened their letters and read some of their correspondence without ever being suspected.[156]

Then, like the proverbial bolt from the blue, came the answer to Britain's prayers for a really good agent inside Russia in the person of Sidney Reilly.[157]

Reilly was another agent acquired by British intelligence during their crash recruiting campaign.[158] At the beginning of World War One Reilly was a captain in the Royal Canadian Flying Corps. He had been born as Sigmund Rosenbaum, the bastard son of a Jewish doctor who had a practice of medicine in a small village near Odessa in Russia. While still a young man, Reilly emigrated from Russia to Brazil but soon had difficulties with the Brazilian authorities. He decided that flight to London was preferable than jail in Rio de Janeiro.

In 1899 Reilly married a pretty young Irish lass by the name of Margaret Reilly Callaghan. Maggie, as she was affectionately called by Reilly, was the widow of a wealthy Englishman. Reilly changed his name from Rosenbaum to Reilly and was known by this name for the rest of his known life.

Reilly was a bit of bounder, as the Brits referred to him, and in Reilly's case, with good reason. British intelligence would later denounce him as a "bloody cad." Reilly left Margaret shortly after their marriage and went to New York. He was in New York when the war broke out. During his short stay in New York he bigamously married a second wife, Nadine. He also managed to get himself appointed as a munitions purchasing agent for the Russian government. Reilly became friendly with Somerset Maugham's supervisor in New York — the chief of the British Intelligence branch there who was masquerading under cover as chief of the British Purchasing Commission, Sir William Wiseman.

In March of 1918 British intelligence co-opted Reilly from the Canadian Flying Corps and sent him to Bolshevik held Russia. In April he arrived in Murmansk but was promptly arrested by the British Royal Navy. Reilly's travel documents were valid and had been issued by the Soviet representative in London, Maxim Litvinov. Unfortunately Litvinov had misspelled Reilly's name on the documents as "Reilli." This typographical error had brought Reilly under suspicion of the Royal Navy who were checking the documents of all travelers in the war zone. Rear Admiral Kemp, the senior British naval officer in the White Sea at the time of Reilly's arrest, after checking with the British intelligence headquarters back in London by wireless, arranged Reilly's release.

Coincidentally, it was on 3 March 1918, that the Soviet government accepted the peace terms dictated by the Germans. Meanwhile, George Hill of British intelligence persuaded his agents to damage the coal mine shafts and destroy the supplies so as to prevent the German-Austrian army from acquiring them.[159]

Reilly supposedly developed a life long hatred for the Soviets during his time in Russia as a British intelligence agent. He also turned out to be, what was considered by his superiors in SIS, a good secret operative and reporting source. All commentators agreed that he was too mysterious about his own operations and too much of an independent agent to ever be considered a good team player. Maybe this should have been considered by his superiors.[160]

Reilly's code designator was ST 1 (Esti-1) but it was some time after he arrived in Russia before SIS received any message bearing that signature. Reilly's methods were unique. He insisted on operating without a senior case officer's direction although SIS desperately wanted to give him operational guidance and, many times, the re-direction of his efforts.

Reilly badly needed professional guidance. According to Eleanor Toye, one of his secretaries, Reilly used to suffer from severe mental delusions — one time he thought he was Jesus Christ.[161] Illustrative of his poor operational judgment, on one mission he used a French operative without coordination with the French intelligence service. This was out of bounds according to the informal agreement among the Allies not to use each other's agents.

This ill-used French operation consisted of a group of ex-Czarist officers whom Reilly employed to assassinate the German Ambassador to Russia, Count Wilhelm von Mirbach, on 6 July 1918. Reilly hoped that the assassination would provoke Germany to reopen hostilities against Russia, which was eagerly desired by Britain. They hoped Russia would act to repulse the Germans and effectively be back in the war against Germany.

Another assassination attempt attributed to Reilly's sub-rosa direction was on Lenin while he was attending the All-Russian Congress at the opera house in Moscow. The attempt failed however because Lenin had been alerted to the plan. The Cheka revenged the attempt on Lenin's life by the arbitrary slaughter of all suspected dissidents. Unfortunately, among the so-called dissidents were a lot of innocent on-lookers who were also indiscriminately rounded up and executed.

Reilly, allegedly fearing capture and execution for his complicity in plotting the attempt on Lenin's life, went underground. Lockhart, Reilly's erstwhile case officer, at least on paper and in theory only, was taken into protective custody by the Cheka. Lockhart was soon released without charges being brought against him, or of being formally accused of participating in the attempt on Lenin's life.

On 31 August 1918 there was a second attempt on Lenin's life but he was only wounded (he was hit by two bullets). As a result, the capricious retaliation by the Cheka intensified. Mobs, secretly organized by the Cheka, supposedly enraged by the assassination attempts on Lenin that had been masterminded by foreign elements according to Cheka propaganda, attacked diplomatic installations. Captain Cromie, the Naval Attache of the British Embassy, was murdered by the frenzied rampaging of these Cheka mobs while defending the old British Embassy building in Petrograd.

Shortly thereafter Bruce Lockhart was rearrested. This time he was in serious danger of being executed. Reilly, leading his usual charmed life, blithely walked through the Cheka dragnet and escaped from Russia across the Baltic Sea to Sweden.

The British government, in retaliation for Lockhart's detention, arrested the Soviet representative in Londom, Maxim Litvinov, as a hostage. Heeding the action of the stern British government, the Soviets had second thoughts about Lockhart's part in the attempt on Lenin's life and released him on 2 October.

Reilly arrived back in London to a hero's reception. He received the British Military Cross for his operations in Russia, even though these same operations had caused the retaliatory bloodbath by the Cheka against their own people. It is estimated that approximately 8000 alleged counter-revolutionaries were summarily executed in Russia because of Reilly's operation.

Reilly, Lockhart and Cromie were gone from Russia and their sub-agents liquidated as part of the Cheka purges. The Russian scene was then barren as far as any useable intelligence flowing into SIS headquarters in London.

The SIS then chose Paul Dukes, an Englishman who was working for the American Relief Administration in Samara, Russia, as their next candidate for spy master in Russia. This cross use of agents among the Allies could not be ignored by the Soviets. They surmised, probably correctly, that there was a joint agreement between Britain and the United States concerning shared intelligence sources.

Dukes had been a music student at the Moscow Conservatoire before the war. His ambition was to be a classical orchestra conductor. He was given the code designator of ST 25 by the SIS.

Dukes entered Russia from Helsinki, Finland. For ten months he sent reliable intelligence back to SIS in London. His reports were well reasoned, from apparently reliable sources, and voluminous. Dukes's Russian was so fluent he was mistaken as a native on many occasions. A group of Russian navy officers who knew Dukes also spoke English, wanted him to recruit British sailors in the Baltic ports.[162]

One of Dukes's covers was that he was supposed to be Joseph Ilytch Firenko, a member of the Red army. In September 1919 several of Dukes's agents were arrested by the real Cheka. As a result, Dukes's cover was blown and he had to escape back across the border into Finland.[163]

When he arrived in London Dukes was given a knighthood. His predecessor, Sidney Reilly, was present at the ceremony to congratulate him. Reilly, at that time, was attached to the British Foreign Office but was still covertly maintained on the SIS rolls, supposedly creating ideas to further harass the embryonic Soviet regime.

The SIS had concluded in September of 1919, mostly on the advice of Reilly, but also based upon their own estimates, that it would be difficult to maintain any viable covert intelligence network in Russia. The consensus was that the potential result was just not worth the expenditure in time, money and agent assets. What they had derived from all their efforts, and very high costs, had not been worth the end result.

After Dukes return from Russia he confirmed this prognosis. This decision was further made unanimous when it was learned of the rapidly improving counterespionage capabilities of the Cheka. Further, covert operations inside Russia were too dangerous for the personal safety of the SIS agents and when they were captured the resulting publicity was bad for the reputation of the British government. Each exposure of British duplicity in the field of clandestine blunders caused the public image of the redoubtable British efficiency to suffer badly. The reputation eagerly fostered by the SIS of its omnipotence suffered the most.

The heroism and tactical success of some of British intelligence's Soviet operations blinded the British leadership to their otherwise strategic failures. The greatest weakness of British intelligence was its failure to obtain and disseminate timely and valid intelligence. The British intelligence personnel performed great feats of attempted covert action operations but they did not have a single representative in Bolshevik controlled Russia to report on conditions and intercept intelligence failed to fill the gap.[164] No doubt Dzerzhinsky's Cheka contributed to this lack of success.

The SIS, again abetted by Reilly's suggestion, decided that there was another method of approaching the Soviet target. They reasoned that anti-Soviet operations could more effectively, and most certainly more safely for the British SIS agents, be carried out through anti-Soviet Russian emigre organizations. They hoped that by generous funding and discreet guidance through unattributable cut-outs they could use the emigres' sympathizers still inside Russia as reporting sources and action agents for sabotage.

The most likely emigre group for the SIS to use was headed by Boris Savinkov, a former Minister of War in the Kerensky government. Savinkov had worked closely with Reilly and Dukes while he was still in the Kerensky government. As the leader, Savinkov claimed more than 250,000 displaced White Russians in his group. He said that many of these emigres had friends and relatives still inside Russia who were willing and eager to help the cause that could result in the overthrow of the hated new Soviet regime.

Savinkov's headquarters were in Paris but he bragged that he personally lead guerrilla forays into Russia from Warsaw and Finland. With his devil-may-care bravado, and claims of a fighting force waiting to be unleashed, Savinkov quickly captured the imaginations of some of Britain's leading policy makers.

Reilly was Savinkov's best promoter. He did everything within his limited power to get more money and supplies for Savinkov's operations from British intelligence. Reilly never was able to deliver all the money and equipment that Savinkov kept insisting upon in larger and larger quantities. In addition, what was probably sound operational policy for the SIS, they did not want to endlessly pour money down the proverbial rat hole without seeing some results for their money. The intelligence product from Savinkov's group was not very impressive. It is axiomatic in western intelligence circles that emigre organizations never do produce on the lofty promises they make to prospective backers.

Reilly's motives in so relentlessly championing Savinkov's cause made the SIS suspect Reilly's real intentions. Was Reilly supporting Savinkov because he honestly believed Savinkov's emigre group could produce worthwhile intelligence from inside Russia? Could it be possible, on the other hand, that Reilly's brokerings for Savinkov were because Reilly was getting kick-backs from Savinkov to line his own pockets? Could Reilly have some ulterior motive, as yet unknown, wondered the SIS?

These suspicions finally made the SIS place their ace spy, Reilly, under surveillance in an attempt to resolve them. The SIS grew even more suspicious when Reilly kept insisting that the SIS give Savinkov more specific requirements regarding exactly what details they needed to complete their intelligence estimates of the Soviet strengths and capabilities.

Detailed requirements of this nature, if they are made known to the Soviets, would reveal exactly where the SIS needed additional information, and, more importantly, where the SIS already had agents in a position to supply the already known information. This, in turn, would enable the Cheka to quickly determine where within their midst was a spy and take steps to eliminate him forthwith. Another axiom in intelligence planning is never to allow the opposition to know one's intelligence requirement that may reveal gaps in coverage. Knowing what information was needed would only enable the enemy to take extra protective measures to deny access to the information.

SIS also had suspicions from the surveillance regarding Reilly's frequent contacts with the members of the London based Soviet Trade Mission. The SIS thought Reilly's friendly social relationship with these Soviets was strange in view of his often expressed hatred of all Soviets. They asked themselves, could Reilly really be in bed, so to speak, with the Soviet government on the one hand and with their enemies, the dissident emigre groups of Savinkov on the other? Further, for all of his supposed good contacts with the Soviets, and his often touted highly placed emigre sources controlled by Savinkov, there was very little intelligence coming through Reilly from either source.

The SIS finally rationalized the strange conduct of their so-called ace spy. They said that either Reilly had come under the influence of the White Russian community, personified by the Savinkov group, or he had gone over to the GPU, the successor to the Cheka, the Soviet secret police!

Reilly's defenders within the SIS suggested another possibility. Reilly, they said, could be playing a dual role of double agent without informing the SIS. He could be trying to develop reporting sources from both groups and bargaining information to the highest bidder, or to the side to which he was most loyal. His supporters claimed his ultimate loyalties, of course, were with the British government and the SIS.

Reilly's detractors were not convinced. They deduced from his life style and the obvious high level of conspicuous consumption of wine, women and lavish parties, that he was more likely lining his pockets with kickbacks from the funds the SIS was paying Savinkov.

Reilly's more loyal supporters in the SIS merely clucked their tongues in sympathetic understanding of Reilly's actions. They tended to excuse Reilly and suggested that, poor chap, he was probably burnt out from his long labors on behalf of Britain and the SIS in Russia. They also reluctantly admitted that Reilly might be unwittingly being used by both Savinkov and the GPU in the internal Russian struggle. It occurred to some of them that Reilly might really be working for the GPU against the SIS. Therefore they reasoned, the relationship with the emigres must be nothing but a cleverly contrived diversion to fool the SIS and probably fill Reilly's and Savinkov's pockets with British money.

In any event, Reilly's intelligence reports from both sources were becoming more and more at odds with what was being reported in Soviet diplomatic cable traffic. These transmissions that were being decoded by the British Government Code and Cypher School (GC&CS) revealed a different story than Reilly's reports. In other words, the light of revelation began to dawn at SIS headquarters.[165]

If Reilly had a Soviet case officer (maybe even Felix Dzerzhinbsky the head of the GPU himself in a case as important as this) he had become tired of creating essentially false but close enough to the truth to be believable intelligence for Reilly to pass to the SIS. The information passed in a double agent operation must be sufficiently real to fool the recipient intelligence service, and keep them placated and interested enough to continue to support the operation, without revealing any information damaging to the donor.

One decisive result of the turmoil brought about by the revolutions in Russia was that the anti-establishment purges had left everyone frightened. Members of the former establishment feared they would be personally detained without charges, their wealth confiscated. They might even be executed as counter-revolutionaries merely on suspicion that they were, or had been, rich or belonged to the intelligencia.

No one felt immune, for example, Trotsky chaired a meeting attended by Russian officers of the regular army together with some Bolshevik commissars. He was polite to the Russian officers, referring to them as "gentlemen" and was courteous to an extreme. He then turned on the Bolshevik commissars and poured forth on them a torrent of abuse much to everyone's amazement. He called the commissars "skulking slackers" and berated them for having their blouses unbuttoned.[166]

The Russian establishment as an on-going viable source of good intelligence for America and Britain was gone forever. The British were forced to develop the art of radio intercept and the breaking of Soviet government encoded cable and wireless traffic to a very high degree as their principal source of intelligence.

Much of the most valuable intelligence was provided to the British, by the Government Code and Cipher School (GC&CS). This was the decrypted traffic of the Soviet government. The Bolsheviks suffered during their first decade in power from two serious cryptographic handicaps. The first was their fear of relying on the relatively sophisticated codes and ciphers which they had inherited from the Czarist regime and their consequent introduction of less secure systems based at first on simple forms of letter substitution. The second handicap was the dispersion of the Czarist code breakers who had given pre-revolutionary Russia the world's lead in cryptoanalysis. Worse still, from the Bolshevik point of view, some of the best code breakers defected to the opposition.[167]

The ability of the British cryptoanalysts to break and read Russian wireless traffic started when a number of Russian cryptographers, the most notable being Felix Feterlain, escaped the chaos of revolutionary Russia for the tranquility of Britain. Feterlain was able to completely reconstruct most of the Czarist diplomatic codes. The new Soviet government in Moscow had merely taken over these codes intact and were using them on a daily basis to communicate with their representatives in foreign capitols. The British took full advantage of this entre' into the innermost thoughts and intentions of the Soviet leaders.

As so often happens in espionage work, the British politicians blew this valuable intercept operation. Maybe an argument can be made for the politicians' case for using intercepted traffic to embarrass the Soviets, but it is difficult to justify to the case officers who are responsible for future intelligence estimates. The momentary gain offered by disclosure of the information obtained by an intercept could hardly offset the potential future worth of an operation that might go on for years giving the SIS a ringside seat at all high level Soviet decisions.

Lloyd George, the British Prime Minister, learned from decrypted Russian radio traffic that the Soviet Trade Delegation in London was subsidizing the influential British newspaper the *Daily Herald* and using its pages for black and grey pro-Soviet, anti-British propaganda.

The Prime Minister believed that it was diplomatically necessary to expose this Soviet propaganda operation. On 15 October 1920 two honorable British dailies, the *Morning Post* and the *Daily Mail*, published the actual clear text messages from the Soviet government in Moscow to the Russian Trade Delegation representatives in London. This story, and texts of the messages, exposed the covert funding of the grey and black propaganda of the Soviets that had been appearing in the *Daily Herald*. There could be no doubt as far as the Soviets were concerned where and how these messages of theirs had been obtained. As a result, the Soviets changed their codes and this valuable source of intelligence was lost. The intercept program of the GC&CS desperately sought to find the key to break into these new Soviet codes.

Most simply explained, in order to "read" or "break" an enciphered or coded message, the jumble of apparently random letters and numbers in the message must be skillfully rearranged and replaced by other letters and numbers until the original message is revealed., This is a tedious, painstaking, frustrating, sometimes impossible task. Breaking the message is made as easy as child's play if the cryptoanalyst has a copy of the code or cipher that was used to originally scramble or encode the message.

Codes and ciphers of the opposition used in breaking the opposition's messages are usually obtained by black bag jobs, or more politely phrased, surreptitious entries. The opposition embassy, consulate, trade mission or wherever the codes and ciphers are securely stored, is illegally entered. The safe or vault where the codes or ciphers are stored is opened by an expert safe cracker and the codes and ciphers are photographed.[168]

Everything must be put back in the safe and in the room just as it was so that the opposition will never know that their codes or ciphers have been copied. If there is the slightest clue that would alert the opposition to the entry, and cause them to suspect that their codes have been compromised, they would stop using them. The whole entry operation, which takes a monumental amount of preparation, money and manpower, would then have been for naught!

The SIS was able to recruit a Volga German named Gregory who was a code clerk in Maxim Litvinov's Soviet Mission in the Estonian capitol of Finland. Gregory was not able to supply GC&CS with the actual codes the Soviets were using. He did, however, give them paraphrased messages which enabled the cryptoanalysts to break and read a considerable amount of Soviet traffic. This traffic revealed that the Soviets were instructing their representatives all over the world to give only lip service to the idea that the Soviets were not going to continue to strive for world domination through the spread of communism. In actuality, the Soviet representatives were specifically instructed to continue with increased vigor and renew their efforts to subvert the free world to communist domination.

This new entre' into the Soviet traffic of coded radio and cable messages, made available through Gregory's defection in place, revealed to the SIS that the Soviet Deputy Foreign Commissar, Maxim Litvinov, was covertly funding the Sinn Fein. The Sinn Fein was the revolutionary Irish Republican movement that was fighting the British for an independent Ireland.

Intercepted Soviet traffic also revealed that they were covertly subsidizing a notoriously violent anti-British group of Indian nationalists. The Comintern (Communist International) was providing these Indian subversives with guerrilla training through the local communist parties.

Possibly the bitterest blow to British pride that came from these intercepts, was when they learned that the Soviets were covertly doing everything they could to actively impede the British Colonial office in the administration of their colonies. The British had always assumed that their "little brown brothers," as Kipling referred to the residents of some British colonies, were supremely happy being under the yoke of British imperialism. It came as a surprise for them to learn that not all was well and quiet in the colonies — that some of the "little brown brothers" were susceptible to communist propaganda and wanted to fight the British for their independence.

On 16 March 1921 Britain signed a trade agreement with the Soviet government. This gave the new communist regime, for the first time, some official recognition in the world community. Part of this British-Soviet agreement bound both parties not to engage in subversive activities against each other.

Litvinov's secret funding of the Sinn Fein in Ireland and the covert support of the nationalists in India was a clear violation of this agreement. Since the Easter Sunday uprising in Dublin, Ireland in 1916 by the Irish Republican revolutionaries, the prospects of the Soviets entering this volatile situation in Ireland was a considerable worry for the British Colonial Office. Both the Prime Minister and the Colonial Office wanted to expose the Soviet meddling in these affairs. This time, however, they were constrained by the GC&CS and the SIS from publishing copies of the clear text of the intercepted messages.

A solution to the knotty problem of whether to reveal in clear text the Soviet cheating on the agreement came in time from another source. The SIS chief in Berlin, Timothy Green, had by "clandestine means" been able to obtain copies of the same documents from the office of the Soviet representative in Berlin. (Even in those days the reference to obtaining something by "clandestine means" usually meant the information had been obtained through a burglary or surreptitious entry.)

The stolen information revealed the same covert Soviet funding, that is, the secret financing of the Sinn Fein and the Indian nationalists. These stolen documents were sent by the British Foreign Office to the Soviets with a strong protest note from the British Prime Minister. The communists, as they are masters at doing, were able to wiggle out of the charges. They denounced the documents as forgeries and claimed they were not to be believed. Regardless, subsequent intercepted traffic revealed that the Soviets intended to keep on sending money to the Sinn Fien and to the Indian nationalists.

After several diplomatic reversals for the Soviets, where there should have been at least some minimal successes, the Soviets finally became suspicious. In the early part of 1923 the Soviets became aware of the possibility that their new diplomatic codes were systematically being broken by the CG&CS. However, for reasons that no one knew, apparently not even the Soviets (unless it was that old bureaucratic devil of the right-hand-not-knowing-what-the-left-hand-is-doing) the Soviet Foreign Ministry, and their embassies overseas, continued to use the obviously compromised codes.

There is no indication that the British government, after learning that the Soviets knew their codes were compromised, were being fed disinformation by the Soviets through the intercepted traffic. The Soviets apparently were not that sophisticated at that time despite their super-sophistication in other spheres of espionage techniques.

This could have been a cat and mouse game but it did not have that appearance. Whatever it was, the British government eventually decided to continue to monitor and read the traffic. Since they were reading the Soviet diplomatic traffic and the Soviets knew they were reading it, they had nothing to lose if they used the intercepted texts to blast the Soviet's game of subversion.

In an ultimatum to the Soviets, the British government told them to either stop their subversive campaigns in Ireland and India or the 1923 Trade Agreement, which was very advantageous to the Soviets, would be canceled.

The Soviets were still addicted to evasion, deceit and trying to wiggle out of the evidence. They claimed that the intercepted radio traffic had been distorted in transmission and therefore misunderstood. These flimsy protests aside, subsequent radio traffic revealed, as before, that the Soviets were instructing their foreign representatives to give only lip service to the British demands. The Comintern representatives in Ireland and Indian were instructed to temporarily suspend their political activity and put their agents on stand-by until things cooled down. Then, after a little while, the Soviets instructed that the subversive campaigns in Ireland and India should be re-activated.

This Soviet placebo, which they knew the GC&CS would read in the intercepted traffic temporarily served its purpose. It placated the British government just long enough for the Soviets to revamp their codes and transmission procedures. The GC&CS could not immediately break into this new code and valuable intelligence was lost until they could master the new Soviet systems.

Soon thereafter, on 15 September 1924, the notorious Zinoviev letter appeared. This diplomatic disaster set any amicable relations there may have been between the two countries back to the beginning and into a state of diplomatic tension. The Zinoviev letter was supposedly an intercepted communication between General Zinoviev, Lenin's designated President of the Third Communist International (Comintern), and the Communist Party of Great Britain (CPGB). Zinoviev outlined in his letter, in very explicit terms, that the Soviet Union's true intentions, regardless of any overt statement to the contrary, was the complete and devastating subjugation of Great Britain to Soviet control.

Very few people ever claimed to have seen this controversial letter. In those days, however, few British officials seemed to have any doubts about its origins or entertain any skepticism about the letter's authenticity. This letter turned out to be a fraud but it still caused the British government a loss of face for having believed it and published it in the first place.

Who the authors of the Zinoviev letter were remains a mystery. Some facts are known which can be used for speculation and certain inferences appropriately drawn. It was Sidney Reilly who originally passed the Zinoviev letter to SIS. One question might be, who in the SIS trusted Reilly sufficiently to accept the letter from him as genuine? After the blow out and the letter proved to be a fraud, the British Prime Minister was embarrassed when he learned that the discredited Reilly had been the source of the letter. It is possible that Trotsky was the prime mover of this operation!

At this point Reilly's credibility was at its lowest. Reilly was already suspected by the SIS of shady dealing because of his incessant advocacy that they should make more use of Boris Savinkov's emigre organization. Moreover, his dissolute and scandalous personal life made him the kind of a bloody chap that did not belong in the old school tie British tradition as a member of the exclusive club of SIS members.

Giving the SIS its due, there was a lot about Reilly that did not set too well either with the average person who knew anything about him. What Reilly was doing with his personal life was simply not in conformance with the security requirements of a secret intelligence organization such as the SIS.

Reilly's bigamous marriage was also a source of embarrassment to the SIS. This was especially true when his first wife continually pestered his SIS colleagues to give her Reilly's address where she claimed he was hiding from her. Equally embarrassing were her incessant demands upon the SIS exchequer for alimony payments. Eventually, after exhausting every reserve of good will toward their fellow spy, the SIS made the decision that Reilly would have to retire.

From the available evidence there is little doubt that Reilly was personally involved in forging the Zinoviev letter. Reilly himself, however, probably had little concept of the serious ripple effect the forgery would cause. Even if he had been concerned about the political repercussions of his forgery he was apparently more interested in promoting the fortunes of Boris Savinkov and his emigre organization. He thought the threat of increased Soviet incursions into Britain, as seemed to be promised by the Zinoviev letter, would cause the SIS to turn to Savinkov and his emigre group for help.

About this time Savinkov claimed to have heard of a large well organized counter-revolutionary group inside the Soviet Union known as "The Trust." This organization was specifically anti-Soviet and known inside Russia under its true name as the Moscow Municipal Credit Association. Supposedly this group was covertly fighting against the Soviets and was extremely effective in gathering intelligence about conditions inside Russia.[169]

It was also claimed that The Trust had the ability to freely move agents in and out of Russia. The Trust was supposedly so powerful in Russia that it posed a genuine threat to the new Soviet regime. It claimed it could practically do anything it wanted to in Russia because so many of its members were actually highly placed officials of the Soviet government. The Trust, according to Boris Savinkov, had paid him the honor of offering to collaborate with his emigre organization.

Savinkov claimed that The Trust had offered him safe conduct to go to Moscow to discuss with the leaders of The Trust the details of future collaboration. During this trip he was also to reactivate his agents who had been lying dormant in Russia.

For posterity to judge, Reilly initially claimed to disbelieve the claims of The Trust and was allegedly adamantly opposed to Savinkov making the trip to Moscow. Savinkov scoffed at Reilly's disbeliefs and swore he believed everything The Trust told him as the absolute truth. Savinkov went to Russia with the agents of The Trust leading him across the border on 10 August 1924.

As far as is known Reilly never saw Savinkov again. Two weeks after Savinkov left Britain the Soviet press announced his arrest, trial and sentence of death. Then, wonder of wonders, the Soviets announced

Savinkov's complete pardon. It appeared that Savinkov, Reilly's alleged bulwark against the Soviets and communism, had defected to the enemy. Probably, more accurately stated. Savinkov had been a very effective double agent from the very beginning.

Savinkov had merely gone home to Russia when his tour of duty in Paris for the Cheka was finished. It is gracious to think that Reilly, with all of his expertise and intuition in sizing up agents and his manifest experience in supposedly being able to read the mind and soul of the Russians, had been so completely hood-winked and duped by Savinkov.[170]

Reilly was apparently incensed because of Savinkov's defection. He tried to affect the facade in public that he had not been double-crossed by Savinkov but had always secretly suspected him of being a Cheka agent. Reilly refused to admit that he had met a Soviet agent who had completely fooled him. Savinkov had turned out to be a Soviet agent that was smarter than Reilly.

Any further test of wills, or who could to what to whom, was rendered moot when the Soviets announced that Savinkov had died in May of 1925. Speculation by historians who studied this matter are divided whether Savinkov committed suicide or was executed by the Cheka. All historians agree that Savinkov did not die in his sleep!

There was no indication throughout the Savinkov affair that The Trust had betrayed Savinkov. The Trust, as far as the gullible British politicians were concerned, was still a legitimate, anti-Soviet organization. Reilly continued to do all he could to perpetuate this belief.

Reilly, for the record, vowed to continue his anti-Soviet crusade without Savinkov's help. He claimed he wanted to finance and lead an armed resistance group against the Soviet government from inside Russia. He pestered Winston Churchill and other prominent politicians for financial and political support. Since the SIS had divorced themselves from Reilly, few people, particularly those in power, would have anything to do with him, much less help him with his foolish scheme.

Reilly still retained one strong link with the SIS through Ernest Boyce, the SIS chief of station in Helsinki, Finland. Boyce had, in a small way, been secretly funding The Trust in exchange for intelligence from inside Russia. Boyce introduced Reilly to The Trust man in Finland, Alexander Yakushev.

Boyce considered Yakushev one of his most trusted, reliable and prolific reporting sources. Yakushev had everything a good case officer, like Boyce, would insist upon in a good agent. He also had everything the opposition would insist their penetration agent must have in order to be acceptable to the British as an agent. That was exactly the case here. Yakushev was a GPU agent under cover working for The Trust. He had been feeding Boyce disinformation throughout their entire relationship of case officer and reporting agent.

Before Yakushev's GPU status was known, Reilly apparently accepted Yakushev for what Boyce believed him to be. Reilly then supposedly became intrigued with the idea of using Yakushev as an agent inside Russia to overthrow the Soviet government. Yakushev, playing his part, told Reilly everything he wanted to hear. Reilly told Yakushev he wanted to meet the leaders of The Trust face to face somewhere outside of Russia to plan for the coup to overthrow the Soviets.

Yakushev protested Reilly's plan and told him that the leaders of The Trust were too highly placed within the Soviet government to leave Russia. Yakushev proposed that if Reilly wanted to meet with the leaders of The Trust and receive approval of his plan, he should travel to Russia where he could meet the leaders at an assuredly safe rendezvous.

Boyce immediately protested the prospect of Reilly chancing such a trip to Russia. He cited the possibility of imminent capture, torture and the consequent possibility that Reilly would reveal SIS secrets. Reilly, however, was adamant and not to be dissuaded.

No one, not even those blessed with perfect hindsight, could rightly say what might have prompted Reilly to accept this risk. Unless, of course, he knew that there was no risk! The question might be phrased, what could have distracted Reilly sufficiently so he was not repelled by the obvious dangers in Yakushev's proposal? Where were his intuitive and constant evaluations of his agent and his operational security? Where was his skepticism of such a blatantly provocative offer? Did not his recent experience with the Cheka agent, Savinkov, fairly shriek out to him that this offer might be a trap?

The answer to all of these questions, at least from the point of view of detached historians, is a resounding, "Reilly must have known the answers before he left?"

Despite the obviousness of the provocation, Reilly accepted Yakushev's offer to visit the leaders of The Trust in Moscow. He intimated to Boyce that he had accepted Yakushev's offer but did not tell him any details which might allow Boyce to stop him — such as the time he planned to leave Helsinki to cross the border into Russia. Boyce continued to remonstrate with Reilly not to go to Russia with Yakushev but apparently Reilly retained his own counsel.

On the night of 25 September 1925, while Boyce was away from Helsinki on consultation in London, Reilly slipped across the Finnish border into Russia. He was led by two guides supplied to him by The Trust.

Mysteriously, two days later, as preplanned should they ever need to exchange bona fides with each other, Boyce received a postcard authentically written by Reilly and sent from Moscow. This card indicated that all was well with the operation and that Reilly was safe. Of course, if Reilly was being tortured he might have been forced to reveal his safety signals and made to write the card.

The following week *Izvestia*, a Soviet newspaper considered to be the official voice of the Communist Party of the Soviet Union, reported that a group of smugglers had been apprehended by the police at the Finnish border. The report stated that three of the smugglers had been shot and killed on the spot while the fourth, a Finnish soldier, had been taken prisoner. It was assumed by Boyce and the SIS, that Reilly had been one of the three that had been shot and killed.

The news of Reilly's death was belatedly published in the London newspaper, *The Times* on 15 December 1925.[171] Most cognoscente complacently felt that the Reilly episode for the SIS had finally been put to rest.[172]

The Trust was later exposed as a gigantic Cheka hoax for combating internal dissention in Russia and for enticing emigre groups outside of Russia to expose themselves and their operations. The chief of the Cheka, Felix Dzerzhinski, decided that The Trust had been too badly compromised over the years to continue as a front for future anti-Soviet provocations and ordered it be terminated.

In 1927 The Trust was demobilized and most of its principals were liquidated as a matter of routine. Dzerzhinski's maxim had always been that only a dead former agent was a good former agent. Alexander Yakushev, who had led Reilly across the Finnish border to his alleged death, saw Dzerzhinski's handwriting on his wall and jumped back into Finland one step ahead of the GPU execution squad. Yakushev eventually showed up in the SIS stable of ex-Cheka agents and told the full story of

Reilly's capture, Significantly, Yakushev could not provide any details of Reillys alleged execution.

The Soviets have always deprecated any reference by the British to Reilly as "Britain's Master Spy." In fact, any such claim was too far from the truth to stand close scrutiny. If Reilly had been a loyal British SIS agent he was too subjective in the assessment of his agents to be even an average spy. He was also too self-centered to be a good team player for an intelligence service to classify as even a good spy. Moreover, in the club-like ambient of the old school tie of the SIS, Reilly would never have been considered "one of them" in order for them to classify him as their "Master Spy!"

Boyce subsequently left the SIS under less than favorable conditions. Boyce was almost cashiered out of the SIS for having allowed The Trust to dupe him for so long. Boyce was asked the ever searching questions after all blown operations: Did not his regular objective assessment of the on-going operation with The Trust indicate to him that the whole operation was too pat to be genuine? Senior officers in the SIS nailed Boyce with the hindsight question: If the operation was going too well was he not suspicious of it being a double agent operation where the other side had done everything in their power to keep it going so well? Conversely, if the operation was not going too well should not Boyce have suspected that it might be a double agent operation and the opposition was throwing up roadblocks to make the operation look real? In any event, Boyce later became a commodities broker and made pots of money so he did not care too much what the SIS thought about his operations with The Trust.

Reilly was rumored to be alive and well, living comfortably in the Soviet Union on a Cheka pension. Later it was reported that he was living in Rio de Janeiro in Brazil on money stolen from the Savinkov operation. He was also reported to be living in London, in a thin disguise, as if he wanted to be apprehended. Like all historical figures who take on a fictional sheen, Reilly became a legend and, as a legend, will probably never die!

On balance there is a compelling possibility that Reilly was really working for the Cheka from the beginning of his career with SIS or, at the very least, he was recruited by the Cheka shortly after he began working for the SIS. Assuming, that Reilly was really all that he and the SIS claim him to be, then we must properly acknowledge his devil-may-care lucky exploits in Russia. He undoubtedly operated with a sophisticated elan in the espionage arena. He was probably much too flamboyant to have survived in any other period of espionage history. His effectiveness in getting good intelligence out of Russia at the beginning of the Bolshevik reign, if true, is much to his credit. He left a rich fictionalized lore of the period and his participation in it.

The overview of some of the high points and problems of British intelligence operations against the Bolsheviks and Soviets during the formative period of the revolution illustrates how Trotsky and Stalin must have perceived the rules of the game of shadow diplomacy. Better said, they perceived the lack of rules of the game as played in the arena of clandestine warfare by the British.

Trotsky, an intellectual and much more of a cosmopolitan than the average Bolshevik leader, could understand the nature of clandestine operations among nations. He perceived it as part of a larger picture of diplomacy and the necessities of conducting foreign relations in a broader sense.

To Stalin, on the other hand, the apparent lack of rules in clandestine interaction appeared to him to be a license, approved by the most civilized and moral nation on earth, Great Britain. Stalin looked upon these operations as granting reciprocal unbridled permission for cheating, conniving and murder. Stalin was never able to appreciate the fine distinction that, even in clandestine warfare, there are principles, ethics and morality — that regardless of appearances of bare knuckle, no-holds-barred, street brawling, clandestine matters among nations, even in non-attributable skirmishes, are always resolved in a "civilized manner."

In the most explicit of terms, for example, 'agent termination' is never synonymous with 'agent liquidation'.

TROTSKY'S ACUMEN

> When Trotsky cuts off an opponents head [with his writings], he holds it up to show that there are no brains in it!
>
> George Bernard Shaw

Any attempt to interpret Trotsky's operational qualifications in running clandestine agents before and during the Bolshevik revolution is shrouded in myth and secrecy. Any commentary can be judged only by subjective measurements. The full truth regarding Trotsky's proficiency in operational matters will probably never be revealed. In addition to the unknowns inherent in any clandestine operation there are the deliberate false trails he laid to confuse and evade the Czar's Okhrana plus his natural inclination to always employ obfuscating tradecraft. Both of these now serve to lead the researcher down false paths and make it difficult for any objective evaluation of the true nature of Trotsky's handling of intelligence operations.

Some false trails were deliberately built into the operations run by Trotsky to downgrade their importance from contemporary political enemies. Other operations he camouflaged with tradecraft in order to be able to categorically deny his participation in nefarious activities for purposes of historical evasion; he wanted the future to beam upon him kindly. Other obfuscations were deliberately built into his operations as a part of an inherent cover so as to be non-attributable, or at least plausibly denied, by the Soviet government.

> There were the pseudo-intellectuals with long hair and dirty fingernails who talked through the night in the crowded coffee houses of the European capitals who thrilled at the idea of being Comintern spies. They used complicated code-names, fantastic recognition signals and were addicted to disguises. They switched their loyalties, all without any genuine espionage relationships, from Lenin to Trotsky, from Trotsky to Bukharin to Stalin. Therefore, there are

> many claimed espionage relationships of Trotsky with no basis in fact.[173]

The Bolshevik's motto might well have been never tell the truth when a lie might serve to confuse or evade the Czar's Okhrana. Judging from the known conspiratorial tradecraft used by his fellow Bolshevik's (Lenin used over twelve known cover names), Trotsky must have been at least on their level of proficiency. He was at least as knowledgeable in using all tricks of the trade to avoid and confuse the opposition and posterity.

In exile, however, Trotsky's conspiratorial acumen was seldom evident. Although he eventually perceived the GPU was under the direct control of Stalin, he trusted its known agents in Alma Ata in 1928 and in Turkey in 1929, when he feared his life was in danger from the exiled Czarist officers who were living in the vicinity.

Incomprehensibly he entrusted his correspondence, with his conspirators who remained inside the Soviet Union, to the GPU in the consular offices in Istanbul. He naively assumed that the GPU would not know his son's name and sought to hide his authorship of correspondence by signing "Leon Sedov." He stupidly underestimated the GPU, snidely writing in 1932 that the GPU knew of his movements only through what they learned from the newspapers.

Another example of security neglect was that everyone knew that Trotsky was partial to Jews, especially leftist Zionist Jews from the Soviet Union. He paid no heed to the warnings of non-Jewish comrades about this chink in his defenses, yet all his aides who were subsequently discovered as GPU/NKVD penetrations of his entourage were Jews. There is no record in Trotsky's or Sedov's files of Jewish or Zionist comrades being investigated before being allowed to join in Trotsky's movement, yet applicants of other origins were subject to suspicions as possible penetrations.[174]

There is one instance in which Trotsky took direct charge in a covert action and propaganda operation to discredit the American mission in Petrogad. This is known as the Kalpashnikov case. Andrew Kalpashnikov was a Russian officer in late 1917 working for the American Red Cross Mission in Petrograd. In the pre-war period he had been an attache in the Czar's Russian Embassy in Washington and, as a result had a pro-American bias. The American Red Cross had sent

Kalpashnikov to the United States to buy ambulances that were needed on the Eastern front by the Russian army. When the Soviets unilaterally left the war, the ambulances were no longer needed by the Russians so they were diverted to the American Red Cross in Romania.

Kalpashnikov had been in Halifax when Trotsky was arrested by the Canadian authorities and was asked to translate for Trotsky. Ever afterwards Trotsky had associated Kalpashnikov personally with that demeaning, very unpleasant, experience and wanted revenge against him.

Colonel Anderson, who was in charge of the American Red Cross unit in Romania cabled the American Embassy in Petrograd asking that the ambulances be shipped to him since they were no longer needed by the Russians. At that time the ambulances were on flat cars in the Petrograd railway yards.

Suddenly, on the eve of their departure to Romania, Trotsky issued an order that the ambulances be detained and Kalpashnikov be arrested. Trotsky claimed that Kalpashnikov planned to deliver the ambulances to the anti-Bolshevik forces on Rostov-on-Don commanded by General Kaledin. Trotsky read a message to the crowd at the Alexandrovsky Theater from Colonel Anderson to the American Ambassador Frances in which it was stated that the ultimate destination of the ambulances was to be Rostov-on-Don.

The Rostov-on-Don destination has been only one of Colonel Anderson's stops as a possible escape route. It had been replaced by a different destination in a subsequent cable which Trotsky either did not see or chose to ignore.

Colonel Robins of the American Red Cross Mission in Petrograd, and an American intelligence officer, assisted in clearing up the matter by appealing to Trotsky whom he knew. Eventually Kalpashnikov and the ambulances were released. It turned out that Trotsky did have Anderson's subsequent telegram when he made the original charge which he then knew to be false. On the other hand, Trotsky had gained considerably in the propaganda arena with his public charges against the Americans and had let the American Red Cross and the Embassy know that he knew what they were doing because he had access to the communications; a not so subtle warning!

How far did Trotsky get involved in the routine espionage and counterespionage intelligence operations in the post-revolutionary phase?

An enigma of Trotsky's life as a revolutionary after his banishment was the fact that he never instituted a counterintelligence organization nor fostered that healthy suspicious mentality among his followers. Between 1917 and 1921 there is considerable evidence of his dominant position in creating the Soviet counterintelligence service. He installed political commissars in every Soviet army unit and commandeered Cheka representatives for joint operations with Red Army intelligence units. Throughout the civil war, Trotsky worked closely with Dzerzhinsky, the leader of the Cheka, for whom he continued to express admiration for even in his writings in exile after he knew that Dzerzhinsky had conspired with Stalin against him.[175]

Even partial acceptance of the low-down and dirty, scheming accusations made by Stalin against Trotsky, puts Trotsky in the forefront of constantly running intelligence operations of various kinds. Supposedly he avoided discovery for his nefarious acts until the great Stalin ferreted him out and exposed him as late as 1936 in the Purge Trials, about twenty years after the fact.

For example, Christian G. Rakovsky, one of the co-defendants with Trotsky in the 1938 Moscow Trials that were staged by Stalin, confessed that he and Trotsky accepted covert financial aid from non-Russian sources during and after the Bolshevik revolution. There was other, more substantial evidence of penetration, double agent and disinformation operations supposedly run by Trotsky, that were corroborated by the confession of Rakovsky. All of these made Trotsky look like a very competent case officer.

In 1924, Rakovsky was appointed the Soviet Ambassador to Great Britain. It was alleged by the GPU that soon after his arrival in London, two British intelligence officers, Captain Armstrong and Captain Lockhart called upon him. They told Rakovsky they knew he was a confidante of Trotsky and offered a mutually beneficial agreement. British intelligence would help the Trotskyite underground in Soviet Russia to overthrow the existing government in return for intelligence information from Trotsky on Soviet intentions and military capabilities. Trotsky is supposed to have pondered this offer for a few days and then instructed Rakovsky to establish relations on those terms with British intelligence. When Rakovsky was later appointed Soviet Ambassador to France, according to Rakovsky's confession at the Purge Trials, Trotsky ordered him to establish a similar relationship with the French government.[176]

Two other alleged agents of Trotsky's, according to the testimony produced at the Purge Trials, were Alex Shestov, an engineer attached to the Soviet Trade Mission to Germany and Sergei Bessonov, a member of the Soviet Berlin Trade Mission. Bessonov was a member of the central Soviet trade mission and had, although stationed in Berlin, a liaison function with over ten different countries. Lyova Sedov, Trotsky's son, arranged with Bessonov to handle Trotsky's secret communications from Russia to Trotsky and from Trotsky to his representatives in the countries Bessonov dealt with in the course of his routine duties. Bessonov would include these letters in the official pouches he sent and received and they were all in secret writing.

Shestov was a member of the Board of Directors of the eastern Siberian Coal Trust. It is alleged that he was able to organize sabotage cells in Siberia through his position. Sedov convinced Shestov to work for German intelligence by providing them with information about Siberia. The Germans were particularly concerned about the new industrial base Stalin was building in far-off Siberia which was out of range of the German bombers. Shestov balked at becoming a spy for Germany but finally relented when Sedov convinced him it was for the good of the Trotskyite cause.[177]

This independent corroboration of Rakovsky, if it can be given any credence at all, leads one to believe that maybe all the charges against Trotsky being a master at clandestine operations were not contrived by Stalin. If Rakovsky's confession was forced form him during torture and therefore unreliable, the other supporting evidence still indicates the Machiavellian behind-the-scenes hand of Trotsky in many, well run, clandestine operations.

Trotsky's Left Opposition from the beginning had functioned on two levels. Openly, on the public platforms, in its own newspapers and organized cadre. It also operated behind the scenes in small clandestine conferences led by Trotsky and his cohorts who planned the underground strategy and tactics. Trotsky built a secret conspiratorial opposition organization. By 1923, this underground apparatus was extensive and powerful — using codes, recognition signals, danger signs and advanced tradecraft techniques. There were secret printing presses throughout Russia with Trotskyite cells in the army, diplomatic corps, Soviet state governments and Party institutions.

In the winter of 1921-22, the former lawyer and leading Trotskyite, Nicolai Kristinsky, was appointed to be the Soviet Ambassador to Germany. It was alleged at the Purge Trials that General Hans von Seeckt, the head of the German Reichswehr, told Krestinsky that Germany would help the Trotskyites because Germany was sympathetic to the aims of the Left Opposition.

Trotsky told Kristinsky that Germany was looking to the west to revenge themselves on France and England and was no enemy of Russia. He told Kristinsky that when he returned to Germany, he would propose to General Seeckt that Germany should provide a regular subsidy for the development of clandestine Trotskyite activities inside Soviet Russia.

Kristinsky named the sum of 250,000 gold marks to General Seeckt who agreed and proposed the quid pro quo that the Trotskyites provide to Germany Soviet secret military information. In addition, Trotsky was to procure Russian visas for German spies to enter the Soviet Union legally. Kristinsky claimed at his trial that Trotsky accepted all these conditions which were then put into effect in 1923.[178]

Seeing through the history of who did assist the Bolshevik revolution is the relentless disclosure of allegations that foreign merchants, bankers and diplomats subsidized the various political factions in pre-revolutionary Russia.

Surprising as it may be, the recurring evidence of cooperation between the ostensibly implacable foes of socialist revolutionary ideals, the capitalist businessmen, and the Bolshevik radicals was an accepted fact of life in Russia in 1917. Trotsky took advantage of any material benefits wherever he could find them, even if they came from the decadent capitalists.

The cooperation between these two extreme groups was, for the most part, clandestine and cloaked by conspiratorial security techniques. Few, except the participants, were aware of this odd relationship. Trotsky, throughout the active revolutionary phase of his life, was an important participant in these covert funding liaisons.

A discernible link between Trotsky and the funding of the Bolshevik cause by the international bankers was through Abram Givatovzo, Trotsky's cousin. Givatovzo was a private banker in Kiev before the Bolshevik revolution. After the revolution he became a leader in the private banking community in Stockholm. From Sweden he continued to channel financial assistance to the Bolsheviks. Subsequently, under

Trotsky's direction, Givotovzo acted as the banking agent for the Soviet government in their currency reform transaction of 1918.

When Trotsky was returning to Russia from New York he had $10,000 from some unexplained origin on his person when he was arrested in Halifax. There is some evidence that this money could have come from German financial sources passed through German intelligence to Trotsky.

During the time Trotsky was the co-leader of the Bolshevik government he lived with great meagerness.[179] "At his own request he received but fifty dollars a month as the highest official in the Soviet government. A psychologist might say that Trotsky had a money complex. He was so afraid of plots to implicate him in something unethical or illegal, that he threw people out of his office when they offered honest and legitimate financial aid to Russia."

Trotsky was quick to denounce, or otherwise silence, any implication of his affiliation with the Germans. Trotsky demanded and got the life of Admiral Stchastny, the savior of the Russian fleet because of possible tie in to the Germans. Admiral Stchastny was executed at Moscow on 22 June 1918. The Admiral had disobeyed the Bolshevik orders to scuttle the Russian fleet before the Germans could capture it. Instead he had saved the fleet by sailing the 168 vessels to a safe harbor at Kronstadt.

The mistake the Admiral made was that he unwittingly gave Trotsky some documents he had received that denounced Trotsky as a German collaborator. Trotsky was the chief and only prosecution witness against the Admiral in a closed trial that reached a unanimous verdict and carried out the execution the same day.[180]

A more forthright descriptive term for Trotsky's operations might be one of subtle penetrations, or covert liaisons, that brought the financial support of foreign bankers to him and the Bolsheviks. Apparently Trotsky himself case officered many of these operations. In another clandestine sphere, through cut-outs, Trotsky directed some very sophisticated propaganda and covert action operations.

Three of Trotsky's own operations are considered as illustrative. These operations could be called either penetrations of the opposition, semi-covert unwitting double agent operations or, as a last report, covert unwitting liaison operations.

The ambassadors of the three principal Allied governments in Russia — France, England and the United States — were out of favor with the Soviets. This lack of essential rapport between the Allied ambassadors and the Soviets resulted mainly from the fact that the countries had never proffered any friendliness toward the Soviets. This was principally due to the fact that the Soviets wanted to take Russia out of the war and the Allied powers desperately wanted Russia to remain in the war.

Also, these three ambassadors had all been accredited to the Czar's government and were disturbed at having to do business with a usurper government. It would have been a serious violation of that diplomatic accreditation if they became friendly with a revolutionary force avowed to overthrow the Czar. In some instances these ambassadors had even taken positive steps to try to defeat the Bolsheviks. There was therefore ample reason for wariness on both sides that prevented any mutual trust or reciprocal acceptance between those ambassadors and the Soviet government after the Bolshevik revolution.

The first of Trotsky's operations involved Raymond Robins. The New York State Lusk Commission reported in 1919 that materials it had seized from the files of the Soviet Information Bureau in New York confirmed that Robins and his wife were closely allied with Bolshevik activities in America. Robins was already deeply committed to a close rapport with the Bolsheviks when, in early December of 1917, he assumed a position of leadership with the American Red Cross unit in Russia under the leadership of Colonel William Boyce Thompson.

Robins's persistent efforts to advance the Bolshevik cause in America gave him a high degree of prestige and acceptance among the Bolsheviks when he arrived in Russia, especially with Trotsky. He apparently arrived indicating a willingness to cooperate and, by inference, ready to agree to whatever Trotsky suggested. Possibly unwittingly, but so closely similar as to be indiscernible, Robins's self-expressed goals and objectives in Russia were the same as Trotsky's.

Robins was determinedly pro-Bolshevik but the French Ambassador, Noulens, who derived his political support from the wealthy French "200 families," was vehemently against Robins. Noulens started a rumor to discredit Robins. He arranged for one of his secretaries to pointedly ask, in the presence of American Ambassador Francis, "Who is the real American Ambassador in Russia, Francis or Robins?"[181]

Karl Radek was Trotsky's deputy in the Press Bureau of the Peoples' Commissariat of Foreign Affairs. Trotsky was the Director of this Commissariat and would have had Radek act as a cut out between him and his cooperative agent, Robins. On one occasion Radek was indiscreet. He told the Austrian Consul General in Moscow that Robins acted as an intermediary between the Bolsheviks in Russia and friends in the United States who were sympathetic to Bolshevism.

Another unwitting double agent, but probably aware of Trotsky's sly hand holding the puppeteer's strings, was Bruce Lockhart, of the British Foreign Service. Lockhart, a genuine British Foreign Service officer, coopted into intelligence work by the Prime Minister, was a Russian speaking, experienced diplomat. He was originally officially sent to Russia by Prime Minister Lloyd George with the full concurrence of the British Foreign Office. In 1917 Lockhart's second appointment to Russia, had been directed solely by Lloyd George without the concurrence, or even knowledge of (until after the fact), of the British Foreign Office.

The second time Lockhart was not sent to Russia in any ordinary diplomatic capacity. He was sent to Russia to be Lloyd George's personal envoy to the Bolshevik government and to establish an unofficial liaison with the Bolsheviks. This was a strange deviation from the usual established channels of protocol, for a Prime Minister to preempt the staid, tradition bound, rigid structure in a matter of British foreign policy.

Maxim Litvinov, the unofficial Soviet representative in London, gave Lockhart a letter of introduction to Trotsky personally, saying that Lockhart understood the Bolshevik position and that he was sympathetic with it. This was tantamount to handing Trotsky an agent already recruited, or at least prone to cooperate, on ideological grounds. Under these circumstances Lockhart could be considered as waiting for an assignment from his case officer, Trotsky.

Lord Milner, a British banking director and hero of the Boer War in South Africa, had pressed Lloyd George to make the Lockhart appointment. Before Lockhart left London he was briefed almost daily by Milner who was exceedingly pro-Bolshevik. In addition to championing the diplomatic recognition of the Bolsheviks by Britain, Milner also tendered financial support to their opponents in Southern Russia; that is, Milner covered all possibilities that may occur in Russia.

New York bankers also promoted financial aid to both the Bolsheviks and the forces fighting them in the south of Russia. The bankers' theory being that regardless of who won the fighting in the south of Russia, they would be in a preferred position for their commercial ventures in the post-revolutionary period.

The French member of the three Allied unofficial representatives to the Bolsheviks was Jacques Sadoul, an old friend of Trotsky's. Sadoul was officially appointed by the French government to be its "unofficial" liaison agent with the Bolsheviks. Sadoul, was an openly self-proclaimed Bolshevik sympathizer. Sadoul would be amenable to satisfy Trotsky's intelligence requirements as a loyal Bolshevik.

Sadoul felt a very strong ideological attraction for the Bolsheviks. In August of 1918 he defected to them, for which he was condemned to death in absentia as a deserter and traitor by the French government. However, this sentence was not carried out after Sadoul returned to France and joined the French Communist Party.[182]

Trotsky could give the Allied powers, through these three unofficial liaisons, exactly the information he wanted them to have. In turn he would have these three fulfill his intelligence requirements by receiving back from them the information he requested in exchange. The exchange of significant information is a must for both sides in any liaison operation in order to keep it "oiled" and running smoothly.

Trotsky had a close personal relationship with each of these men. He had definite intelligence objectives that he wanted to obtain from the governments of each of them.

Another liaison operation probably case officered by Trotsky was the Bolshevik relationship with the resident German intelligence officer in Russia. This German used the name Bauer, Bayermeister and Bayer. Although not listed as the chief of German intelligence, probably for cover purposes, he was most active and listed as the second in command of the Nachrichten (intelligence) Bureau, of the German General Staff. His nominal chief was listed as Major Luberts. The United States High Commissioner for the Near East, Rear Admiral Mark L. Bristol, wrote on 31 January 1921:

> The alliance with the German military [intelligence] agent [Rudolph Bauer] and the militarized Moscow [Bolshevik]

> government [Trotsky] is well known, as is the economic and financial aid given by the Germans to the Bolsheviks.[183]

Jacob H. Rubin was the self-admitted Bolshevik agent during his testimony in 1921 before the House Committee on Foreign Affairs when he said, " I have been with the American Red Cross at Odessa. I was there when the Red army took possession of Odessa. At that time I was favorably inclined toward the Soviet government because I was a socialist and had been a member of the party for twenty years. I must admit that to a certain extent I helped form the Soviet government in Odessa." Rubin also admitted to the House Committee that he had been arrested as a Bolshevik spy by the government of South Russia that was fighting the Bolsheviks.

Rubin was a banker and the president of Rubin Brothers and Company of 19 West 34th Street in New York City. In 1917 Rubin was also associated with the Union Bank of Milwaukee and the Provident Loan Society of New York. In 1920 the United States Department of State received a message from its Consul in Odessa stating that Rubin had formally expressed a desire to remain in Odessa with the Bolsheviks rather than return to the United States.

Another American agent that came under Trotsky's control was Robert Minor. Minor was a cartoonist for an American newspaper, *The Philadelphia Public Ledger*. Minor went to Russia in 1918 to report on the Bolshevik revolution. While in Russia he joined the Bolshevik Bureau of International Revolutionary Propaganda. This Bureau was controlled by Trotsky's Peoples' Commissariat for Foreign Affairs. Once again there is evidence of Trotsky's direction in exploiting an agent.

Trotsky was Director General of the Bureau of International Revolutionary Propaganda. Coworkers on the Bureau with Minor were the unofficial French liaison to the Bolshevik government, Jacques Sadoul; John Reed, the American journalist of whom more later; Reed's wife, Louise Bryant, who later became the wife of the United States Ambassador to Moscow, William C. Bullitt;[184] another American, Albert Rhys, and Williams a self-proclaimed communist and Bolshevik adherent. (In 1917 Williams attempted to return to the United States to establish the Soviet Information Bureau but permission was denied to him by the Department of State.)

In November of 1918, Minor and several other comrades from Trotsky's Bureau, went to Germany to distribute pro-Bolshevik propaganda. Minor was subsequently arrested in Paris by the French police and was turned over to the American military authorities. Minor's father was a judge and minor political figure in Texas. It was alleged that the father brought political pressure to bear on the American federal government for his son's release, Minor was released and returned to the United States where he toured America on the lecture circuit extolling and promoting the wonders of Bolshevik communism.

What a wonderful, non-attributable political action agent for the Bolshevik government! An American citizen, the embodiment of the all American boy, touring his native land praising the virtues and accomplishments of communist Russia to his credulous fellow Americans. Trotsky was, at the very least as chief of the Propaganda Bureau, giving operational guidance to this operation.

Albert Rhys Williams, also from Trotsky's Bureau, did not stop trying to establish a Soviet Information Bureau in the United States. In 1918 the Soviets appointed Alexander Nyberg, an alias of Santeri Nuorteva, to establish a Soviet Bureau in the United States. This American Bureau would be under the guidance of Trotsky's Bureau in Russia. Nyberg (Nuorteva) had immigrated to the United States from Finland in January of 1921. Unfortunately, after the trouble of establishing him as the head of the Soviet Information Bureau in the United States, he was not very effective.

In 1921 Trotsky replaced Nyberg (Nuorteva) as Director of the Soviet Bureau with Ludwig C.A.K. Martens. Trotsky had finally succeeded in placing one of his most trusted men in America. Whatever opposition there may have been to Trotsky's handling of this operation was ignored in the new found harmony existing between Lenin and Trotsky who were working smoothly in concert.[185]

The use of party activists in espionage operations is initially hampered by the security factor that many, if not all, already have dossiers in the police files. Trotsky was against mixing Soviet diplomatic or official personnel with espionage activities. In this instance Trotsky over came his scruples and used Martens. If there were Soviet diplomats working in America at the time, he might have preferred to take a chance with them. He was well aware that Martens was known to the Bureau of

Information and would not be accorded diplomatic immunity if caught spying.[186]

Martens promoted trade between American businesses and Soviet commerce despite the fact the United States had an embargo against such trade. Although trade may have been the ostensible justification for the creation of the Soviet Information Bureau, Martens at the same time was providing covert financial support to the newly formed CPUSA. This covert funding to a domestic communist party was part of Trotsky's program to support continuous international revolution and the world spread of communism.

Trotsky's artful hand as senior case officer is never more obvious than in the area of fostering the international spread of communism. The Comintern relied to a large extent on natives of the country in which they worked. All resident directors should be Russia but the senior representatives should never work in the country of their birth; a German must never work in Germany, a Britain must never work in Britain. This policy would provide an alibi in the event of compromise. It could be denied that the Soviet Union had anything to do with the agent but it could be suggested very strongly that, for example, the German caught in France was spying for his homeland of Germany and so on.[187]

Some of the funds for the Soviet Information Bureau's expenses in the United States came from local American contributors. Most of the funds, however, were supplied by couriers carrying loot from the Soviet government to the United States.

Among those couriers was an American journalist and writer, John Reed, who was being used as a courier by the GPU. Reed was secretly smuggling diamonds and jewelry from the Soviet Union into the United States for sale on local markets. These funds were to help finance the CPUSA and Soviet espionage activities in the United States and Latin America.

A roster of personnel employed by Trotsky's Soviet Information Bureau in New York was compiled by the British Scotland Yard. This was part of a joint operation with the United States government against the common enemy of communism, at the request of the Lusk Committee of the State of New York. The list of employees of the Soviet Information Bureau included, among others, the name of Julius Hammer. The British report stated, "In Julius Hammer, Martens has a real Bolshevik and ardent left wing adherent who came to the United States not long ago from

Russia. He was one of the organizers of the left wing movement in New York and speaks on the same platform with the left wing [John] Reed...."

The Hammer family has had close ties with the Soviets since 1917. Armand Hammer is currently the Chairman of Occidental Petroleum Corporation and still has close associations and business relationships with the Soviet Union. Jacob, the grandfather of Julius and Armand, was born in Russia.

At the 1919 convention that gave birth to the CPUSA, Julius Hammer served with Bertramn Wolfe and Benjamin Gitlow on the steering committee that formed the CPUSA. In 1920 Julius Hammer was sentenced to three and one-half to fifteen years in Sing Sing prison in New York for criminal abortion. Lenin commented at the time, that although Julius was imprisoned on the charge of practicing illegal abortions, he was in fact, imprisoned for being a communist.

A former aide to Colonel House, Kenneth Durant, was assistant Secretary of Trotsky's Soviet Information Bureau in America. Colonel House was the chief aide, and close personal friend, of President Wilson. Durant was identified as a Soviet courier for the transfer of funds from the Soviet government to Soviet agents in the United States. Durant was the press secretary and publisher of *Soviet Russia*, the unofficial organ of the Soviet Information Bureau.

From 1923 to 1940 Durant was the office manager of the Soviet news agency "Tass" in the United States. J. Edgar Hoover, the Director of the Federal Bureau of Investigation, described Durant as "at all times...particularly active in the interests of Martens and the Soviet government."

John Reed was a prolific writer, a witting propagandist for the Bolshevik cause and a member of Trotsky's Bureau. Reed contributed to the communist oriented American magazine *Masses* and wrote what is the best known book in English about the Bolshevik revolution, *Ten Days That Shook the World*. Lenin wrote the introduction to *Ten Days...* and commented that "Here is a book I would like to see published in a million copies and translated into all languages." Needless to say *Ten Days...* is very laudatory of the Bolshevik cause.

After the Bolshevik revolution, Reed became the American member of the Executive Committee of the Communist Third International; commonly known as the "Comintern." He died of typhus in Russia in

1920 and is buried in the Kremlin lying near to Lenin. (Reed is the only American ever accorded this dubious honor.)

Reed was characterized by his biographer as a "spokesman of the Bolshevik in the United States." Enigmatically much of Reed's income came from the ultra conservative publication *Metropolitan* which was owned by right wing capitalist Harry Payne Whitney. Reed also received funds, while living in Petrograd, from the Bolshevik sympathizer, Raymond Robins, while Robins was serving in Russia with the American Red Cross.

Meanwhile, on the American homefront, patriots were girding to combat the rising threat of Trotsky's "Red Menace." The organization United Americans was formed for this purpose in 1920. Its goals were to fight the teachings of socialists, communists, the International Workers of the World (Wobblies), all Russian organizations and radical farmers' societies. Membership in the United Americans was limited to citizens of the United States and the stated objective of the group was to have five million members fighting these radical organizations.

In *The New York Times* of March 1920 a detailed account of a planned Soviet Red invasion of the United States was printed. In order to make the story more sensational *The New York Times* reported that the invasion would cost the Reds more than $20 million dollars. The Soviets, it was alleged, had obtained this $20 million "from the robbery and murder of the Russian nobility."

This story came from the United Americans. At the same time they leaked this invasion story, they decided to bolster the fright quotient of the scare. They revealed the existence of a large number of home grown American subversives. These United States citizens were supposedly ready to pounce on unsuspecting Americans in their homes while they were asleep and murder them in the name of communism. The United Americans leaked the "precise" number of these domestic subversive communists to *The New York Times*. Also,they claimed there were exactly 3,465,000 Americans who were members of their organization.

There is a quagmire of innuendo and suspicion concerning the part that American commercial and financial communities exercised in affecting the intelligence coming out of Russia. The bankers were accused of playing some manipulative role in fostering the beginning and outcome of the Bolshevik revolution. The answer is, there was some minor role played by certain bankers interested in a Bolshevik victory but it was on an individual basis and not a profession-wide organized effort.

There is an equally disturbing allegation that international Jewry led a conspiracy that fomented and manipulated the Bolshevik revolution to their own ends. Particularly there have been many historical allegations that the Bolshevik revolution was the result of a cabal's conspiracy of Jewish bankers, mostly Wall street but also European Jewish bankers sharing equal credit or blame.

Winston Churchill, the British statesman, wrote that the international and atheistical Jews had a great role in the creation of the Bolshevik revolution. Churchill calls the international Jews a "sinister conspiracy" that emerged from the populations of the countries where Jews had been persecuted. Churchill traced this Jewish conspiracy back in the history of leftist subversion, implicating Trotsky, as a Jew. He added, Bela Kun of Hungary, Rosa Luxemburg of Germany and Emma Goldman of the United States; all admitted strong communists and, also coincidentally, all Jews.

There is a United States Department of State document dated 13 March 1918 entitled, "Bolshevism and Judaism." that states that Max Warburg, was a Jewish banker from Hamburg, Germany. Warburg was closely affiliated with the New York banking firm of Kuhn, Loeb and Company, that helped finance the Bolshevik revolution. They allegedly, financed Trotsky solely because he was a Jew. It was later proved that Warburg had acted as an individual, only for personal commercial purposes. No hard evidence of a Jewish conspiracy to support Bolshevism, or Trotsky as a Jew, can be found. Money was given to Trotsky because he was in a position to help with Warburg's commercial aspirations. The fact that Trotsky was a Jew was irrelevant.

In sum, and on balance, there does not appear to be any concrete or plausible evidence that the Jews were involved in the Bolshevik revolution because they were Jews. There were many people of all creeds, races, professions and persuasions involved in the Bolshevik revolution. The fact that some of them happened to be Jewish is about all that can be cited to support the so-called Jewish conspiracy.

The purpose in even bringing up the Jewish allegation is that we are following the Trotsky saga wherever it goes. The story reflected this issue of Jewish bankers, particularly American Jewish bankers, financing the

Bolshevik revolution because one of the Bolshevik leaders, Trotsky, was a Jew.

In the summer of 1927, Trotsky made an all out bid for power in the Soviet Union. He was voted down in a referendum of all Bolshevik Party members with a count of 740,000 to 4000. In *My Life* Trotsky described the hectic conspiratorial activity following the vote — the meetings and the printing of pamphlets and strike notices all planned for an uprising to overthrow the Soviet government on 7 November 1927, the Tenth Anniversary of the Bolshevik Revolution.

Trotsky's attempt failed. His propaganda seemed to have no effect on the general population. The small group of Trotsky supporters were dispersed without bloodshed and the leaders were arrested. Trotsky was expelled from the Bolshevik Party and later sent into exile.[188]

TROTSKY IS BANISHED

> Prison reform will not work until we send a better class of people there!
>
> Anonymous

> The historian is accountable for the reputations of the mighty dead whom he conjures up and portrays. If he makes a mistake, if he repeats slander on those who are blameless or holds up profligates or schemers to admiration, he not only commits an evil action; he poisons and misleads the public mind.
>
> Albert Mathiez

Government archives are filled with weighty tomes concerning the turmoil that was Russia giving birth to the Soviet Union. It is a classical Cain and Abel fratricide. The Mensheviks, Bolsheviks and political opportunists of all stripes were striving to outwit, out scheme and out kill each other in their drive for power in the vacuum left by the overthrow of the Czar.

Trotsky and Lenin were the architects, catalysts and prime movers of the Bolshevik revolution that brought communism to Russia (renamed the Soviet Union). This cataclysmic upheaval changed Russia forever from a paternal, but ruthless and despotic autocracy, into a socialist dictatorship without compassion for human rights. Later, by comparison with the utopia the Soviets had promised to the Russian people, the days of the Czar were remembered as kindly and benevolent by those who lived under both systems.

Unfortunately, Soviet-manipulated history has eradicated Trotsky's real contribution to the success of the Bolshevik revolution, portraying Stalin in their history as Lenin's principal partner during the revolution and the legitimate heir to Lenin's mantle of supreme leadership.

Stalin's hatred of Trotsky started at the beginning of the Bolsheviks' first revolutionary stirrings. Trotsky had returned from exile in 1917 with an international reputation as a Russian revolutionary — laurels won during the great conflicts of 1905 as a leader and competent propagandist. World War One, with its divisive counter forces, and the human suffering that it brought to Russia, had undermined the Czar's regime. The Czar's enemies within Russia gave the tottering Czarist government a push until it collapsed of its own ineptitude, corruption and brutality. Russian, for the first time in its history, found itself in a position to be receptive to the ideals of the socialist process.

The Russian people revelled in their new found freedom from Czarist rule, however transitory, and particularly after the exhaustion of a major war. The people were eager for a political solution that would assure the continuation of this new freedom.

Preaching immediate social revolution, Lenin, as the chief of the Bolshevik party, formed an alliance with the independent Trotsky, a loner but a fiery orator and an accomplished organizer. Stalin was, at this time, a relatively obscure party worker, certainly not in a position comparable to Trotsky or Lenin. Stalin was a peasant, who spoke Russia with a country accent and was considered a bumpkin by the more sophisticated urban Bolsheviks.

Two men emerged from the morass that resulted from eight months of the Kerensky provisional government, who had the will and the power to rig the world stage for the stormy events that followed. Lenin as the instigator and the philosophical pundit and Trotsky as the tried and proven insurrectionary leader, both grasped boldly for power. Stalin, intellectually far behind these two leaders, was massed with the rest of the party followers. Stalin was characterized as one who always played it safe until he could gauge which way the wind was going to blow., The power that had been Russia was lying unclaimed in the rubble of the war's aftermath. Lenin and Trotsky, in an almost bloodless coup, seized the power and with tyrannical zeal established the Soviet dictatorship on the remains of the Kerensky provisional government.

Stalin, sensing a sinecure, left the comfortable anonymity of the bureaucratic masses and enlisted in Lenin's relatively safe work of political theory and government administration. Trotsky, always an activist, on the other hand, organized the Red army from the destitute, derelict soldiers exhausted from the war and whipped them into shape as the New Red army. Trotsky's leadership enabled the Red army to win a bloody civil war over the White forces led by professional Czarist generals. Trotsky for his battlefield triumphs was acclaimed a hero and revered as a ranking star in the communist organization.

When Lenin died in 1924, Stalin, who as the ranking bureaucrat present, adroitly maneuvered Trotsky's friends into powerless positions. He exploited alliances with Trotsky's enemies who would gladly conspire against Trotsky. He intrigued with anyone who would cooperate to help gain his goal of isolating Trotsky from all bases of power.

The eminent British historian and contemporary statesman during Trotsky's era of prominence, Winston Churchill, has a different idea where the emphasis for Trotsky's failure to assume Lenin's mantle of power may be,

> ....he might well have been made the Dictator of Russia by the armed forces, but for one fatal mistake. He was a Jew. He was still a Jew. Nothing could get over that. Hard fortune when you have deserted your family, repudiated your race, spat upon the religion of your fathers, and lapped Jew and Gentile in a common malignity, to be bilked of so great a prize for so narrow minded a reason.[189]

On 12 January 1928 the GPU informed Trotsky that on 16 January he and his family must leave Moscow. They were to be banished to Alma-Ata in the Turkestan area of Russia pursuant to a deportation order from the Soviet Central Committee. The ostensible reason for the deportation, according to the proclamation the senior GPU officer in charge of the detail read to Trotsky, was his counter-revolutionary opposition to the Bolshevik revolution.

When the morning of the sixteenth dawned, Natalia had a high temperature but she was determined not to show any illness that might be misconstrued by the GPU escort as a weakness. Thousands of Trotsky's well wishers and friends clogged the street in front of the house of Boloborsdov, the former Commissar of the Interior. The Trotskys had moved to Boloborsdov's house after being forced to vacate their apartment in the Kremlin.

Another crowd, estimated at several thousand, demonstrated at the railroad station against Trotsky's deportation. The crowd surged across the railroad tracks preventing the train from moving. The GPU escorts could do nothing in the face of such a determined opposition from the people. They had to return the Trotskys to Boloborsdov's house for the night.

The next morning the GPU was determined that no similar demonstration would deter them from deporting Trotsky. Trotsky, this time, refused to leave without a fight and barricaded himself in his room at Boloborsdov's house. The GPU smashed through the glass of the door, forcibly dressed Trotsky in his street clothes, coat and scarf and carried him bodily out of the house.[190]

Trotsky and his family were thrown on to a train and kept locked in their compartment until the train was well out of the marshaling yards of Moscow. Thus, in ignominy was Trotsky, one of the most powerful instigators of the Bolshevik revolution, expelled from Moscow by Stalin.

Two hundred and fifty kilometers from the end of the railroad terminus, and about 4000 kilometers from Moscow, Alma-Ata, when the Trotskys arrived in winter, was a snowy wasteland, desolate and foreboding. There was no electricity, only a few roads that were really nothing more than wagon tracks in the mud, no drainage, no municipal services such as garbage collection and only hovels for houses.

The Trotskys were housed in a four room apartment in a ramshackle, tumble-down building. The wind and rain blew through the cracks and broken boards and the floor sagged at a precipitous angle. Natalia quickly set about trying to make the hovel more comfortable for Trotsky and the children. Trotsky was soon immersed in his writing and oblivious to his surroundings. Natalia worked tirelessly to keep their squalid rooms clean and tidy for his peace of mind.

The GPU kept Trotsky effectively isolated by harassing anyone who appeared to befriend him. Trotsky's correspondence both ways was heavily censured. In addition to the meager financial allowance granted to him by Stalin, he received some monies from his writings, some editing sent by literary friends and an occasional piece of translation work.

Despite GPU efforts to keep Trotsky isolated, their bulletins to Moscow revealed that the number of Trotsky partisans visiting him increased daily. Trotsky sympathizers visited Alma Ata in crowds and their visits took on the character of a pilgrimage of the faithful to a shrine.[191]

Trotsky's health began to worsen with recurrent attacks of fever, gout and colitis. The censorship of his correspondence became more stringent and Trotsky began to suffer from isolation paranoia.

On 16 December 1928 Volynski, a special representative of the GPU, arrived in Alma-Ata from Moscow. He carried a warning to Trotsky that he must cease all political activity or he would be further isolated. Trotsky naturally refused. After conferring with Stalin, on 20 January 1929, Volynski returned with an order that the Trotsky family were to be deported from Soviet Russia forever.[192]

Trotsky, Natalia and their son Lyova left Alma-Ata on 22 January 1929. A tractor pulled their automobile through the snow. When the tractor bogged down, the GPU insisted on the group proceeding by sleigh around the deeper snow drifts and then back into the automobiles until they reached the railhead at Frunze. There the GPU told Trotsky that he was being deported to Constantinople (Istanbul).[193]

Trotsky, Natalia and Lyova boarded the cargo ship "Ilyich" at Odessa. The ship was towed by an ice breaker for the sixty miles through the frozen sea before it could get under way in open water. The Trotsky's arrived in Istanbul on 10 February 1929.

When Trotsky arrived in Istanbul he was received like visiting royalty. Foreign correspondents lined up to interview him and international headlines reported his every move.[194]

In Constantinople Trotsky and his family were initially housed in a wing of the Soviet consulate. They were shown every courtesy by the consulate staff and accorded the respect they were entitled to by virtue of their previous position in the revolutionary hierarchy. Fokin, the chief of the GPU in the Soviet Consulate officially welcomed Trotsky.

Trotsky was concerned if he remained in the Soviet consulate that the GPU officials there might confiscate his precious research papers. He had little recourse but to remain at the consulate until other private and secure living quarters were found. If he left the physical security of the consulate to live freely in Istanbul, he feared revenge against him personally by former Czarists, and possibly against Natalia and his son. This revenge could come from the large group of emigre White Russians who had fought in the civil war and now lived in Istanbul who were hostile to Trotsky.[195]

Stalin had given Trotsky a grant of $1500 dollars for his resettlement, so his choice of living quarters were severely proscribed. Natalia later reported that the amount given to Trotsky was $5000. Stalinists who sought to denigrate Trotsky claimed that he got away with millions of dollars from the Soviet treasury.

The Trotskys decided that, regardless of the threat from the White Russian emigres, they would leave the safety of the consulate. They moved first to a furnished apartment at 29 Isset Pasha street in a suburb of Istanbul.[196]

Natalia continued to search for a better house in which her beloved Trotsky would be happier and more comfortable to continue his writing. She finally found a dilapidated house to rent on the island of Prinkipo in the sea of Marmara, a short boat ride from Istanbul.

The house on Prinkipo was owned by a bankrupt Turkish pasha who was anxious to rent it, without being too particular about the references of the tenants. Prinkipo was an hour and a half boat ride from the harbor at Istanbul. The island had a few wealthy summer residents but it was normally inhabited only by a few peasant fishermen and shepherds tending their flocks.[197] The Trotskys moved to Prinkipo with the intention of remaining there only until they could find a permanent haven for their exile some other place in the world.

There was never any difficulty with the Turkish authorities although many other countries were anti-Trotsky. In the Turkish struggle for independence from Greece, Kemal Pasha had received arms from the Soviet Union which were delivered through an agency of Trotsky's when he was Commissar of War. Trotsky said, " Fellow soldiers don't forget such things. That was why Kemal Pasha did not lock me up despite the pressure from Stalin."[198]

The President of Germany had indicated in a well received speech before the Reichstag, the German parliament, that Trotsky would be welcome to reside in Germany. When Trotsky applied for a German visa, the conditions demanded for his residence in Germany were unacceptable to him. The German government thereupon withdrew the offer and refused him asylum under any conditions.

The Norwegian government cited the high cost of providing protection for Trotsky if he were allowed to settle there as their reason for refusing to grant him a residence permit. France merely stated that, as far as they were concerned, their order expelling Trotsky from France during World War One was still in effect and Trotsky would not be allowed to enter France. The Labour government in Britain denied Trotsky's visa application without any comment.

Trotsky appealed to many other countries and, in a similar fashion with the flimsiest of pretexts, each of them denied his request for a residence visa or temporary asylum.

He did not bother to apply to the United States. He had been privately told that it would be useless. The United States would never grant him permission to even enter the country much less consider any permanent residence status. Trotsky thus came to realize that for him, this earth was a "planet without a visa," which he took as the title for the last chapter in his autobiography. Governments everywhere apparently still feared Trotsky and some even rejoiced in his plight.

Natalia set to work and soon transformed the crumbling shell of the house on Prinkipo into a liveable place for Trotsky to work.[199] The house eventually, through her persistence, assumed a pleasant homey atmosphere and a cheery ambiance. Amid the dust of many years, Trotsky's papers and books pushed for space against the dingy marble and shoddy gilt of the house's former princely grandeur. The household, however, remained in a constant state of barely organized confusion despite Natalia's earnest efforts to bring order and neatness.[200]

From Prinkipo Trotsky published *The Bulletin of the Opposition*, which he filled with commentary on Soviet and communist policies. Soviet officials who traveled abroad read it and carried it back to friends. It was reported that Stalin carefully studied every copy. The paper was well informed about conditions inside of the Soviet Union and Trotsky was not a critic to be ignored. *The Bulletin* enlightened Stalin better than the reports of his own political police on the moods and hopes of the opposition.[201]

With the exception of one move to Prinkipo's only hotel, the Savoy, when a fire damaged a part of the house, the Trotsky family settled down to live in their new home.[202] The family in 1932 consisted of Trotsky and Natalia. Lyova had returned to Germany in February of 1931 to resume his engineering studies. There were also the combination secretary/guards: Jan Frankel from Prague, Pierre Frank from France, Otto Scheussler from Leipzig, and, the one who was to become Trotsky's confidant and closest friend, Jean van Heijenoort from Paris.

There was also a Russian typist, Maria Lliuishna Pevsner who had an apartment on Istanbul and came over by ship each morning from the mainland and returned each evening. If the sea was rough she would spend the night on the island at the Savoy hotel. In June of 1938, an

American Trotskyite, Sara Jacobs, who knew Russian, offered to work at Prinkipo without pay. Maria Lliuishna Pevsner left on 18 June which saved the financially strapped group her salary.[203]

The only other resident was Trotsky's grandson, Seva Volkov, who had been born in 1926 to Zina, Trotsky's older daughter by his first wife, Alexandra. Zina left Prinkipo in October 1931 for Berlin. She left in order to be with Trotsky's son, Lyova, and to receive medical treatment; she was suffering from a recurrent mental condition.

Sentries provided by the Turkish government patrolled the grounds outside the house. Inside the house, male secretaries tried to do double duty as guards. Regardless of the security precautions taken by the Trotsky household to protect him, a Trojan Horse in the guise of a GPU penetration agent was a constant threat.

Trotsky believed in his own infallibility. He claimed that the force of his ideas and his personality would convert anyone who might have originally been opposed to his philosophies or intended to harm him. Trotsky was a supreme egotist who honestly believed that he could win over the most dedicated, deeply committed, opposition agent Stalin could send against him. Trotsky never realized that he was mortal until that late afternoon of the twentieth of August 1940 in Mexico City, when it was too late!

Max Eastman, an American Trotskyite, visited Prinkipo and wrote this account:

> Lenin warned against Trotsky's 'excessive self-confidence' but when the 'cocksuredness breaks down' he is non-plussed. He does not know how to cherish a doubt, how to speculate... and to confer is out of the question.[204]

Friends in Berlin had warned Trotsky that one Valentine Olberg was a GPU agent being sent to Prinkipo by Stalin to spy upon him. Fortunately this warning was heeded by Trotsky. When Olberg appeared at Prinkipo offering his services, Trotsky politely but firmly refused his offer.[205]

Jacob Frank, also known as Graef, arrived in Prinkipo on 29 January 1929 and remained for almost five months as one of Trotsky's trusted secretaries. Frank was also a GPU agent sent to Prinkipo to spy on Trotsky. Frank had been recommended to Trotsky as a dependable person by, Raissa Adler (maiden name Timo Feievna), the wife of Alfred Adler, the noted Viennese psychologist.

Frank was a Lithuanian Jew, a member of the Austrian Communist Party and had worked in the Soviet Commercial Office in Vienna. He did not try to hide any of his past communist affiliations from Trotsky for two reasons. First, the GPU knew Trotsky's penchant for taking opposition agents under his care, like the average person might take in stray cats. The GPU calculated on taking advantage of Trotsky's tremendous conceit in this regard. Secondly, the very warm recommendation of Frank by Raissa Adler was truly sincere. The GPU believed Trotsky would accept her word over any gossip about Frank's past associations with communism and the GPU.

In addition to Olberg and Frank the GPU recruited spies from the local peasants and two of the most successful GPU spies who penetrated the Trotsky household — the two Sobolevicius brothers, Abraham and Ruvin, both Lithuanian Jews and both fervent communists.

The GPU had established the practice of recruiting agents to work in western Europe from among Russian speaking Jews who were native to countries that bordered on Russia. Some of the best examples of this practice were the Sobolevicius brothers, Jacob Frank and, later in the Trotsky saga, Mark Zborowski.[206]

Ruvin Sobolevicius used the name Roman Well as an alias to penetrate the Trotsky household. Before he came to Prinkipo he had infiltrated many Trotskyite organizations in Germany. He succeeded in these penetrations mainly because of the recommendation from Jacob Frank. The two spies served as cross references for each other in a web of mutual recommendations. They exchanged information on how best to burrow their way deeper into the confidence and trust of a Trotskyite group.[207]

Abraham Sobolevicius used the name Abraham Senin as an alias. He freely admitted that his wife was a Soviet citizen and a member of the CPUSSR. His wife, he readily acknowledged, was employed at the Soviet Trade Mission in Berlin. The association of Senin's wife with the Soviet communist party appealed to Trotsky as making Senin another political waif to be converted to his own ideology. It was also another warning sign that, in his supreme egotism and conceit, Trotsky chose to ignore.[208]

The Sobolevicius brothers had risen rapidly in German Trotskyite organizations before going to Prinkipo. In 1931 they both visited Trotsky at Prinkipo. Later Abraham Sobolevicius (as Abraham Senin) spent considerable time in Russia briefing the GPU how best to take reprisals against the relatives of Russian Trotskyites who were still living in Europe. Rubin Sobolevicius (as Roman Well) spent the first months of the Spanish Civil War shuttling between Toulouse and Barcelona, reporting to the GPU on the basic communist loyalties or Trotskyite sentiments of members of the International Brigades.[209]

A Trotskyite sympathizer who secretly visited Trotsky on Prinkipo was Yakov Blyumkin (aka Blumkin). In 1918 Blyumkin was a Cheka agent who helped assassinate the German Ambassador to Russia, Count Mirbach. Trotsky had subsequently interceded for him and he was grateful to Trotsky for having saved his life. Blyumkin had risen in the Cheka/GPU ranks to become the "illegal resident" in Istanbul.[210] He agreed to transmit a message from Trotsky to Karl Radek inside the Soviet Union. A GPU penetration within Trotsky's entourage reported Bylumkin's visit to Prinkipo.

Lisa Gorskaya, a GPU agent in Istanbul was instructed to "get close" to Blyumkin and seduce him to find out all she could about him and his relationship with Trotsky.[211] Blyumkin trusted Gorskaya and confided to her the help he had given Trotsky which she duly reported to Leonid Eitingon, the senior GPU officer in the area. Blyumkin was arrested and as he was being taken away he realized he had been duped. He shouted to his former mistress who was watching his arrest, "Lisa, you have betrayed me!" Later, just before Bkyumkin was executed his last words were "Long live Trotsky!"

The self-destructive indifference to security that Trotsky permitted to exist at the start of his exile at Prinkipo set the pattern for the future. This woefully inadequate regimen consisted of inefficient night watchmen serving as guards outside his houses (instead of using professional soldiers) and inexperienced (even though deeply devoted) amateur secretaries as guards inside the houses. Granted that all of these guards were prepared to risk their lives in Trotsky's defense, they were no match for the professional infiltrating GPU agents.[212]

Trotsky's physical defenses at Prinkipo, and every other place he lived including Coyoacan in Mexico, were clearly ineffectual and amateurish. However, even had his physical protection been the strongest and the most effective, he would still have been at the mercy of the infiltrating penetration agent who could apparently (and did) worm his way into the very heart of Trotsky's trusted entourage.

TROTSKY IN EUROPE

> The worst fault a person can have is to be conscious of none!"
>
> Anonymous

> History does nothing, possesses no enormous wealth, fights no battles. It is rather man, the real, living man, who does everything, possesses, fights. It is not 'History' as if she was a person apart, who uses man as a means to work out her purposes, but history itself is nothing but the activity of men pursuing their purposes.
>
> Karl Marx

Trotsky, still in Prinkipo, was invited at the end of 1932, to speak to the Danish Social Democrats in Copenhagen. This was to be in celebration of the fifteenth anniversary of the Russian October revolution. When Trotsky indicated a desire to leave for Copenhagen in November of 1932, the Turkish authorities gave him and Natalia Turkish passports for aliens. Trotsky and his family had been deprived of Soviet citizenship by a decree of 20 February 1932. They used these same Turkish passports, the only ones they had, when they entered Norway and Mexico.[213]

On 14 November Trotsky, disguised as a Mr. Sedov, left Istanbul with Natalia and three of his secretary/guards; Jan Frankel, Pierre Frank and Otto Schuessler. Trotsky's disguise and alias were quickly penetrated by the GPU. In order to embarrass Trotsky and hold him up to world ridicule the GPU broadcast to the world his travel route and the alias he was using.

Many people, goaded by Stalin's derogatory propaganda against Trotsky, still feared him. As a consequence, his journey received an undue amount of odious publicity. He was harassed by the police and hostile crowds at every stop along his route. He finally arrived in

Copenhagen on the 23rd of November, unharmed, undaunted and still in zesty spirits.

In Copenhagen the Danish Royal family snubbed Trotsky because they were related to the Russian Czar's family that had been executed by the Bolsheviks. Trotsky had been accused at the time of the Romanov's massacre of having master-minded the killings when, in fact, he had nothing to do with them. These charges that Trotsky had murdered the Romanovs were shrilly repeated in the communist press during the entire time he was in Copenhagen.

Communist party newspapers loyal to Stalin accused Trotsky of having sabotaged the Bolshevik October revolution which further exacerbated leftist public opinion against him. This concerted propaganda barrage demanded the people of Copenhagen use physical violence to prevent Trotsky from speaking at the anniversary program. These "popular demonstrations" against Trotsky were orchestrated by the GPU and staffed by local communist party activists.

Trotsky moved several times in Copenhagen to avoid the hostile crowds that were constantly demonstrating in front of his lodgings. The newshounds, and crowds bent on violence, followed him everywhere. They quickly discovered the new hiding places his guards moved him to on an almost daily basis.

Trotsky's speech was eventually given without incident. Friends had been apprehensive that the anti-Trotsky mobs would disrupt the speech but it proceeded peaceably and was generally well received by the audience.

In July of 1933 the new Daladier government of France revoked the 1916 order expelling Trotsky from France and granted him a visa. Trotsky left Prinkipo on an Italian ship with Natalia and Max Schactman, an American Trotskyite who had been expelled from the CPUSA in 1928 for Trotskyite deviations. Also traveling with Trotsky were three secretary/guards; Sarah Weber, Rudolf Klement and Jean van Heijenoort, a French communist.

On 9 October 1938 Trotsky went with Natalia for a rest cure in the Pyrenees and even paid a visit to the holy shrine at Lourdes. Possibly he hoped that even a non-Christian communist, albeit an anti-Stalin and good Trotskyite communist, might find some solace from his lumbago with a miracle at the Grotto of Lourdes!

Trotsky and Natalia lived in country houses in France for almost a year without secretaries or guards. They were content being alone and, in later years, would often reminisce about the happy times they had at Domesne.

On 1 December 1934 Trotsky learned from a radio broadcast that his old friend Sergie Kirov had been murdered by one Nikolayev.[214] At first the news releases from Moscow reported that the counter-revolutionary White Guards were suspected of having masterminded Kirov's assassination. Then the master propagandist, Joseph Stalin, took over the flow of disinformation. Without any evidence Stalin accused Trotsky, Zinoviev and Kamenov of directing Kirov's murder.

Eventually Nikolayev and fourteen others were executed for Kirov's murder but the blame was firmly attributed to Trotsky as the master planner. In the minds of the Russian people, and communists around the world, Stalin's accusations against Trotsky made him the real culprit. Subsequently, Stalin condemned thousands of Russians to death who were only suspected of harboring sympathies for Trotskyism. Stalin used Trotsky's alleged connection with the Kirov murder to send thousands of innocent people, all falsely accused of complicity in Kirov's killing solely because they were alleged to be Trotskyites.

Kirov's assassination on 1 December 1934, took place in the Smolny Institute, which was the Party headquarters in Leningrad. The assassin was Leonid Nikolaev, a disaffected young communist. It was a badly bungled affair. Nikolaev was only the hatchet man for the GPU. The GPU planned the assassination but since it was supposed to be a random killing by a disaffected youth they were not blamed.[215]

Zinoviev was sentenced to ten years in prison at hard labor and Kamenov to five years for Kirov's murder. Trotsky feared, and future events were to prove him correct, that Stalin had offered Zinoviev and Kamenov a deal for their lives and a promise not to torture their families. The truth was that Stalin had allowed them to live if they implicated Trotsky as the leader of the terrorist's counter-revolutionary plot and chief plotter of Kirov's murder.

Trotsky appreciated, possibly for the first time, the enormity of Stalin's threat against him. He realized his life was in imminent danger as were the lives of all his family. Stalin had subtly announced this ultimatum to all those with the ability to read the party totems. By his elimination of

Zinoviev and Kamenov Stalin had declared unrestricted warfare on Trotskyism and its leader wherever it or he could be found.

Trotsky was also convinced that his supporters, and everyone who had ever befriended him (that is, those who dared share his political philosophy or had not been unequivocal in loudly accepting Stalin's political views), would now be accused of the same conspiracy against Stalin and the Soviet government as was Trotsky.[216]

In those cases where a specific charge could be verified, the perfidy of Stalin's charge could be proved. A hotel in Copenhagen where three of the defendants testified to having met Trotsky, had ceased to exist many years before the date of the alleged appointments. The authorities of an airport near Oslo, where one of the defendants claimed to have landed a German plane for a meeting with Trotsky, testified that at the time indicated (and during several months before and after the alleged date) no foreign planes had landed at the airport.[217]

Trotsky realized that his family, wherever they could be found, would be persecuted and executed by Stalin's minions. He deeply lamented the fate of his younger son, his first wife, his two sons-in-law and his grandchildren (except Seva who was with him) and his remaining daughter. Those relatives in the Soviet Union, Trotsky knew, were doomed without any hope of escape from Stalin's revenge.

Trotsky had already seen his family suffer from Stalin's edicts. Zina (Zinaida) and Nina, Trotsky's daughters by his first wife, Alexandra, had both suffered from severe consumption when they were young and this condition had persisted through their early adulthood. Most of the time Nina was sicker than Zina and it was Zina who tenderly nursed her sister through many a bout of a wracking coughing attacks of consumption. Nina died at the relatively young age of twenty-six years, much to Zina's dismay.

This experience of caring for her dying sister, to whom she was deeply devoted, had a profound effect on Zina's mental stability. At the same time, Zina's husband, Platon Volkov, had been deported to Siberia in 1928 by Stalin's orders. Taking upon herself her sister's grief, Zina also grieved for Nina's husband, Man Nevelson, who had been sent to Siberia about the same time as Platon Volkov. Zina attributed part of the cause of her sister's early death to her worry about Nevelson's ultimate fate in Siberia.

Zina had been allowed to leave the Soviet Union with her five year old son, Seva (Vsievolod, Sieva or Seva) in January of 1931. This was not a clean escape, free from mental anguish. Zina had been forced, as a condition of her freedom, to leave her daughter behind in Russia with her mother, Trotsky's first wife, Alexandra. At the time, Alexandra was already caring for Nina's two children so one more did not add that much to her burden.

Zina had spent less than a year at Prinkipo which was, for her, a period of great emotional tension with Trotsky. For his part, Trotsky could not understand Zina's irrational behavior and he did not try to sympathize with her. Zina resented Natalia's assuming her role as Trotsky's wife in place of her mother, Alexandra. All of this further exacerbated her relationship with Trotsky.

Natalia, for her part, always maintained in later years that she went out of her way to be considerate of Zina and tried to show her all the affection she could as a step-mother. Zina did occasionally warm to Natalia's overtures of compassion but deep down she could not help but reveal her bitter resentment. This constant female bickering between the two, made Trotsky nervous and irritable. His work slowed and almost stopped.

Unable to stand the hostile atmosphere of the household, Trotsky reacted to the tense situation and thought he saw a way to ease Zina out of the house. Zina's irrationality was apparent, even to Zina herself, and Trotsky persuaded her to accept psychiatric counseling in Berlin where she could live with her half-brother, Lyova. Zina reluctantly agreed and promised to undertake the proffered treatment in Berlin.

Once in Berlin, however, Zina rejected any overtures of friendship from Lyova. Gradually, without friends and eluding all efforts to get her into treatment, she became more introverted and morose.

In February of 1932 Stalin announced that Trotsky and all his family had been officially expelled from Russia. The decree stated that Trotsky had been deprived of his Soviet citizenship. The actual effect of this order was to also take Soviet citizenship away from all of Trotsky's family as a matter of derivative law. Zina keenly felt the hopelessness of being a stateless person and drew more into herself, depressed and isolated from reality.

Seva Volkov, Zina's son, had remained at Prinkipo with Trotsky and Natalya. As soon as Zina was settled in her own apartment in Berlin she begged Trotsky to send her son to her. Trotsky was concerned about Zina's irrational behavior. He doubted that, in her distraught condition, she would be able to properly care for his grandson.

Trotsky was understandably reluctant to send the boy to live with Zina who by then was experiencing hallucinations and probably not to be trusted with the boy's care. Regardless, Zina, after much wheedling, finally prevailed upon Trotsky, who was inclined to give in to Zina anyway, and Seva Volkov was sent to Berlin to live with his mother.

This incident manifests Trotsky's character, so deeply hidden on most occasions, of letting sentiment and emotional attachment for his favorite daughter overcome his objective rationality. He was fully aware that it would be better for Seva's well-being to keep the boy at Prinkipo with him where he could be properly looked after and lovingly cared for by Natalia. In trying to placate his pleading, emotionally and mentally maladjusted daughter, he allowed Seva to join Zina in Berlin in December 1932.

Trotsky felt guilty about Zina's condition. He could not escape the thought that he was to blame for her illness. Zina had always worshipped Trotsky and Trotsky knew that he did not show Zina the depth of the love he truly felt for her. Trotsky honestly loved Zina with a strong paternal love. In Trotsky's defense, Zina's irrationality and temper tantrums did not make her the easiest person to express demonstrative affection for. Conversely, Zina conveniently blamed Trotsky's indifference to her as the root cause of her psychiatric problems.

Trotsky's worst fears about Zina and Seva's safety were soon confirmed. Shortly after Seva arrived in Berlin to be with his mother, she committed suicide. Zina killed herself by inhaling gas on 5 January 1933. She was found dead at two o'clock in the afternoon.[218]

Alexandra wrote Trotsky from the Soviet Union blaming him for Zina's death. Alexandra claimed that, by forcing Zina to reveal her innermost thoughts during psychiatric counseling, he had compelled her to reveal emotions which she desperately wanted to keep as her own secret's. Therefore, according to Alexandra, Trotsky had caused Zina to commit suicide. Alexandra closed her letter to Trotsky with a most prophetic comment, "Our children are doomed!" she wrote.

Charles Yale Harrison, an American Trotskyite, had the temerity to criticize the Soviet Union for driving Zina to suicide. He was promptly expelled from the John Reed Clubs, accused of being "one of the star hired men of the yellow Hearst press" and denounced for echoing the same line as Archduke Cyril, Pope Pius, Alexander Kerensky, United States Congressman Hamilton Fish and the leaders of the Second International[219]

Custody of Seva Volkov, alone in Berlin after Zina's suicide, was assumed by her brother, Lyova Sedov, and Jeanne Martin. Lyova had telephoned to Trotsky at Prinkipo to report the news of Zina's death. He had also been able to get through to his brother, Sergei, in Moscow with the same depressing news. Lyova had asked Sergei to pass the news to Alexandra. This incidentally was the last time the brothers, Trotsky's sons, spoke with each other.

In April of 1935, Trotsky learned that his first wife and Zina's mother, Alexandra, had been deported to Siberia and that Sergei's wife, Lelia, had disappeared. Lelia's disappearance was the first of many such episodes where Trotsky's relatives and former friends simply "disappeared."

Seva Volkov, whom Trotsky looked upon as the promise of his future, continued to live in Berlin with Lyova and Jeanne Martin. In March of 1935 they all moved to Paris to escape the Nazi persecution of Jews, then beginning with a vengeance in Germany.

In France, Trotsky was increasingly concerned that the French government would finally accede to Stalin's continuous, and increasingly urgent, demands that he be expelled from France and repatriated to the Soviet Union. In anticipation of such a possibility, Trotsky asked for and was granted asylum in Norway. Internal considerations that had previously made the Norwegian government refuse to accept Trotsky's request for asylum had changed. A Socialist government in Norway, more sympathetic to Trotsky's political philosophy, had come to power through recent elections.

Walter Held, (real name Heinz Epe) a German Trotskyite, who had taken refuge in Norway, arranged for some Norwegian friends to ask the government of Norway to issue a visa for Trotsky. On 8 June 1935 Trotsky was notified the visa had been granted. It was a visa valid for six months only.[220] At the same time, malicious Soviet propaganda reported that in late 1935 Trotsky had sent an eight page letter to Karl Radek outlining his secret agreements with Germany and Japan that would take effect when he returned to power in the Soviet Union with their help. Supposedly these agreements ceded the Ukraine to Germany and assured

Japan that the Soviet Union would not object to Japan's conquering of China.[221]

Trotsky and Natalia arrived in Oslo on 18 June 1935. Konrad Knudson, a socialist newspaperman and his wife, extended the hospitality of their house to them. Their new residence was in Vexhall, a village near Honnefoss, about sixty miles from Oslo. Exhausted and weakened by their flight from France, Zina's tragic suicide and the depressing worry about the fate of their loved ones in the Soviet Union, Trotsky entered a hospital in Oslo while Natalia rested at home. Trotsky could not afford a prolonged period of recuperation. Regardless of how much he needed the rest, he had to leave the hospital before he had fully recovered.

Domestic tranquility settled on the Trotsky entourage while they were living in Knudsen's house — on 7 August 1935 Trotsky announced to the world that he had finished what many consider his most incisive work about events in Russia, *The Revolution Betrayed*; Erwin Wolfe, one of Trotsky's secretaries, married Knudsen's daughter; Natalia was happy with her sewing, and everyone in the group was happy that Trotsky was happy.

Then, on 5 August 1936 when the storm troopers of Major Quisling's pro-Nazi party, the National Sammling, broke into Knudsen's house. They ransacked the rooms where the Trotskys were staying. When they left it was discovered that many of Trotsky's important papers documenting Stalin's biography had disappeared. The only reasonable conclusion was that the storm troopers had stolen them on the order of some one in authority.

A week later, a radio broadcast from the Soviet Union announced that Zinoviev, Kamenov, Trotsky and fourteen other defendants were to stand trial in Moscow.[222] These defendants were accused of attempting the assassination of Stalin.[223] All of them were accused of conspiracy to overthrow the Soviet government and, for good measure, for treason against the Russian motherland.

> The terrorism of Stalin rests upon a universal system of spies and informers. The Moscow trials show us what it [the Soviet Union] has made of justice. Sometimes its place is taken by summary death sentences, sometimes — which is worse — it works behind an apparent correctness, but its verdicts, concocted beforehand, are the outcome of false

> witness and *fake confessions*. [Emphasis in the original.] This last point is worth dwelling upon. Nothing is more significant nor more abominable. The accused of Moscow *confessed* the crimes charged against them which they had not committed. They formally acknowledged the truth of the facts proved to be [by other independent evidence] substantially false.[224]

This was the first of the so-called "Purge Trials." Then came the second wave of rigged trials when Stalin desperately tried to eliminate the NKVD officers who had been his instruments for the first series of purges and who knew the secrets of extorted confessions and planted evidence.[225]

Trotsky was specifically charged in the indictment as the main defendant and his son Lyova, who was living in Paris, was charged as his father's principal collaborator and co-conspirator. Another count in Stalin's charge against Trotsky was that he allegedly worked in close collaboration with the Nazi Gestapo and of sending sabotage teams from Norway into the Soviet Union.

Trotsky could not believe the grandiose falsity nor the complete lack of evidence of the charges. Where would he ever get the resources to send sabotage teams into the Soviet Union and why, after a lifetime of anti-Nazi philosophy, would he ever think of working peacefully and in harmony with the likes of the Gestapo? He could only marvel at Stalin's ingenuity in concocting such fabrications. He knew, even though the charges against him were absurd, that this was Stalin's attempt to discredit him further in the minds of socialist and communist world opinion.

Trotsky immediately began to prepare his defense. From 19 through 24 August Trotsky read every newspaper and listened to every radio broadcast of the judicial farce going on in Moscow that was trying him "en absentia." Despite his continuing ill-health, and bouts of lumbago that partially immobilized him, Trotsky worked diligently trying to marshal evidence in rebuttal of the charges.

His efforts at refutation were immediately frustrated. On the first day of the trial, the Norwegian government demanded that Trotsky sign a document that he would not interfere in the politics of another country. This was obviously a devious reference to the Soviet Union and the purge trials. This gag on Trotsky was enforced by Trygve Lie, the Norwegian

Minister of Justice. Lie was alleged to be Stalin's stooge who would obediently carry out any anti-Trotsky order from Stalin.

Trygve Lie was no stranger to Trotsky. Lie had visited Moscow in 1921 and, according to Trotsky, had been identified as an agent of the Comintern since its earliest days. Lie, immediately upon the conclusion of the Moscow trials, submitted a demand to Trotsky that he voluntarily sign a statement agreeing that his mail be censored and that he stop writing articles about current political affairs which could involve Soviet affairs.

Trotsky refused, seeing in Lie's demands nothing more than an attempt by Stalin to gag him. Trotsky demanded an open trial in Norway to establish his innocence of the Moscow charges. He hoped, if he were allowed to be tried in Norway, that he could ridicule Stalin's judicial mockery for what it was and to expose Lie's complicity with the GPU.

There is a document, little known except to academic researchers entitled, *American Aspects of the Assassination of Leon Trotsky,* "Hearings Before the Committee on Un-American Activities, House of Representatives, July-August 1946. Specifically Testimony of Louis F. Budenz Hearings, 22 November 1946." This document makes clear the ostensible justification used by the Norwegian government to try and throttle Trotsky.

In an attempt to supply spurious justification for Lie, who was acting ostensibly for the Norwegian government, Stalin turned to the ingenuity of his GPU agents in the United States.[226]

In his testimony before the House Committee, Louis F. Budenz, former editor of the communist newspaper in New York *The Daily Worker*, and admitted Comintern/GPU espionage agent, gave one such reason. Budenz claimed that the groundwork for the GPU pressure on the Norwegian government to oust Trotsky was based upon some confidential information he supplied to Roberts, his GPU case officer in the underground espionage apparat of the CPUSA. Roberts has been identified as the senior GPU undercover agent in the United States at that time. He was Dr. Gregory Rabinovich, whose cover was the representative of the Soviet Red Cross in the United States.

Budenz had reported to Roberts that a prominent left-wing pacifist, Reverend A.J. Muste, had visited Trotsky in his Norwegian exile home in June of 1936. Allegedly, Trotsky told Muste he had plans for the violent overthrow of the Soviet government and the assassination of

Stalin. This information eventually reached Budenz who passed it on to Roberts.[227]

The Reverend Muste, when interviewed by the House Committee, categorically denied that any such discussion with Trotsky had ever taken place. Budenz's report, however flawed and regardless of its lack of verification, was turned over to Roberts and subsequently given credence in Moscow by Stalin,. From there it was a short trip for the information to reach Stalin's stooge, Trygve Lie, who used it to justify his harassment of Trotsky in the name of the Norwegian government.

Attesting to the use of this false information against Trotsky was Roberts's subsequent commendation to Budenz. Roberts acknowledged Budenz's aid by saying, "This information had been of great value in private representations made by the Soviets in Norway..." i.e., that Trotsky was planning the violent overthrow of the Soviet government from his exile in Norway.

Trygve Lie visited Trotsky several times during Trotsky's house arrest. These meetings were never characterized by an excessive amount of courtesy nor restraint by either party. Trotsky hated Lie, accusing him of having all the vices and none of the virtues of a bourgeois bureaucrat and of being a toady and stooge of Stalin. Lie would retaliate by telling Trotsky that Norway had committed a grave error in granting him asylum and that he should be expelled.

Despite Lie's order for Trotsky's house arrest, he was allowed to appear in open court as a witness against the Quisling storm troopers who had broken into Knudsen's house and stolen his papers. Trotsky hoped he could use the trial to launch a refutation of Stalin's charges or, at least, to generate a protest against Lie's persecutorial harassment.

At the trial it became clear that the judge and prosecutor were acting as Lie's agents and they treated Trotsky like the defendant rather than a complaining witness. Lie then ordered Trotsky to cease all comments on the trials in Moscow and again Trotsky refused. Lie next ordered all of Trotsky's secretaries, who were not natives of Norway, deported. This meant, for all practical purposes, that all of Trotsky's secretaries would be deported.

Lie, not yet satisfied with the vehemence of his harassment, decided to make Trotsky's simple house arrest a virtual prison. He installed his own military police as guards inside Knudsen's house so that the other residents could not talk with Trotsky nor he with them. Lie, on 2

September 1936 ordered Trotsky and Natalia moved to a small house south of Oslo.

On 11 December 1936, Trotsky again was called to appear as a witness at the Quisling robbers' trial. This time Lie was taking no chances that Trotsky would be able to place his case before the press through the forum of the open court. Lie ordered the court emptied of all spectators and members of the press.

To illustrate the type of charges Trotsky faced that were essentially defenseless, Yagoda, the chief of the GPU, made up allegations against Trotsky that were complete fabrications. In 1927 Yagoda reported to Stalin that one Stroilov, a GPU provocation agent, had helped some Trotskyites get a supply of paper for their printing press [hectograph] and then testified against them. Stalin told Yagoda to claim this pliable agent also worked for General Wrangel, the counter-revolutionary White Army leader. The agent was to claim, without any foundation in fact, that the Trotskyites collaborated with General Wrangel.[228]

Trotsky was therefore effectively prevented from defending himself against charges Stalin made against him at the Moscow Trials. He deplored being forced to live in squalid isolation with no recourse to justice. He was deprived of any communications with friends with whom he could, at least, argue or exchange banter. At this point in his exile, Trotsky was at his lowest state of mind and utterly dejected.

Trotsky eagerly sought escape, asylum or solace of any kind from any place in the world. The only person who apparently heard his plaintive plea was Diego Rivera. A world renowned muralist of ultra-left sympathies, Rivera had aspirations to be acknowledged as the world's foremost communist muralist. A native son of Mexico, his recognized artistic ability had brought him world admiration which automatically made him a celebrity to be honored in his own country. Rivera was able to persuade the Mexican president, Lazaro Cardenas, to grant Trotsky asylum in Mexico.

Lie was delighted to have a plausible excuse to be finally rid of the Trotsky nuisance. Probably only after receiving Stalin's concurrence, Lie agreed to allow Trotsky to leave Norway for Mexico.

Winston Churchill spoke eloquently of Trotsky'a fate in Norway,

> ....the once triumphant Trotsky, whose frown meted death to thousands, sits disconsolate — a skin of malice stranded

> for a time on the shores of the Black Sea and now washed up on the shores of Mexico.[229]

On 19 December 1936 Trotsky and Natalia, without personal guards or secretaries, but with a Norwegian police escort thoughtfully provided by Lie, set sail on the Norwegian tanker S.S. Ruth bound for Tampico, Mexico.

Trotsky thus left Europe never to return, while his assassin, and the assassin's helpers, were rehearsing skills of the fine art of murder in the Spanish Civil War preparatory to confronting him in Mexico.

THE SPANISH CIVIL WAR

The only guide to a man is his conscience... It is very imprudent to walk through life without this shield, because we are so often mocked by the failure of our hopes and the upsetting of our calculations; but with this shield, however the fates may play, we march always in the ranks of honor.

Vernon Walters, Former Deputy Director of Central Intelligence Agency

Many people have character who have nothing else!"

Don Herrold

The Spanish Civil War bears some mention in a book about the Trotsky assassination because it was in the killing fields of this war that Trotsky's future executioners appeared on the scene and honed their annihilation skills.

At the start of the Spanish Civil War the two opposing philosophies were lucid and clear. For the true believer on each side there was no middle ground. There was no murky area where rational argument might be convincing and intervene to stop brother from murdering brother. There was no international referee who might at least order a pause for reconsideration in an attempt to halt or delay the impending carnage. There were only diametrically opposed, black and white, polarized right and wrong, with absolutely no room for compromise, for each dedicated opponent.

Each side was convinced that theirs was the righteous quest and the enemy was diabolic, blood thirsty, ill-informed and they were all charlatans without principles bent on a course politically disastrous for Spain.

Lenin characterized Spain as the "little Russia of Europe." At the time he made the analogy it was long before the Civil War of 1936 had erupted. Lenin was referring to the semi-feudal, semi-capitalistic economy of Spain with its general social and political climate remarkably similar to pre-revolutionary Russia. During the period of the Spanish Civil war the similarities became even more pronounced. Due to the aid given by the Soviet government to the Republicans, the Soviet GPU was able to exercise considerable control over the people of Spain. Maybe Lenin, if he had still been alive, would have found this aspect of complete GPU police control even more strikingly similar!

The magnitude, and the brazen manner, that Stalin's GPU considered themselves to be in complete control of the situation in Republican Spain was nakedly expressed in a report of *Pravda*. On 17 December 1936 the official newspaper of the Soviet government, quoting a statement from the Communist Party of the Soviet Union reported:

> So far Catalonia [a province of Spain] is concerned, the cleaning up of the Trotskyites and anarcho-Syndicalist elements has already begun and it will be carried out with the same degree of energy as it was within the Soviet Union.

Workers' strikes, declining productivity, soaring inflation and the inability of the government to cope with the resulting emergencies gave way to factionalism and resulting chaos. The communists relentlessly attacked the administration and managed to gain several seats in the Cortes, the elected parliament of Spain.

In pre-revolutionary Spain, the communists agitated for workers' strikes and incited revolt against all authority. The government forces reacted violently against the communist agitation and ruthlessly slaughtered thousands of otherwise innocent Spanish laborers who belonged to the communist dominated trade unions.

Martial law was declared by the government and throughout Spain some 35,000 striking socialists, anarchists and communists were jailed for defying the government's order to return to work. It appeared that the country was on a course that could only be resolved by open revolt.

Fascist pro-government forces were assigned to quell the strikes. They began a pacification program in the rural areas, well away from the population centers, so as to avoid alarming the general public to the extent of their brutality. These forces began a systematic series of beatings and killings of leftist labor leaders and popular pro-revolutionary politicians. Anti-government Workers' Defense Leagues sprang up in the countryside to militantly defend themselves against the brutal persecutions.

The communist groups burned the Roman Catholic Churches in Spain; burned them right to the ground so they could never be used again. This was partly because of the widely held anti-Catholic sentiment of most of the leftist elements in Spain but, a more practical reason was the fascists used the churches as munitions storehouses and places to cache weapons. The fascists hoped the buildings would be immune from attack by the Republican peasant forces. The reverence they had for the holy ground on which the churches had been built and their peasant superstition that anyone harming a church would be cursed, would protect the buildings.

The fascists also used the Catholic churches as a network of safe houses, or sanctuaries, along their routes of escape from territory held by the Republicans. The Republicans soon learned of this usage and it only added to their determination to rid Spain of all Roman Catholic churches and rid the nation of the Catholic priests who instigated these treasonous acts.

Summary executions of leftist leaders by the fascists brought swift and equally lethal retaliation from the communists. In July of 1936 the fascists stalked out of the Cortes, the elected Spanish parliament, and General Francisco Franco and his troops, revolted against the remaining disorganized, but still the officially recognized and legally elected, government of Spain. The Spanish Civil War had begun![230]

The Republican army was fundamentally a free lance effort energized by the communists, socialists and the leftist labor unions. Each political group marshaled their individual military units, all deeply distrustful of each other and each communicating with the other only when absolutely necessary.

The Spanish Civil War was essentially a war without a front. The leftists, socialists and communists were trying to control the legally constituted Republican government of Spain with its headquarters in Madrid. The rightists, nobility, landlords, the Catholic Church and most of the permanent military force were supporting Franco's insurgent government. Both sides were more concerned with rallying popular support from the urban population and the rural peasants than with the actual fighting. Each side was convinced that Franco's revolt would fail if it did not quickly succeed. Both sides were equally convinced that neither side could win a war of any duration without substantial assistance from outside Spain.

Hitler contributed from Facist Germany to the Franco Nationalist forces, an entire air force including 10,000 pilots, mechanics, and weapons specialists. He received Spanish territory in the Baleric Islands from Franco in return. Mussolini of Italy "volunteered" 75,000 Italian ground troops and technical specialist to Franco in return for a base in Africa.

France contributed neutrality which translated into a route for smugglers of both sides, a holding area for refugees for both sides and a few corrupt French government officials who took money from both sides. Pierre Cot, the French Minister of Aviation was accused of revealing to Moscow secret French military techniques. He did not sue for libel against the newspaper who printed this story which only gave it more credence. In 1953 Pierre Cot was awarded the Stalin Peace prize.

Stalin belatedly contributed to the Republican side, with about 1000 pilots, planes and tanks. Most ominously Stalin contributed some 2000 "advisors," mostly GPU officers. These advisors in turn, ultimately contributed to the preparation of those selected from the Spanish Republican ranks who would later carry out the assassination of Trotsky in Mexico.

Under Vyacheslav Menzhinsky and Grigoryevich Yagoda, successor chiefs of the GPU/NKVD, the operations against Trotsky had been limited to surveillance and harassment of him personally. They also directed penetrations of Trotsky's entourage and, probably most important, the disestablishment by Judas leaders and propaganda of the Trotskyite organization. Under their successor, Nikolai Yezhov, the NKVD embarked on a policy of assassination of the Trotskyite leaders. In December of 1936 Yezhov created the Administration of Special Tasks to be in charge of assassinations and kidnappings.

The Soviet government wrongly believed that the Republicans would quickly defeat the Nationalist forces of General Franco and were slow to react when the Nationalists evidenced military strength. In an effort to strengthen their presence, the Soviets sent one of their most able diplomats, Marcel Rosenberg, on 27 August 1936 to be the Soviet Ambassador to the Republican government. The head of the Soviet military mission was General Jan Berzin and General Lazar Stern. General Stern used the cover name of General Emilio Kleber with a Canadian passport and a cover story to match. Stern was to be the "Saviour of Madrid" and not allow Franco's forces to capture the city.

Virtually everything that was done by the Republicans during the Spanish Civil War was controlled by the Soviet advisors and their agents. Louis Fischer, an American journalist whose views reflected the official Stalinist line, reported:

> There is a powerful hatred of communism developing in Spain. The communists [Soviets] made many enemies by their irritating methods to control important [Spanish] institutions.[231]

True Spanish patriots who were sincerely fighting for a better Spain, and those who may have allied themselves with the leftists only as a matter of expediency, were quickly expelled from the Republican forces by the Soviet advisors. Any Trotskyites, Sindicalists, or any non-conforming or non-communist groups or individuals that did not hew to the strict Soviet/communist line were ostracized from the Republican ranks. Those who displayed any reluctance to leave willingly were dealt with by summary punishment which usually included trumped up charges, jail time and execution.

Constancia de la Mora, a trusted Stalinist was in charge of censorship for the Spanish Republican government. News from Republican Spain was filtered through communist sieves and rigidly controlled. The result was that the outside world heard the Republican story only from the communist angle. The stories of the suppressed and persecuted anti-communist majority of Republicans never were permitted past the censors. The calculated news releases succeeded in reducing the whole Spanish tragedy to a black and red pattern of a life and death struggle between fascism and democracy.[232]

The Republican communist coalition fighting Franco, who had aid from the other fascist powers, appealed to Stalin for military aid. The Soviets were loud in their denunciation of Franco and what he stood for as a fascist. They were equally vociferous in their praise of the valiant fighting men of the Republican forces. The military aid from Stalin, however, was too little too late to have any decisive effect on the outcome of the war.

All Soviet intelligence and covert action operations in Spain, which in the normal course of events meant almost all Republican intelligence operations, were monitored for approval by the veteran GPU officer, Alexander Orlov. He was known also by various street names of Shved (Swede), Nikolsky, Lyova.[233]

Orlov was born in Russia as Leon Feldbin.[234] He was twenty-one years of age when he joined the Bolshevik Party at the time of the 1917 revolution. He had a natural dissembling nature, just what was needed to quickly rise through the party ranks to a position of authority in the Cheka (the Soviet Union's police and intelligence organization). He married a Russian woman and they had a daughter named Vera, whom Orlov adored. Vera's well being was the principal reason Orlov subsequently defected from the Cheka to the west.

Orlov was obviously always planning ahead, either for future GPU operations in the States or hedging his bet against the possibility of a defection to the West when he initiated a trade program with the United States in 1932. He succeeded in obtaining an invitation to spend nearly three months in America as the guest of General Motors. Orlov's visit also laid the groundwork for the later appointment of Gusev, one of Orlov's principal aides, as resident GPU agent in the States. Gusev was to become largely responsible for the Soviet penetrations of United States military forces and the scientific research establishments.[235]

Before his defection Orlov was working in Spain, but before that, in 1931 he had been assigned to England. There he worked with another GPU officer named William Fisher. Years later in the United States this William Fisher became famous as Colonel Rudolph Ivanovich Abel, a deep cover Soviet agent operating in the New York City area.

The tie-ins of persons and the inter-mixing of operations in the ultra-complex web of Soviet intelligence all bears an ultimate relationship to Trotsky and his assassination. The cast of characters of these labyrinthine plots sometimes remains the same and only the scene changes. It is recalled that William Fisher, for example, was first in contact with Trotsky in the office of *Iskra* in London in 1905. A slight digression produces interesting sidelights of the Trotsky drama and eventually leads back to the principal story.

At the time of his arrest in New York in August of 1957 Abel was a Colonel in the KGB, a successor agency to the GPU. He had resided in Canada and United States under false identities for about ten years before his arrest by the FBI. Abel had organized and directed an extensive network of deep cover Soviet agents in the United States while using the covers of a free lance artist and photographer under the alias of Emil R. Golfus.

Abel might have continued to operate this Soviet espionage network in America for many years except his deputy, Rhino Hayhanen, defected to the CIA. Hayhanen had been Abel's assistant from the beginning of 1957 but was a lazy worker, a womanizer and a drunk. He had stolen KGB expense money from Abel's operational petty cash fund and was generally considered by Abel to be a dangerously insecure operational liability. Abel arranged for Hayhanen to be recalled to Moscow so he could be rid of the bumbling, incompetent security threat to his otherwise smoothly functioning operations.

On his way to Moscow Hayhanen had deduced the fate that awaited him there and decided defection to the CIA station in Paris was preferable to the hangman's noose in Moscow. Hayhanen was able to give the CIA in Paris enough information about Abel's activities in New York so that the FBI could locate him and arrest him.

Abel's American lawyer, James R. Donovan, argued that Abel should not be executed. He theorized that some day an American spy might be arrested in the Soviet Union and it would be useful to have Abel in reserve as an exchange chip.[236] Four years later, Gary Powers, flying a super-secret, U-2 reconnaissance plane over the Soviet Union was shot down and arrested. Abel was subsequently swapped for the safe return of Gary Powers.[237]

Returning the narrative to Orlov, during his assignment in England in 1931 he directed the recruitment of Kim Philby and Donald Maclean.[238] Philby was destined to become one of the most senior officers in the British intelligence service while continuing to secretly work for the GPU. While in liaison with the CIA in Washington, Philby passed United States secrets to the Soviets concerning British-United States joint intelligence operations, military cooperation agreements, atomic energy and bombs, and economic partnership pacts. Maclean was also a fount of information for the Soviets. As a ranking officer of the British Embassy in Washington, Maclean passed to the Soviets American atomic secrets plus other scientific and technical information of monumental importance to the Soviets and extremely detrimental to the security of the United States.

In probably one of the least rational comments of the era, British intelligence stated, as a justification for the Philby, Maclean, Burgess debacle:

> It is better that there should be a few Burgess and Maclean cases than that our nation should be like a McCarthyite police state with elaborate checks on people's movements.[239]

Donald Maclean was a student at Cambridge University in England and already a member of the British Communist Party when Orlov recruited him. Maclean's father had been a solicitor (lawyer) and the Chairman of the British Labour Party which may help explain why Donald choose a government career. Proverbially, as good dedicated spies often do, Maclean rose rapidly in the British Foreign Service. Between 1944 and 1948 he was Counselor in the Senior Branch of the British Foreign Service and later became head of the American Department, while continuing to secretly pass all information that came to him to the Soviets.

Information regarding Maclean's spying for the Soviets had been given to the British government by two Soviet defectors. Igor Gouzenk, the code clerk in the Soviet Embassy in Ottawa, Canada, and Konstantin Volkov, the former senior Soviet intelligence officer in Istanbul, Turkey. They both provided evidence to British counter-intelligence of Maclean's espionage for the Soviets long before any definitive action was taken by the British government to curtail his activities and stop his treachery.

Eventually, armed with the evidence of Maclean's traitorous conduct, MI-5, the British counterpart of the American FBI in charge of internal security, began closing in preparatory to arresting Maclean. On 25 May 1951 Maclean secretly fled England to the Soviet Union accompanied by his homosexual friend and fellow Foreign Service officer, Guy Burgess.

Burgess had also been a Soviet spy of long duration in the British Foreign Service.[240] At the time of his flight to Moscow with Maclean he had just returned from Washington where he had served as Second Secretary in the British Embassy. He had secretly passed to the Soviets all information that he was able to collect in his official position about the United States, which was considerable.

Maclean and Burgess had been warned about their impending arrest and given enough time to evade the arresting officers from MI-5 by the infamous "Third Man," Kim Philby. Philby had been the senior liaison officer with the FBI, CIA and NSA for all of British intelligence in the British Embassy in Washington. This was a position roughly comparable to being an ambassador of intelligence matters for his country accredited to the United States intelligence community and privy to all secrets that a trusted ally would enjoy.

In the mid-1930s Philby had covered the Spanish Civil War as a reporter for the London *Times* which had been his cover for his real work as a Soviet espionage agent.

Philby adopted as his cover, the role of an ultra-fascist reporter thoroughly on the side of General Franco. Assisting him in this rightist, fascist role was his mistress Bunny Doble, the divorced wife of Sir Anthony Lindsay-Hogg. Doble described herself as a "passionate Royalist." Doble was unwitting of Philby's dual role and accepted him as a fellow fascist.[241]

Before Maclean and Burgess fled, Philby himself had been under suspicion by MI-5 as a possible Soviet agent. After the two Foreign Service officers escaped the MI-5 dragnet, the suspicions regarding Philby's complicity increased. He was directly suspected of having been the person who had tipped them off. The authorities and the English press reasoned that, in addition to the two escapees, there must have been a "Third Man" who leaked to them the immediacy of their arrests. Speculation abounded that the infamous "Third Man" must have been a member of British intelligence and by a process of elimination wise men reasoned that it must be Philby. There was not enough hard evidence to convict Philby and the case dragged on!

Philby toughed it out through several lengthy and hostile interrogations by MI-5 and MI-6, England's foreign intelligence organization, roughly equivalent to the CIA. Privately Philby eventually recognized that the accumulation of evidence against him had become overwhelming. On 23 January 1963, twelve years after his treachery had first been suspected, Philby also fled to the Soviet Union.

Subsequently in his book *My Silent War*, published from his self-imposed exile in Moscow, Philby admitted his perfidy of many years as a Soviet agent. He bragged of working inside the most sensitive echelons of British intelligence while passing everything to his Soviet case officers. Philby died in Moscow on Wednesday, 11 May 1988, unrepentant, still claiming that the Soviet Union had been his "home" and "homeland" since the 1930s.[242]

These high level recruitments into Britain's intelligence organizations, proved Orlov's astuteness and ability as a case officer. Both Philby and Maclean became extraordinarily valuable agents to the Soviets to the detriment of the American and British governments.

Returning to Orlov, Spain and Trotsky, in 1936 Orlov was chosen by Grigoryevich Yagoda, the chief of the GPU, on direct orders from Stalin, to be in charge of all Soviet intelligence operations in Republican Spain. Orlov was given the power to coordinate all activities of the Spanish Communist Party, all Spanish Comintern operations and the operational direction of the GPU in Spain.

Orlov made certain that only trusted communist comrades were placed as commissars in the International Brigades. He used informants throughout his vast network to ferret out disloyalty among the GPU and to report back to him any criticisms against the Soviet leadership.

Stalin feared that the Partido Obrero Unificacion Marxista (POUM) which was considered to be the Trotskyite political party in Spain, might give Trotsky himself a base of operations in Spain. The Izquierda Comunista (IC) was a Trotskyite party that had split from the Partido Comunista de Espana (PCE). It was also directed by Andres Nin. The POUM was actually a fusion of the IC with the Bloc Obrero y Campesino (BOC) (Workers and Peasants Bloc).[243]

Andres Nin, the co-founder of the POUM had been Trotsky's secretary in Moscow and was Minister of Justice in the Catalan government until arrested by the communists in December of 1936.[244]

Orlov directed the execution of Andres Nin, because he was the leader of the POUM.[245] Nin had been an ardent Trotskyite and, before his conversion, had also been a trusted member of the Soviet Comintern. After he was arrested, Nin mysteriously vanished from the GPU jail where he had been detained.

On 23 October 1937, the Chief of the Republican Barcelona police, Lt. Colonel Burillo, reported the excuse for Nin's arrest. He claimed that the POUM and Nin had been extensively engaged in espionage for the fascists and sabotaged military operations at the fighting front.[246]

Nin's body was discovered only after a specially convened delegation of inquiry was sent by the British Parliament to learn what had happened to Nin.

The people of Spain retained a passion for justice in spite of the GPU's heavy handed methods of dispensing punishment. A populist campaign, that seemed to spring up over night, clamored for information about Nin. Banners appeared on balconies and lamp posts asking the question," Where is Nin?"

Nin had been taken by the GPU to its prison in Alcala de Henares. There he had been brutally tortured, flayed alive, interrogated and finally shot on orders of the GPU by a firing squad consisting of ten German soldiers from the International Brigade.

Irwin Wolf, another of Trotsky's former secretaries, was also kidnapped and executed in Spain. In addition, most of the other POUM leaders were also executed, about seven or eight in all. The POUM lawyer, Benito Pabon, became so terrified for his life that he fled to the Philippines.[247]

In Catalonia the Trotskyites tried to withstand the pressure from Orlov's inquisitorial purge of all non-Stalin communist factions. Orlov was insistent and set out to provoke the Trotskyites in Barcelona, the capitol of Catalonia, into premature violent revolt. The Trotskyites stormed the streets and rioted. Orlov used this public insurrection that he had contrived as an excuse to declare that the Trotskyites threatened the stability of Catalonia's law and order and must be curtailed. He used the massive power of the GPU to quell the rioting and, in the process, eliminated most of the Trotskyite faithful by executions.

In September of 1936, Juan Negrin was the Finance Minister of the Republican government. He suggested to Arthur Stashevsky, the Soviet Economic Attache in Madrid, that all of the Spanish Republican government's gold reserves should be moved to the Soviet Union for safe keeping.

This suggestion was sent immediately to Stalin who replied by cable that Negrin's request should be honored as expeditiously as possible. Orlov was put in charge of arranging for this shipment of gold to the Soviet Union. As he was conducting the crating and loading of the gold, he received the following cabled message: (Paraphrased)

> Absolutely secret. This message must be decoded by Shved personally. [Shved was Orlov's pseudonym]. Imperative you arrange movement to Russia of Spanish gold by Soviet steamship. Utmost secrecy essential. Give no (repeat) no receipt. Imply receipt must await verification gold received Moscow State Bank. I hold you personally responsible for success this most important venture.
>
> Ivan Vasilyvich [Stalin's pseudonym]

At Orlov's suggestion, Negrin gave him fake American credentials. This was to avoid suspicion if he was questioned by unwitting Republican officials at the loading docks. With these credentials Orlov could claim that he was preparing the gold for shipment to America for safe keeping. In order to make the masquerade more believable, Negrin had prepared these fake credentials to identify Orlov as Mr. Blackstone, a representative of the Bank of America.

The total worth of the gold, without attempting to estimate the numismatic value of the ancient gold coins included in the hoard, was at least $518 million United States dollars. Stalin is alleged to have said, when the Republican government's gold arrived in Moscow that, "The Spaniards will never see their gold again as they will never see their own ears!"

When Maurice Thorez, the French communist party leader, appeared before the Soviet Politburo in Moscow in mid-September of 1936, he proposed the creation of an international brigade of socialist fighting men to help the Republicans fight the war in Spain. The non-attributability of such an international non-communist fighting force without any direct connection to, or ostensible commitment with, the Soviet Union appealed to Stalin.

This scheme of Thorez would give Stalin the option to plausibly deny any Soviet intervention in the Spanish Civil War if he were accused of doing so. It would also, on the other hand, give his position as leader of the communist world credibility because the Comintern agitprop mechanism would broadcast his true sponsorship of the International Brigades to the initiated.

As unappealing as front line fighting, or rear echelon GPU treachery, was in the Spanish Civil War, volunteers from all over the world wanted to join the Republicans who were fighting the evils of fascism. The Soviets were therefore able to appeal for help to the socialist and communist young men from all parts of the world and from all walks of life, to come to Spain and be a part of the "good fight." It was thus that the International Brigades were created!

The communists in the United States contributed their own unit to the International Brigades called, the "Abraham Lincoln Battalion." The Battalion was made up, in a small part at least, of well intentioned liberals, idealistically fighting against fascism. Many, if not most, of the other men in the Battalion were avowed communist party members, or, at the very least, what in those days were called "fellow travelers."

Men who volunteered to fight in Spain with the International Brigades were not necessarily informed of its sponsorship. Therefore, evidence that an American fought with the Abraham Lincoln Battalion cannot be used to prove, that he was, at that time or subsequently, a member of the CPUSA or a communist sympathizer.[248]

After the Spanish Civil War the actual idealists of the Abraham Lincoln Battalion were tarred with the same brush as the die-hard communists who fought with the Battalion. After World War Two during the days of McCarthyism and the witch hunts of the Federal and State governments who were seeking to root out communists, these innocent anti-fascists were also pilloried. It becamc a common denominator that, service with the Republicans in Spain, ipso facto, automatically made one a communist.

The Veterans of the Abraham Lincoln Brigade (VALB) had their headquarters at 125 West 45th Street in New York City. Whereas some of those who joined the International Brigades may not have been communist, there is considerable evidence to the contrary regarding the VALB. "The record shows that the VALB faithfully parroted the Kremlin line for a period of fifteen years and that most of its leaders were members of the CPUSA."[249]

When Stalin approved the concept of using military units composed of communists and anti-fascist volunteers, the XI International Brigade was the first one created. At the time it was formed only one brigade was in existence. Yiddish was the "linga franca" for the Americans in the Abraham Lincoln Battalion as, of course, was English. The largest percentage of volunteers in the Battalion were Jewish. Also, when the soldiers from other International components gathered with the men of the Battalion, Yiddish was the common language.

The passports of each volunteer in the Battalion was taken from him by Phil Bart, a newly appointed commissar of the Battalion. Bart told the men that this was only a precautionary measure to prevent the passports from being lost in battle. Bart promised that the passports would be returned to the men after they had safely returned from the front.

The ultimate destination of these passports, the end use, was to "paper" illegal Soviet agents. The GPU was able to use these genuine passports for documenting many of its illegal deep cover agents.

In 1940 when Leon Trotsky was assassinated in Mexico, the Spanish action agent of the GPU was using a Canadian passport in the name of Frank Jacson. The passport had originally been issued to a member of the International Brigade named Tony Babich. The GPU technical section had changed the name and photograph in the passport. The spurious Canadian passport had originally been issued to Babich, a Canadian citizen who had been born in Yugoslavia. Babich had volunteered for the International Brigade, in the Canadian MacKenzie-Papineau Battalion, and had been either killed in the fighting or murdered by the GPU.

Another of the passports taken from a member of the International Brigade during the Spanish Civil War by Soviet intelligence surfaced in 1946. This use of this passport for illegal documentation was exposed by the defection of Igor Gouzenko, the GRU code clerk in the Military Attache's office in the Soviet Embassy in Ottawa, Canada.[250]

The Royal Canadian Commission, in assessing the damage to Canadian security done by the Soviet military spy ring, revealed the use of this illegal documentation. The passport had been taken from a Canadian citizen in the International Brigade and issued to a "very important" illegal Soviet agent operating in the Los Angeles area of California. This deep cover agent was using the passport of Ignacy Witczak, a Polish immigrant to Canada who worked as a farmer and shoemaker. The real Witczak had emigrated from Poland in 1930. He had surrendered his Canadian passport to the Soviets in Spain in 1937 when he became a member of the Mackenzie-Papineau Battalion, the Canadian command in the International Brigade.

When Witczak left Spain to return to Canada he was told by his Battalion's commissar that the trucks carrying the Battalion's bedding and the personal effects, including the passports had been bombed and everything destroyed. In fact these stolen passports were seen personally in the GPU headquarters in Moscow by Walter G. Krivitsky. Krivitsky was the GPU chief in Europe and the principal arms purchaser for the International Brigades. He saw these passports in Moscow during one of his visits to the GPU headquarters long after they were supposed to have been destroyed.

Before the Spanish Civil War was over, more than 500,000 human beings had been killed. The communists had effectively wrested control of all elements of the Republican government from the duly elected Spanish leaders by the end of the first year's fighting. The Republican leader, Largo Caballero, a moderate liberal, had been ousted from the leadership of the Republic. With Caballero gone, control of the Spanish Republican War Department had been virtually taken over by the Soviet advisors.

These Soviet advisors, with the omniscient help of the GPU and its terrorist methods, had threatened to cut off the supply of arms and military supplies to the Republicans unless they cooperated. The Soviets insisted, that unless they were given complete control over the disposition of the arms, and their tactical deployment, the arms would remain in the warehouses under strict Soviet control.

Dr. Juan Negrin had been praised by the international Comintern propaganda as a worthy successor to Caballero. Negrin had been widely publicized inside Spain by the GPU controlled media when he assumed control of the Republican government from Largo Caballero. Negrin openly began to cooperate with the Soviet advisors and they soon permeated every echelon of his government, even more so than before.

The outside world received only a blatantly one-sided mish-mash of what was actually happening inside Republican, Soviet-controlled, Spain. The part that was publicized internationally was only that very small part that remained after subjective editing favorable to the Soviet communist position.

Trotskyites were systematically denigrated in the focused news from Spain that was allowed to pass through the Soviet censors. The barbarisms of the Soviet advisors and the GPU were ignored or slanted as having been fabricated by the fascist press. The Trotskyites were routinely blamed for the worst atrocities.

The gross over-simplification of the roles of the antagonists in Spain that were reported by the communist controlled news sources led the world to believe in the neat separation of the conflict. The world was lead to believe that the war was between vicious depraved fascists, with Hitler and Mussolini aiding Franco, on the one side. According to the Soviet media releases, the Republican side was the underdog, lovingly paternal and helped by the benevolent Stalin, who graciously and compassionately was assisting the downtrodden of Spain who were the true forces of democracy.

The facts, of course, were different. For example, military aid had been donated to the Republicans by Mexico's President Lazaro Cardenas long before any aid had been forthcoming to Spain from Stalin. This Mexican gift was largely unknown to the rest of the world. The Comintern propaganda networks chose not to feature this story, reserving all their verbiage for when Stalin came through with some aid from the Soviet Union.

When belatedly, heavily restricted and tightly controlled supplies stared to trickle into Spain from the Soviet Union, they were on a strict cash basis, with money up front or the weapons were not delivered. Soviet aid to the Republicans was never in the form of a gift, as had been Mexico's contribution and most of the other munitions sent to Spain by other countries. Moreover, the Soviets demanded that the price include the salary and expenses of the GPU advisors and GRU, the Soviet military intelligence organization technicians sent by the Soviets to instruct the Republicans in the use of the equipment.

The exorbitant cash payments in advance demanded by Stalin had a devastating political consequence on the Republicans' ability to win the war. Control of these Soviet arms shipments, doled out parsimoniously in driblets at marked-up prices, gave the Soviets a decisive edge. The Soviets extracted demands of bribery and blackmail from the Republican government officials and military leaders for the actual delivery of these weapons even after the excessive price had been paid.[251]

Soviet supplies were stored under GPU control in Soviet warehouses. In many cases the military materials were given out only as rewards in support for favorable treatment to the Soviets by a Republican official or they were withheld from those Republican officials who were out of favor with the Soviets. Those Republican military officers who the GPU wanted discredited in the popular opinion of the Spanish people usually lost their battles because their equipment situation had been tampered with by the Soviets.

Largo Caballero had tried to explain to the cynically pro-Soviet Spanish communists how the withholding of crucial supplies at critical junctures in the military campaigns many times saved Franco from defeat. The prolongation of the civil war by the Soviets' arbitrary cutting off of supplies at propitious times to save Franco from losing a battle was deliberate, according to Caballero. The Soviets had to be assured of complete tactical control over the Republican fighting forces before they would release military supplies to the Republican troops.

A further proof of Stalin's hypocrisy was shown by his covert support of Franco. Throughout the Civil War, Stalin continued to send Russian oil by way of fascist Italy to Franco to fuel his planes and tanks. These were the same planes and tanks that were being used in the fighting to kill Stalin's own advisors and the soldiers he claimed were his brother Republicans, including those fighters from the International Brigades.

Regardless of Stalin's duplicity in the war in Spain, his consuming vendetta against Trotsky and the Trotskyite disciples, remained paramount. Stalin's eradication of all aspects of Trotskyism, and other opponents of his own brand of Marxist philosophy in Spain, was thorough, diabolically clever and brutal. The GPU deliberately fostered a systematic persecution of Spanish labor leaders, government and military men and women sympathetic to Trotskyism in any form. This type of indiscriminate purging was bound to also include victims of personal vendettas and grudges. On the whole, the Soviet policy of rounding-up these individuals and summarily disposing of them, did more damage to the social and moral fabric of Republican Spain, and robbed it of its will to fight, than did all of Franco's legions.

The GPU had secret prisons in Spain to incarcerate the victims of these purges from which few, if any, once committed, ever returned. All trials within these prisons were conducted without any record, without benefit of defense counsel or rights of the accused. The only verdicts of these trials were the ones the Soviet advisors dictated to the judges. In addition to routine prison sentences, there were many verdicts of execution which were swiftly carried out.

In September of 1937 a detailed account of the GPU barbarisms in Spain routinely practiced against the Trotskyites, the Republican armed forces in general and specifically against the Spanish people was given to the American reading public by Anita Brennan. She wrote a lengthy, well-documented study that was published in the *Modern Quarterly*. Apparently not many people paid attention to the truths contained in this article. The average American continued to believe in the myth that the benevolent, paternal Stalin was really helping the Republican cause out of the goodness of his heart.

As the purge trials in Moscow continued through 1936 and '37, Orlov could read the ominous omens that he too was vulnerable. The purge trials eradicated many of the old line communist party members everywhere. The GPU stalwarts who had been Orlov's contemporaries in the old years were some of the principal defendants. The number and variety of ruses used by the current GPU leaders to try to lure Orlov back to Moscow for what he was certain would be his execution, were many and persistent. Orlov continued to refuse to return to the Soviet Union and eventually decided to defect to freedom in the west.

Orlov and his family managed to escape from Europe to Canada aboard the S.S. Montclair. When they finally reached a safe haven in Canada, Orlov wrote a letter to Stalin. In the letter Orlov threatened to publish for the world to read all the details of Stalin's treacheries if he, his family or his relatives still in the Soviet Union, were harmed.

As the senior GPU officer in Spain, Orlov had an assistant whose name was Leonid Eitingon. This man would later be the senior GPU case officer directing the Trotsky assassination in Mexico. Caridad Mercader, the mother of Trotsky's assassin, Ramon Mercader, was a close personal friend of Orlov's in Spain. She was also the mistress of Leonid Eitingon. Caridad and Eitingon were often guests at Orlov's plush hotel suite. Much of the dinner conversation there had to do with the subject of assassination in general and much of this information was later employed by Eitingon in the assassination of Trotsky.[252]

Even Ramon Mercader, the future assassin of Trotsky, served in the Spanish Civil War. He was in one of Orlov's special commando units in Spain. He carried out sabotage and execution missions behind Franco's line on targets ordered by Orlov. All of the training for these special assignments was excellent preparation for the supreme task that was ahead of him in Mexico. At that time, however, no one, not even Mercader himself, knew that he would be chosen as a last resort to murder Trotsky.

As the death knell struck for Republican Spain, the GPU and the Comintern propaganda networks throughout the world began to lay the foundation for world opinion to accept the idea of having Spanish Republican refugees among them as exiles. Committees in countries around the world, ostensibly led by well meaning liberals but actually dominated by controlled GPU agents, sprang up. In many instances these committees were subsidized by Comintern financing. The United States and Mexico were targeted by the GPU as of special importance in establishing these Spanish Republican refugee committees.

From the beginning of the Spanish Civil War Mexico had conformed to the world mandate against extending relations with fascist Franco and offered no diplomatic recognition to him as leader, nor to his government, even after he won the civil war. As a result, Mexico was overtly sympathetic to the Spanish Republican refugees and even permitted the establishment of a Spanish Republican diplomatic mission in Mexico after the civil war.

Some of the Republican refugees who went to Mexico found legitimate work of an open and honest nature. Others, with a more sinister motive and the need to find work as cover employment for their intelligence activities, were given ostensible working status with the Soviet representative in Mexico City. Later, when diplomatic relations were established between Mexico and the Soviet Union, some of these Republican refugees found sanctuary employment with the Soviet Information Bulletin and the Soviet Economic office. Even Trotsky would eventually find sanctuary in Mexico.

MARK ZBOROWSKI[2]

> Mark Zborowski is a perfect informer and fingerman: he commits no thefts but points out where the documents can be stolen; he commits no murder, but informs on the whereabouts of people condemned by [GPU] kangaroo courts. Zborowski understands what his reports to the Soviets imply; he knows what is in store for those he shadows.
>
> David Dallin

In order to properly understand the scope of Stalin's operation against Trotsky, it is appropriate that Mark Zborowski's role be examined. As David Dallin puts it, "We do not know if Zborowski met Jacson [Mercader, Trotsky's assassin] but we may be certain that his information was essential for the operation [of murdering Trotsky]."[253]

Mark Zborowski was born in the Russian Ukraine town of Uman on 21 January 1908. He became a communist at a youthful age after the Bolshevik revolution and later was assigned as a party agitator to Lodz, Poland in 1921. In Poland, pursuing his communist assignment, he was arrested by the police as a suspected Soviet agent.

He was released from prison after a few months and wisely decided to flee Poland for the relative security of France. In France he studied medicine and philosophy at the Universities of Rouen and Grenoble. In his sworn testimony before a United States Congressional Committee he admitted that he was recruited as an espionage agent into Soviet intelligence by the GPU while studying at Grenoble.

In Paris the GPU assigned Zborowski to work in the Union for the Repatriation to Russia with offices at 12 Rue de Hoci in the center of Paris. His cover was the administrative secretary of that organization but in reality he was in charge of the operation. The Union was a sting

---

[2] His original name was Mordka Zborowski (Sborowski, Zborowfski and Zborsky). His GPU/NKVD code name was Mark.

operation run by the GPU to entice former Imperial Russian military officers, and other expatriates, back to Russia. The purpose of this operation was to accelerate the disintegration of the anti-communist Russian emigre organizations in Europe by literally stealing away their members.

The Union's staff was made up entirely of GPU officers working under the guise of homesick sentimental Russian exiles, yearning to return to Mother Russia. These supposed homesick men tried to convince the former Czarist military officers in Paris that they should share this deep-rooted love for the old country and return home. As soon as the dupes set foot back on the soil of Mother Russia they were promptly seized by the GPU. Many were executed as traitors to the new Soviet regime, others were sent to prison while a few took oaths to serve loyally in the new Soviet army.

The GPU cadre posing as the fostering nucleus of the Union for the Repatriation to Russia also served as a ready reservoir of man power for the GPU to use for assassinations and kidnappings. These action-oriented agents were poised for instant dispatch to any trouble spot in Europe where the GPU deemed they were needed.

Zborowski's principal task, after his work at the Union as a Judas goat, was to infiltrate Trotskyite organizations wherever they existed in Europe. The nominal international headquarters of the Trotskyite movement, called The Fourth International, was in Paris. There, Leon Sedov, Trotsky's son (called Lyova by the family) was in charge of organizational activities. Sedov was also the managing director of the Fourth International's newspaper, the *Bulletin of the Opposition*.

A noted Russian historian comments on the Fourth International as follows:

> Trotskyism as a well-defined ideological trend has never existed. At various stages of the revolution, Trotsky attracted varying groups of supporters... they never formed a coherent trend or "ism." On the other hand, some historians exaggerate the significance of Trotskyism, arguing that it has always been a well-established anti-Lenin system of views. Mevedev disagrees with both views, believing that most of Trotsky's assertions were erroneous and demagoguery incapable of being the basis for a political philosophy.[254]

Zborowski, under the nom de plume of "Eitenne" contributed anti-Stalinist, anti-Comintern and pro-Trotsky articles to the *Bulletin*. He wrote these pieces to please Sedov and convince him that he was a loyal, bona fide Trotskyite. This reenforced his cover and prevented anyone from thinking he might be a GPU spy trying to infiltrate Sedov's operations.

Through diligent obsequiousness, some little talent as a writer and very good operational guidance from his GPU case officer, Zborowski became Sedov's chief assistant and alter ego. In this position Zborowski had access to all of Trotsky's papers, the plans for his international opposition to Stalin and the details of Trotsky's agent cadre within the Soviet Union.

Zborowski gave copies of this information to his GPU case officer who was working out of the Soviet Embassy in Paris, for subsequent transmittal to Moscow. Allegedly Stalin read every one of Zborowski's reports. By his own admission, Zborowski met his case officer two to three times each week in a park opposite the Paris Gare d'Austerlitz railway station near where he had an apartment. He would pass his reports to the Soviet case officer in a rolled-up newspaper which he exchanged for a similar rolled-up newspaper that his case officer was carrying.

In July of 1936 when the Spanish Civil War broke out, Trotsky's Fourth International convened a war council in Paris. At that time Trotsky had been expelled from France and was living in Norway under the watchful eye of Trygve Lie. Trotskyites from all over the world arrived in France to attend this important meeting. The meeting site had to be kept a secret, not only to thwart the French police but to evade the prying eyes of the GPU.

These prudent precautions may have preserved the secrecy of the meeting from the French police and newspapermen but the GPU knew exactly where the meeting was held and when and who was in attendance. The GPU continued to receive detailed information about the delegates, the agenda and the conclusions reached at each session. The organizational escort officer assigned by Sedov himself to oversee the meeting and make the delegates comfortable, was Mark Zborowski. He served as a conduit for every bit of information about the meeting directly to the GPU.

Not everyone in the Paris Trotskyite headquarters was taken in by the obsequious Zborowski. The widow of Ignace Reiss, the former GPU senior officer who was murdered by the GPU in Switzerland with

Zborowski's active assistance, was not fooled. Mrs. Else Bernaut, Reiss's widow, recalls that many in the Trotsky movement in Paris took an instant dislike and distrust of Zborowski. In her excellent book, *Our Own People,* which she wrote under the name of Mrs. Elizabeth Poretsky, she cites several instances where members of the Fourth International warned Trotsky and Sedov against trusting Zborowski.

Henryk Sneevilent, a Dutch communist and member of the Fourth International, who was later murdered by the Nazis in occupied Holland, told the leaders of the Fourth International, "There is an agent [inside the Fourth International] and it is that little Polish Jew, Eitenne [Zborowski's nom de plume]."

In 1936 Trotsky's archives had been transferred for safekeeping from Sedov's office to the International Institute of Social Studies in Paris located at 7 Rue Michelet. The Institute's director was reputed to be an anti-communist, bona fide socialist and political writer named Boris Nicolaevsky. Lola (Lilia) Estrine, the future wife of David Dallin, the noted anti-communist author, was the private secretary to Nicolaevsky at the Institute. Estrine was also a frequent advisor to Sedov, a close personal friend of Mark Zborowski and suspected of bieng a GPU agent.

Zborowski, in addition to his work for Sedov also worked part time for the Institute of Social Studies. He convinced Sedov that Trotsky's papers and research material would be safer from the prying eyes of the GPU if they were stored at the Institute.

After Trotsky's archives had been transferred to the Institute, Zborowski told his GPU case officer that Trotsky's valuable research papers were stored at the Institute under very lax security conditions. He outlined a method for the theft of Trotsky's papers and suggested to his GPU case officer that the papers were literally there for the taking.

On the night of 7 November the GPU broke into the Institute and stole the fifteen crates of Trotsky's most confidential papers. The French police commented that only a foreign professional burglar could have committed this crime because the thieves used a blow torch to cut through the iron bars on the windows. This type of burglary equipment had never before been used by French criminals.

The GPU choose this date of 7 November, the anniversary of the Bolshevik revolution, to acquire Trotsky's papers as an anniversary present for Stalin, so the story goes. The GPU tried to make their surreptitious entry of the Institute look as if it were a bungled burglary.

They vandalized the interior of the Institute but took no money or other valuables that were lying about in the open showcases.

Considering Zborowski's easily deduced complicity in this theft, he rightly feared that suspicion for the crime would fall upon him. At the time of the theft, Zborowski avoided being exposed and condemned as the thief for three reasons. He was too well ensconced as Sedov's assistant and personal friend and therefore came under Sedov's protection. Secondly, he hid a small amount of what he claimed were the most important of Trotsky's documents in his apartment while the bulk of them were sent to the Institute for storage. He later produced these documents and claimed that he kept them because he was afraid they might be stolen from the Institute. Lastly, Lola Estrine, a co-worker at the *Bulletin* office and the secretary to Boris Nicolaevsky, the director at the Institute, vouched for Zborowski's innocence.

Lola Estrine had helped Zborowski pack and move Trotsky's precious papers from the office of the Fourth International to the Institute. She also kept some of what she claimed were the more important documents in her apartment rather than trust the whole shipment to storage at the Institute. She categorically vouched for Zborowski's lack of complicity in the theft, as he also swore to her innocence. Their joint and mutual protestations of innocence seemed to be confirmed when they both returned to Sedov those documents they had kept in their apartments.

Lola (also known as Lilia) Estrine is an enigmatic character who deserves special consideration. Her life was a series of paradoxes that beggared rational explanations. Although there is no proof that Estrine was a GPU agent, in the true sense of the word, it can just as obdurately be argued that there is so much coincidence and parallel activities of known GPU agents in her life that the allegation could not be categorically disproved either!

Estrine referred to herself and Zborowski as the "Siamese twins," meaning that neither would undertake to do anything without the other. The papers of Trotsky and Sedov contain dozens of folders about the two, accusing them of being Stalin's spies. However these files admittedly contain more suspicions than actual truth. Trotsky preferred not to believe them and drew Zborowski and Estrine even closer into his confidence.[255]

Later, after Sedov's death in February of 1939, the two of them seemed to vie with each other in praising each other and excusing any misfeasance of the other regarding Sedov's death in their letters to Trotsky in Mexico. Their joint letters to Trotsky were signed "E" or "ET" for "Eitienne (Zborowski's pen name) or they were sometime signed with his street alias "E. Tienov" and Estrine signed her letters as "Lola," "Lil," "Paulson" (her street alias) or just plain "P."

Lilia (Lola) Estrine had been born in Liepaja, Latvia in 1898 as Lola Ginsburg. She moved to Moscow in 1914 where she studied law. She emigrated to Berlin in 1923 where she married Estrine, a local communist party functionary. They moved to Paris in 1933 but she always maintained her Soviet citizenship by renewing her passport annually. She was permitted by the Soviets to do this even after she was well-known as a Trotsky adherent.

At the time she worked for Sedov at the *Bulletin* office in 1935 she and her husband lived in a small apartment that was crowded with relatives. There was her brother, Ralph Ginsberg, who professed to be a medical doctor; and Ralph's wife, Fanny Trachtenberg (known to her Russian friends as Tranchenka), who also claimed to be a medical doctor.[256]

Estrine's work for Sedov at the *Bulletin* office was voluntary and she received no remuneration. Establishing herself first as a volunteer proof reader, then as a researcher and finally as Sedov's private secretary where she had access to everything that went on in the entire Trotskyite movement. (After Sedov's death Estrine wrote many compassionate letters to Natalia Trotsky in Mexico and always included piously worded barbs against Jeanne Martin, the late Sedov's mistress. Jeanne Martin openly charged that Estrine and Zborowski were traitors to Trotsky and were implicated in Sedov's death.[257]

Lilia Estrine was one of those people who apparently knew everyone in Paris that was worth knowing, including the GPU agents, Jack Soble and Dr. Robert Soblen, the former Sobolevicious brothers who had infiltrated Trotsky's home in Prinkipo.[258]

Later Estrine married David Dallin, the author of many books about the Soviet Union. including an intensive study of Soviet intelligence activities entitled, *Soviet Espionage*. It was rumored that, for reasons unknown to the rank and file of Soviet espionage, the leaders of the GPU had helped Dallin with source material and anecdotes for his book. (Dallin's sources may have been unofficial and his own independent penetration of the GPU.)

There is a persistent rumor that Dallin had a considerable amount of additional information about Soviet espionage that had been given to him by his GPU sources that was not included in his book. This material called "the D papers" (Dallin papers) had been hidden either by him before his death or hidden by Estrine after his death. There were a few of Dallin's unpublished papers found after his death but they had been purged and clipped of any sensitive material that might embarrass the Soviet intelligence system.

After Estrine married Dallin, they emigrated to the United States. Dallin entered the United States on a Honduran passport issued to him in Berlin on 1 July 1933. Dallin at the time was a Soviet citizen. Honduran citizenship was arranged by the Soviets for other emigres, among them known Soviet espionage agent Valentin Olberg.[259]

After the Dallins were settled they sponsored Zborowski, his wife and son for immigration into the United States. Zborowski had married Rivka (Regina) Levi in 1938. Rivka had come to Paris with her parents from Lodz but the parents continued on to their dream destination in Palestine leaving Rivka in Paris married to Zborowski. Rivka was a confirmed Stalinist and knew of her husband's espionage activities for the GPU. Their only child, a son named George, was born in Paris.

Jack Soble, one of the former Sobolevicious brothers, became Zborowski's Soviet case officer in the United States. Both David and Lilia Dallin (Estrine) knew of this professional clandestine relationship between Zborowski and Soble in the United States.

Back in Paris, at the time of the burglary of Trotsky's archives from the Institute, Zborowski remained above suspicion of complicity in the theft. There were many other odd situations that developed at the *Bulletin* office that were suspicious but Zborowski always managed to be on the uninvolved periphery of the happenings.

Odd incidents and unusual accidents persistently badgered the *Bulletin* working rooms and the personnel which adversely affected the efficient production of the Fourth International. Somehow these suspicious happenings always seemed to be resolved in a manner favorable to a position that gave the GPU the advantage and a set-back for the Trotskyites but no adversity could ever be clearly attributed to Zborowski.

Leon Sedov was incensed by the Moscow trials when they began in 1936. He compiled his famous "Red Book" that denounced Stalin's crimes with a considerable amount of corroborating evidence of Stalin's perfidy and Trotsky's innocence. Stalin thereupon proclaimed to the GPU that Sedov must be eliminated and he became second on the list of prime targets for the GPU assassination teams, right behind his father.

Stalin's emissary was Sergi Efron, who came to Paris especially to organize the hit squad that had Sedov as its principal target. Preparatory to their executionary role the team had to gather all the information about Sedov so they could determine where to strike at his weakest point. Even though this assassination team already received daily reports of most of Sedov's activities from Zborowski, they needed surveillance coverage of those periods when Sedov was not with Zborowski.

One of the members of the surveillance team that Sergie Efron organized was Francois Rossi (aka Roland Abbiate). Rossi had a detailed drawing of the layout of Trotsky's house in Coyoacan, Mexico on his person when questioned by the police.[260]

Efron operated under cover of a political refugee at Zborowski's Union for the Repatriation to Russia. He organized the surveillance team with Dmitry Smirensky, aka Maurice Rollin and Varia Chadek. Under various pretexts, Smirensky had been recruiting surveillance agents for sometime. He usually tested the new recruits, after a suitable period of training under his tutelage, with other assignments before incorporating them with the surveillance team.

The team, as finally organized that surveiled Sedov, was composed of yet another Stalinist named Pozniakov; a printer and photographer and detective, Pierre Ducomet; Vadim Koindratiev and Pierre Schwarzanberg (both Vadim and Pierre were officers of the Union of Repatriation to Russia), and Roland Abbiate also known as Francisco Rossi. The only woman on the surveillance team was Renata Steiner.

Efron's team tried several times to lure Sedov into a trap in order to execute him on the spot. Sedov, apparently unmindful and oblivious to the frantic scrabbling of the team in trying to ambush him, somehow in his innocence managed to elude them and squeak out of the traps.

For example, in 1936 Efron urgently sent Renata Steiner to Juan les Pins in southern France. She was told to stay at the Villa Marie Pension and cultivate the friendship of a young couple who would be staying there. The couple were Sedov and his common-law-wife, Jeanne Martin.

The GPU's tap on Sedov's telephone produced the information that the couple would be staying at this pension. The surveillance was established preparatory to putting the finishing touches on the plan for

Sedov's murder. The blissful couple, unaware they had just managed to foil the GPU's crack death squad, left the pension before the killers could get organized to strike.

Later another failed attempt on Sedov's life occurred in January of 1937 in the town of Mulhause. Sedov had arranged a secret meeting in Mulhause with a lawyer who was handling a libel action brought by his father, Leon Trotsky, against several pro-Stalinists newspapers.

The situation in Mulhause was almost perfect for Sedov's murder but the actual hit was called off. The GPU surveillance team and hit squad, ordinarily did a better job than this. The reason the murder had to be cancelled in Mulhause was because in the excitement of getting everything ready for the execution no one thought of preparing a secure escape route. Therefore the killing had to be aborted. Sedov and Jeanne Martin returned to Paris, again unaware that they had once again narrowly escaped death.

The Efron surveillance group next established themselves in a base house at No. 28 Rue Lacretelle in preparation for the murder of Sedov. This house shared a common wall with the house next door at No. 26 Rue Lacretelle where Sedov and his common-law-wife Jeanne Martin lived. On several occasions Zborowski tried to lure Sedov to a place where the GPU could kidnap him but either Sedov inadvertently failed to keep the appointment or he missed the kidnap rendezvous by a piece of good luck. Sedov's escapes, however were not because of any lack of trying to do him in on Zborowski's part. Zborowski was typically very successful in other execution type operations. For instance, Ignace Reiss, a defected former GRU officer was murdered by the GPU hit squad which included Rossi and Renata Steiner on 4 September 1937 in Lausanne, Switzerland, and Zborowski was alleged to be the fingerman.[261] Walter Krivitsky, a defected senior GPU officer was almost kidnapped by the GPU at the railway station in Marseilles. He narrowly escaped and the fingerman was alleged to be Zborowski.[262] Erwin Wolf, Trotsky's personal secretary, was expelled from Norway before Trotsky was forced to leave. Wolf went to Spain where he was promptly murdered by the GPU and Zborowski was again alleged to be the fingerman. In July of 1938 the headless corpse of Rudolph Klement, a German communist and one of Trotsky's personal secretaries, was found floating in the Seine river in Paris. Zborowski was alleged to be the stalking horse and arranged for Kelment's kidnapping and murder by the GPU.

In early 1937 Sedov printed an admonition to Stalin so all the world could better judge who might be responsible if anything happened to him. Sedov summed up the many strange things that had happened to him recently and added, "I have in no way lost my moral balance nor my confidence in life. I am therefore in no way inclined to commit suicide or to disappear. If something should happen to me, the cause must be searched for on Stalin's side and no where else."

On 15 January 1937 the on-set of Sedov's abdominal pains were alarmingly sudden. Their persistence caused Estrine, who was helping to care for Sedov, to call her sister-in-law, Fanny Tractenberg, who was alleged to be a doctor. Tractenberg diagnosed Sedov's pain as a mild attack of appendicitis. She put Sedov on a mild rice diet. He recovered by 20 January and for several days complained of no stomach pains.

Inexplicably on 8 February violent abdominal seizures returned. Ice compresses on the stomach wall were prescribed but did nothing to relieve the pain. On 9 February 1938 Trachtenberg brought her husband, who was also reputed to be a doctor, to examine Sedov. Their joint recommendation was to remove him to a private sanitarium, the Mirabeau Clinic on Rue Narcisse Diaz in Paris. Somehow, Estrine managed to get the money for Sedov's hospitalization. (She always seemed to be able to come up with large sums of money when they were needed.)

Zborowski told Jeanne Martin, that the Mirabeau Clinic was chosen because it was not known to the Soviets and therefore Sedov would be safe there. Further, said Zborowski, he had registered Sedov at the Mirabeau Clinic under the alias of Martin, his citizenship as French and his profession as civil engineer.

Supposedly, according to Zborowski, only five people knew where Sedov had been taken; Jeanne Martin, whom he had just told; Lilia Estrine; her sister-in-law, Fanny Tractenberg; Trachtenberg's husband; and Zborowski. Years later when testifying before a Congressional Committee, Zborowski admitted that at the same time he had summoned the ambulance to take Sedov to the Mirabeau Clinic, he had informed his GPU case officer of the ailing Sedov's destination.

The truth was, instead of being a hospital unknown to the Soviets, the Mirabeau Clinic was a hospital used almost exclusively by the Russian emigres in Paris and well known to the GPU. The Mirabeau Clinic was used by the Union for the Reparation to Russia to entice ailing emigres who went there to return to the Russian motherland.

The Mirabeau Clinic preached an evangelical brand of medicine that caused a psychological uplift for the patient bent upon returning to his beloved Russia. Tractenberg later said she chose the Mirabeau Clinic for Sedov for two reasons: she knew Sedov was a Russian exile and thought he would be more comfortable among Russians. The second reason was that Zborowski had told her that Sedov must be taken to the Mirabeau Clinic.

Zborowski's guidance, under GPU tutelage, can best be appreciated when the credentials of the staff of the Mirabeau Clinic are understood. Dr. Zhirmunsky was well known to the French police as a Stalinist sympathizer. He had originally come from Russia in 1928 with considerable money, enough to buy and equip this hospital in Paris. The cosmopolitan doctor lived the good life considerably above his financial station and there was no plausible explanation of where the funds came from for his lavish living; certainly not from the ostensible income from the Mirabeau Clinic.

Dr. Simkov, who was originally from Kiev, held a medical degree from Germany. He became the medical director of the Mirabeau Clinic in 1931. Simkov maintained an active membership in the Mechnikov Medical Society, an avidly pro-Soviet professional medical association. Dr. Simkov was the initial surgeon on Sedov's case and knew Sedov's true first name despite the alleged precautions taken of registering him under the alias of Martin. The actual operation on Sedov was performed by Dr. Marceal Thalheimer, who was apparently a legitimate French doctor.

The nurse assigned to Sedov was Helena Eismond (nee Regina), who was originally from Leningrad. Esmond was specifically assigned by the clinic to Sedov's case and was therefore with him more than any other person at the clinic.

Dr. Thalheimer removed Sedov's appendix which was normal and not inflamed. This meant that the stomach cramps he had suffered had to have been caused by some foreign substance that was irritating his intestines. Sedov showed remarkable progress toward recovery for the first four days after the operation. Zborowski took it upon himself, without consultation with Sedov or Jeanne Martin, to dismiss Sedov's night nurse. He later explained that he thought Sedov's condition was so improved that he no longer needed round-the-clock nursing.

Suddenly, during the night of 13/14 February 1938, Sedov went into a delirium state, with a raging fever. One story put forth by the Clinic had Sedov tearing off his nightshirt and wandering the halls naked, mumbling to himself and alternatively shouting obscenities in Russian until he finally collapsed in exhaustion. The next day, on 16 February, Sedov died. Apparently no official of the clinic saw him or heard him until his body was found the next morning where he fell. (If any official of the clinic heard or saw Sedov, they made no record of it. However, there had apparently been some fellow patients who reported the tearing off of clothes and the shouting who strangely took no further action to call a clinic attendant.)

Jeanne Martin persisted in her investigation of Sedov's death and the Mirabeau gave her a slightly altered but equally vague version of what happened to Sedov. They claimed that Sedov alone and unattended, jumped out of bed, ran into the adjoining room, picked up an orange, peeled it and ate it. Then, according to their story, he fell to the floor in a coma or a faint. The clinic orderlies carried him back to his own room.

Jeanne Martin questioned the patient in the room where the orange eating episode was alleged to have taken place. This patient, apparently not briefed on what story to tell, told Martin that he had not seen anything of Sedov or was not aware of anything even approximating the orange eating incident as described to her by the clinic officials.

Jeanne Martin was suspicious of two men who had been patients in another adjoining room next to Sedov's but on the side away from the room of the alleged orange eating incident. These two men had left the clinic and the officials of the clinic refused to give Jeanne Martin any information about them.

By this time Jeanne Martin was convinced the clinic had not been truthful in relating what had happened to Sedov on the night he died. She was convinced that Sedov had either been poisoned or beaten up. During her brief visit with Sedov the morning after the incident on 14 February, Sedov had whispered some words to her which she had dismissed as disjointed ramblings. Sedov had whispered to her, "You know what they did to me last night?" Jeanne Martin planned to question Sedov further about this statement but had been diverted and never had the opportunity before his death to ask him what he had meant.

Martin knew that the night before his relapse, Sedov had been too debilitated and enervated from the operation to walk or even leave his bed much less being able to "jump" out of bed as the clinic claimed. Martin persisted that Sedov could not have left his bed without the help of at least one or two strong men or without a very strong stimulant.

Dr. Thalheimer found severe bruises on the scar of Sedov's operation which he attributed to Sedov's fall. He asked Jeanne Martin if her husband used drugs or had ever threatened to commit suicide? Dr. Thalheimer's immediate diagnosis was that someone had forced a massive dose of drugs into Sedov's body which caused the dramatic reversal of his otherwise good recovery. The doctor tried surgical intervention but Sedov's intestines were in complete paralysis. At 11:00 o'clock on the morning of 16 February 1938 Sedov died.

Estrine and Zborowski assured Trotsky in Coyoacan by letter of the professional competence of the staff at Mirabeau. They further assured him that there could not have been any harm done to Sedov by anyone connected with the clinic. They wrote detailed accounts of imaginary illnesses that had beset Sedov throughout 1937 and '38. They wrote that Sedov was too debilitated to convalesce much less recuperate, particularly after the second operation.

In their letters to Trotsky, Estrine and Zborowski always managed to get in a devastating, snide jab about Jeanne Martin who was, at the time, imploring the French authorities to investigate the brazen murder of her husband. Estrine and Zborowski implied to Trotsky that Jeanne Martin had become psychiatrically demented by Sedov's death. They characterized her as "neurasthenic with a lunatic imagination of foul play... and [she] sees the GPU everywhere."

Estrine and Zborowski were more convincing to Trotsky than either Jeanne Martin or Trotsky's French comrades who were likewise convinced that Sedov had been murdered by the GPU despite what Estrine and Zborowski contended. Trotsky, however, accepted the pap from Estrine and Zborowski that his son had died of natural causes as a result of post-operative complications. Consequently, Trotsky relegated Jeanne Martin to a position of distrust. He no longer accepted her as one of his trusted inner circle or as Sedov's widow.

Before Trotsky had accepted Estrine's and Zborowski's version of Sedov's death, he wrote with his usual eloquence of Sedov's death:

> The first and natural assumption was poison. To gain access to Leon, to his clothes, his food, presented no great difficulty

> for Stalin's agents. The art of poisoning has been developed to an exceptional degree. It is true, the secrets of the art are beyond the reach of ordinary mortals. But the poisoners of the GPU have access to everything. It is fully possible that there now exists a poison which cannot be traced after death even by the most thorough analysis. Or, did they kill him without the aid of chemistry?

The suspicious nature of Sedov's death caused a strong searching look to be focused again on Zborowski by the Paris Trotskyites. Realizing that he had probably outlived his usefulness to the GPU in Paris, Zborowski began writing to Trotsky in Mexico. He requested a transfer to Coyoacan to assist "the master there" with Russian translations and general secretarial duties.

Trotsky had learned of the suspicions of the Paris Trotskyites concerning Zborowski's and Estrine's possible complicity in Sedov's death. He therefore replied to Zborowski's letters with caution and restraint. Even though he continued to deal openly and, on the surface at least, cooperatively with Estrine and Zborowski, he was enough of a survivalist to decline Zborowski's offer to be his secretary in Coyoacan.

By now there was a consensus among the Paris Trotskyites that Estrine and Zborowski were NKVD spies. Henry Sneevilet, (true name Marling), the Dutch Trotskyite leader, referred to Zborowski as that "filthy Polish Jew who is a spy for the GPU."[263]

Alexander Orlov, a former GPU senior case officer who defected to the west, wrote to Trotsky stating that, before he left Russia in 1936, he learned that the NKVD (GPU) had succeeded in infiltrating a spy into Sedov's entourage. Orlov's letter, written on 27 December 1938, contained this passage, "I understand that Stalin was doing his best in order to corner Trotsky and assassinate him and I understood that through this man, Stalin might introduce under the guise of a guard or secretary, an assassin into Trotsky's household."

Orlov later commented, "While working in Spain during the Civil War, I used to come on business to France, and there I did my best to find out the identity of the agent from the Chief of the NKVD [GPU] in Paris, Nikolai Smirnov.[264] I found out that this agent had become the closest friend of Trotsky's son, Leon Sedov, and that he was in correspondence with Trotsky himself. I found out that his first name was Mark. I did not know at that time his name was Zborowski."[265]

Any possibility that Trotsky might accept Zborowski into his personal coterie in Coyoacan was partially negated by this curious letter that Trotsky received from Orlov who unfortunately did not sign his true name. Orlov wrote in the letter to Trotsky that he had obtained the information from General Lushkov, a Soviet GPU officer who defected to the Japanese in 1938.

On 27 December 1938, Orlov's letter, signed with an alias Trotsky did not recognize, described Zborowski's position as Sedov's right hand man. Orlov also wrote that the agent provocateur was named Mark, that he wrote using the pen name of "Eitienne" and that he had been active in the Society for the Repatriation to Russia. Orlov warned Trotsky that a Spaniard might be the assassin who would come in the guise of a Trotskyite.

Lilia Estrine happened to be in Coyoacan with Trotsky when Orlov's letter arrived. (She arrived by bus from the United States border in Mexico City on 29 May 1939.) Trotsky asked her opinion about the spy described in the Orlov letter. Estrine, of course, recognized the spy Orlov described as being Mark Zborowski but refused to betray him to Trotsky. Instead she resorted to dissimulating evasion.

Estrine later testified before a Congressional Committee (this time under oath and with the threat of a perjury indictment) that she said to Trotsky, "That [the Orlov letter] is certainly a dirty job of the NKVD [GPU] who want to deprive you of your few collaborators that you have left in France. And, at the same time he [Trotsky] had received another letter from another unnamed agent, telling him that a woman, meaning me, was coming to visit him and would poison him. So we both laughed, 'See how they work? They want that you should break with the only people that are left to you in France!' And we both laughed and he decided that it was not to be taken seriously but that it [Orlov's letter] had been a hoax of the NKVD [GPU]!"

Orlov tried to contact Trotsky directly to warn him about Zborowski. He went to the office of the Fourth International but he distrusted the appearance of the receptionist there and never made contact. Orlov in the letter had requested Trotsky to place an advertisement in the Trotskyite newspaper stating that the editorial office had received a letter from "Stein." Trotsky placed such an advertisement and asked the writer of the letter to contact a man named Martin at the Trotskyite offices. Orlov never did.

On several occasions Orlov tried to telephone Trotsky in Mexico but was rebuffed by Trotsky's protective phalanx of secretaries who answered the telephone and insisted on knowing the details of what he wanted to speak about with Trotsky. Orlov finally gave up trying to contact Trotsky directly and warn him about Zborowski.

Orlov's failure to personally warn Trotsky, plus Estrine's diverting ploy that Orlov's message was probably an NKVD (GPU) disinformation plot, left Trotsky without final proof of Zborowski's treachery. Fortunately, there were enough other suspicious circumstances so that Trotsky was not, under any circumstances, going to extend an invitation to Zborowski to join his Coyoacan extended family.

Trotsky was essentially helpless in this situation. The Paris NKVD station kept sending him anonymous letters denouncing other genuine Trotskyites, as Soviet agents. No doubt Zborowski helped write some of these letters. Regardless, Trotsky was confronted by a number of adverse reports about his closest allies and associates. To credit these denunciations would have wrecked his movement. In the end, Trotsky disregarded them all.[266]

Later, in 1954, David Dallin invited Orlov to his apartment in New York. He mentioned that his wife, Lilia, would be there. Orlov looked forward to meeting Lilia, which would make it a social occasion which the Orlovs, since his defection, had been avoiding. At the meeting the subject came up about Mark. Lilia Dallin quickly entered the conversation by saying that she and her husband had arranged Mark's entry into the United States. Orlov was astounded and reported the conversation to his contact in the FBI which led to Mark Zborowski's interrogation by the Bureau.[267]

After Zborowski was detained, Jack Soble, who was his NKVD case officer in America became concerned that Zborowski would tell the FBI about him. Soble wrote a letter to the Moscow NKVD Center expressing these concerns and gave the letter to Boris Morris to deliver to his senior NKVD handler, Vasilli Zubilin. Instead Morris, an FBI penetration of the Soble network, gave the letter to his FBI case officer. This episode led eventually to Soble's arrest and imprisonment for espionage.

Both Lilia and David Dallin denied before the investigators of the Congressional Committee looking into Zborowski's past, that they knew of his work for the GPU. Many of the Congressional investigators were skeptical of this feigned innocence in view of the obvious, and very close, relationship between Zborowski and Lilia Dallin.[268]

When Victor Kravchenko defected from Soviet intelligence in 1944, Zborowski wormed his way into a traitorous friendship with him. Zborowski helped Kravchenko write his book, *I Chose Freedom.* Zborowski turned over the pages of Kravchenko's manuscript as they were written to Soble, his NKVD case officer. Kravchenko's second book was about NKVD's use of terror. He cited, for example, the call his host in California had received while he was a guest. A man told the host that if he loved his small daughter he would get rid of the foreigner [Kravchenko] who was visiting him at that very moment.

Zborowski worked in the United States as a professor of anthropology at Columbia and Harvard Universities. Orlov's information that Zborowski was a Soviet agent eventually enabled the FBI to arrest him. The United States Attorney for New York accused Zborowski of having served the cause of the Soviet Union, international communism and the Soviet intelligence service most of his entire life. Zborowski, unfortunately, could not be tried in the United States for the criminal acts he committed against Trotsky and Sedov in France and Mexico.

The FBI questioned Zborowski about the work he had done in the United States for his Soviet GPU case officer, Jack Soble. One source reported:

> In 1945 Zborowski was told by his NKVD superiors to shift from Russian emigres to the American scene.... Only a fraction of Zborowski's deeds have thus far been revealed, especially those of recent years. But from the known facts it is clear that Zborowski has been a loyal and important Soviet agent for many years. The MVD [NKVD] is not likely to let agents of this caliber stray.[269]

During his interrogation, Zborowski was less than truthful regarding aspects of his life in America. Zborowski claimed that the NKVD took the initiative in contacting him after he came to the United States, which is wrong.

On 21 April 1958, Zborowski was indicted for telling lies to the FBI and was brought before the United States District Court on charges of perjury.[270] Zborowski was convicted on 8 December 1958 and sentenced by Judge John M. Cashin to the maximum sentence possible of five years at hard labor. In 1962 Zborowski appealed and was again convicted with an extended sentence of three years and eleven months additional.

Zborowski was eventually released from prison and went to work for the Mount Zion Medical Center in San Francisco as a member of the Medical Anthropology Department. He and his wife moved to a fashionable suburb in San Francisco. He published several books including *Life Is With People* and *Cultural Components of Response to Pain*. Zborowski's principal field of study was research, testing the human tolerance of extreme pain. The world famous anthropologist, who had unwittingly befriended him before she knew he was a Soviet agent, Margaret Mead, wrote a forward to one of his books. Zborowsku died in San Francisco on 12 May 1991.

TROTSKY AND THE AMERICAN COMMUNIST PARTY

> The phenomena of conversion is peculiarly inherent in revolutionary movements. At some point, the revolutionary crosses a threshold of belief, faith and action. Some edge across it slowly; some hurl themselves impulsively. But for all, there is a moment of decision, a step from one world into another.
>
> Theodore Draper *American Communism and the Soviet Union*, p.364.

Communism has been all things to a great number of essentially ignorant intellectuals and a host of pseudo-liberals since it germinated in America some one hundred and fifty years ago.

The first recorded Marxian Socialists in the United States were German immigrants who came to America in 1848 after the repression of their socialist revolution in Germany. In 1864 the International Workingmen's Association, commonly known in leftist circles as the First International, was founded in London by Karl Marx. An American counterpart was formed five years later in 1869.

The conspiratorial form of the First International, the doctrines of violence, the utter disregard for authority, the idealistic search for liberty, equality and fraternity through rebellion, destruction and chaos were all principles that these new American members were forced to adopt on faith. Unfortunately, the worst of these tocsins remained as the catechism for the Communist Party of the United States of America (CPUSA) which evolved in a circumventional way from this inauspicious beginning.

Sometimes overlooked in citing the origins of the pro-communist left in the United States is Louis C. Fraina. Born in Italy in 1894 he grew up in the Bowery on the east side of New York. His rags to revolutionary path started traditionally with selling newspapers on the streets of the Bowery and reading voraciously the literature of radicalism and revolt.

Fraina's first act of genuine radicalism was to break with the Catholic Church of his parents and forebears. He then began to write for an agnostic publication spewing forth all the venomous myths of time immemorial against the Catholic Church. His articles came to the attention of one of the foremost newspaper columnists of the day, Arthur Brisbane, who got Fraina a job as a cub reporter on the *New York Journal*. Fraina soon became a self-taught, highly gifted, intellectual agitator.[271]

There was little Russian influence in the left wing, or socialist communist movement, in America before 1917. Even after the Bolshevik revolution there was relatively little Bolshevik influence. This was a two-way street, the American left wingers knew very little about the Bolshevik communists, or the revolutionary movement inside Russia, and the Bolsheviks knew even less about the left wing movement in the United States.

Toward the end of 1915 Lenin made tentative overtures toward the American left by sending to the United States his beautiful roving ambassadress of good will, Madame Comrade Alexandra Mikhailovna Kollontay. She came ostensibly on a lecture tour, appealing to the down-trodden workers of America whom she tried to convince that they were the victims of capitalist exploitation.

The workers in America, at that time and forever since, were definitely better off in all respects when compared with the Soviet drudges she was exhorting the American workers to emulate. Her message, needless to say, was not received enthusiastically. Madame Kollontay also advocated sexual permissiveness and free love for the workers and that was not at all well received by the American workers' wives.

The man who arranged Kollontay's American lecture tour was Ludwig Lore, a German born, orthodox Marxist. Lore was the assistant editor of the leftist newspaper *New Yorker Volkszeitung*.

In early 1917 Leon Trotsky and his family came to the United States. Trotsky was already known in leftist circles as one of the leaders of the 1905 Russian revolution,. He had escaped from Russia and come to New York as a political exile badly in need of a place to recuperate. The *New York Times* reported that Trotsky had represented several Jewish newspapers in Petrograd and Kiev. This report was probably Trotsky's self-generated subterfuge and cover. It was safer for him to pass himself off as a harmless Jewish refugee journalist rather than to be known as a notorious international revolutionary.

Trotsky wrote several articles for the Jewish newspaper *Novy Mir* in New York while he lived there with Natalia and their two sons in a three-room, cold-water flat in the Bronx. During the two and one-half months he was in New York he also wrote editorials for Lore's *New Yorker Volkszeitung* and gave some thirty-five lectures at $10 per speech. Sometimes Trotsky, Bukharin and Lore would speak from the same platform. Lore, incidentally, became Trotsky's closest American friend, which association was to have fateful consequences for Lore after Stalin came to power.

In the winter of 1917 about twenty prominent left wing socialists met in Ludwig Lore's home in Brooklyn. There was Madame Kollontay, Bukharin, Fraina, Lore, Trotsky and about fifteen others. Trotsky spoke at this meeting so eloquently that he captivated the entire group. A member of the audience said Trotsky could have led them directly into armed revolt against the United States government at that moment. Those present agreed that Trotsky literally talked himself into the momentary command of the American left wing with his speech. It was often said later that if Trotsky had remained in the United States, he would have become the leader of the CPUSA.

Meanwhile, as if to create the ambient for duplicity, conniving and underhandedness that later was to characterize the CPUSA movement, Madame Kollontay was secrctly tattling back to Lenin everything that Trotsky and Bukharin were doing in America.

Trotsky cvidently offended Kollantay who later wrote to Lenin that " Trotsky's arrival strengthened the right wing of our meetings." This letter provoked the last of those personal outbursts of Lenin's against Trotsky, "What a swine that Trotsky is!" - which were afterwards so freely exploited by Trotsky's enemies, especially Stalin.[272]

Trotsky, apart from his sporadic employment with the *Novy Mir* and his lecturing at $10 a speech, spent most of his time in the New York Public Library. He was studying the American economy and trying to learn what made Americans think and act like Americans.

Whatever plans may have eventually developed for Trotsky to lead the American communists in revolt against the United States government, were cut short on 15 February 1917 when Nicholas II, Czar of All Russia, abdicated. Trotsky, and the other Russian exiles in the United States who were planning to play leading roles in the American left wing, packed their bags to return home to Russia.

Trotsky went back to Russia to become the Commissar for Foreign Affairs and later the Commissar of War; Bukharin was put in charge of the Communist party newspaper *Pravda* and Madame Kollontay became Commissar for Social Welfare. These positions required considerable talent and intellect. If this array of talented people had remained in United States to assist the embryonic stirrings of the American left wing movement, it might have developed into a much stronger political force than it eventually did.

The CPUSA never gave credit to these early organizers, as it never doubted its own claim to divine birth. Most of the domestic communists in the United States seemed to believe that the Bolshevik revolution on 7 November 1917 had spontaneously created the American communist movement. One of its members said, "At one time or another the Party [CPUSA] has committed every crime in the book, up to and including the elaborately schemed murder which ended the life of Leon Trotsky."[273]

Most American left wingers at the time did not know what was going on inside Russia and were confused by the conflicting stories that came out of their ideological homeland. For example, Fraina immediately called for a second revolution in Russia after the success of the initial Bolshevik revolution. Fraina wanted a revolution more in line with Trotsky's philosophical image than in Lenin's. Fraina felt that Lenin's leadership had caused confusion in the socialist world as to what exactly the revolution in Russia meant to the socialist movements in other parts of the world. Did it mean social change in one country, Russia, which was Lenin's basic philosophy or did it also include something of Trotsky's philosophy of continuous world revolution?

The general effect of the Bolshevik revolution in the United States was to hypnotize most of the American left wingers. What was the secret of the success of the Bolshevik revolution, they asked? The Bolsheviks won because they were more revolutionary than anyone else, was what the majority of American left wingers answered. The principal lesson to be learned, others claimed, was that a small dedicated group could seize power only if they had enough genuine revolutionary zeal and slavishly followed a pristine-pure doctrine of Marxist communism.

Lenin and Trotsky, their names inseparable to most Americans, were the idols of the day. The fact that Trotsky was personally known in the United States to many of the radical socialists made him even more popular. Every left winger could dream of emulating Trotsky's revolutionary success and overthrowing the established government if they only followed the pure Marxist communist doctrine that had inspired Trotsky.

The first American pro-Soviet organization was the Friends of the Russian Revolution which was founded in 1917. Friends was organized ostensibly to support the Soviet peace proposal to cease, unilaterally if necessary, hostilities against Germany. Other, more doctrinaire left wingers, formed the American Bolshevik Bureau of Information in early 1918. Fraina was named the director of this latter group.

Generally speaking, both groups believed the Soviet propaganda that the Bolsheviks in Russia were the equivalent of militant American trade unionists. The average American communist had on blinders and believed this fabrication. He did not realize at the time that what the Bolsheviks had wrought in creating the new Soviet worker-automaton was the exact opposite of what the free American trade unionist stood for.

A new ultra-left magazine *The Revolutionary Age* with Fraina as editor and John Reed and Ludwig Lore as contributing reporters was first published in the United States in 1919.

When Lenin's "A Letter to the American Workingman" was published, it went to an extra effort to address the American trade unionists as "revolutionary equals." The American workers were not, and never wanted to be, "revolutionary equals." There was, however, a new breed of revolutionary worker, oriented to the left, slowly developing on the American scene.

In 1918 Michael Borodin, an early Comintern/GPU agent tried to smuggle money into the United States together with some propaganda material to finance the embryonic CPUSA.[274] Carl Sandburg, future poet laureat of the United States and Lincoln biographer, was in Oslo at the time covering the Finnish revolution for the Enterprize Association newspaper. Borodin gave Sandburg $10,000 and a copy of Lenin's "A Letter to American Workingmen" to smuggle into the United States. Sandburg himself reported the operation to the United States Customs authorities who seized the money and propaganda.[275]

Ludwig Lore, and the older socialist radicals who ante-dated the Bolshevik revolution, were in their mid-forties. These old timers could no longer believe in the dream of a quick revolution in the United States. There was a new breed emerging though, exemplified by the likes of William Weinstone and Jay Lovestone. Both of these men had been born in Russia and had been student leaders together at the Intercollegiate Socialist Society at the College of the City of New York. They were ready to take over the leadership of the American communist movement from the old guard.

Communism in its own right, instead of radicalized socialism was gradually coming into its own. On 19 April 1919 the first publication in the United States with the word "communist" in the title, *The New York Communist*, was published with John Reed as its editor.

About this time in Moscow, on 4 March 1919 to be exact, the First Congress of the Third International was voted into existence. It was to be known popularly as the Comintern (short for Communist International). The Comintern is international in about the same sense that the American Express Company is international; that is, it has branches in many foreign countries. However, its policies and purges and leadership have been from the start determined wholly by the internal and foreign interests of the Soviet state.[276]

The principal aim of the new Comintern was to help create and support communist parties outside of Russia. Gregory Zinoviev was elected president of the Comintern. Despite Trotsky's strenuous efforts to give the Comintern directorate an international flavor, virtually all of its staff were Russians.

In January of that same year of 1919, a courier from Moscow arrived in the United States. A 1919 telegram was intercepted that the German communists transmitted to Moscow by clandestine radio. This telegram, when deciphered read that, "Guralski arrived here [presumably Berlin] with money on his way to America." The message continued, "My name is now James."[277]

The courier carried instructions for Ludwig Martens, a Soviet agent, to organize an official American agency to represent the Soviet government in America.

> Properly disciplined, the professional communist [in the United States] must always be ready to be at the command of any of the *men of the mist* who may present himself. Out of nowhere, as though he became visible through a heavy fog, one of these men will suddenly float into the life of a comrade, introduced by some superior Party member. Then will begin furtive meetings, and the comrade will be required

> to get certain information in the special field or community or group in which he lives or labors. Occasionally he may be given a vague hint of why the information is necessary, if that will expedite his getting it, but often he has not the slightest notion unless he can guess for himself.[278]

Martens, a Russian by birth, had come to the United States as a representative of a large Russian steel firm. He had developed many good contacts with American capitalists and among the staff of *Novy Mir*. The name of Martens's new organization which opened for business in April of 1919 was the Russian Soviet Government Information Bureau.

The Bureau was staffed almost exclusively with Americans with no Russians observable on the premises. There was no diplomatic pouch between the Bureau and Moscow. The only communications with Russia was by telegrams and couriers traveling on slow steamships. There was great confusion, even within the Bureau itself, regarding exactly what it was supposed to do. Martens preferred to emphasize commercial matters but the leaders of the pro-Soviet groups in the United States felt the new Bureau should militantly spearhead the American proletarian revolution.

In later years the Soviet diplomatic missions in Washington, and the consular offices in other cities, drew a sharp distinction between their own espionage activities and the clandestine operations of the CPUSA. There has always been a strict operational dividing line that compartmented their respective jurisdictions so that a flap by one would not necessarily impede or contaminate the continuing clandestine activities of the other.

This efficient division of work has also been applied to the Soviet Union's propaganda effort. The two traditional networks are directed by, and report back to, two remarkable Kremlin creations: the USSR Foreign Office and the Comintern. (The Comintern until 1943 and the Cominform from 1947 to 1956.) The heads of these two departments are subject to the direction from the AGITPROP department which, in turn, is controlled by the Presidium of the Central Committee of the CPUSSR.[279]

The early American communist movement did not appear to be worried about compartmentation or tactical separation from the quasi-official Soviet Bureau. They believed that the American communist revolution was just around the corner, so why worry about security.

Regardless of how those working in a quasi-official status for the Soviet government in Martens's Information Bureau, and those Americans working for the communist organization might be separated in the public mind, they were really brothers under the skin. Both groups were publicly striving toward the common goal of the overthrow of the United States government and world domination by the Comintern.

In the summer of 1919 elements of the factionalized left wing socialist movements and the incipient communist nuclei, gathered in Chicago to regularize their status as orthodox political parties. They also had to determine their degree of allegiance to the principles of the Bolshevik revolution.

The conservative element of the left mobilized for their convention, as the official Socialist Party, at the Machinists' Hall in Chicago at 113 South Ashland Avenue on 30 August 1919. The radical and moderate left convened in a separate convention. They met at the headquarters of the Russian Federation at Smolny Hall at 1221 Blue Island Avenue in Chicago on 1 September 1919 to form a pure communist party.

At 6:00 p.m. on Sunday, 31 August 1919, John Reed, Benjamin Gitlow, and eighty other delegates from twenty-one states convened their meeting of the Socialist Party. They sang the communist anthem, the "Internationale" in the billiard room on the first floor of the Machinists' Hall. Subsequently, on 2 September in the IWW Hall at 119 Thrope Street in Chicago, where the conservative convention had moved from the Machinists' Hall, they formed the Communist Labor Party. They considered themselves full-fledged communists and unconditionally pledged their loyalty to the Third International in Moscow.

Meanwhile, on 1 September back at Smolny Hall, the more radical convention of the Communist Party of America opened with 128 delegates. Under pictures of Trotsky, Lenin and Marx they had a ten piece band that played the "Internationale." It was the dawn of a new millennium, they thought, and merriment reigned supreme — until the Anarchist Squad of the Chicago Police Department raided them.

Also observing this birth of a segment of the future CPUSA was Jacob Spolansky, of the United States Department of Justice's, Bureau of Investigation.[280] Trotsky's old friend, Louis C. Fraina, gave the opening speech at the inaugural meeting of what was to become the CPUSA.

These two factions of the foundling CPUSA continued to vie with each other and joust for the preeminent position in the minds of the American left. These two groups also fought over certification (accreditation) from Moscow – the ennobling pat on the head from the

Comintern to be designated as the official Communist Party of the United States of America.

Ultimately, preeminence would mean very little. Four months after the communist movement was launched, with the two factions still in contention, the entire communist movement was driven underground. As Theodore Draper wrote in his authoritative book, *Roots of American Communism*:

> The wave of strikes in the late 1919 on the American labor scene was large and boded ill for the fledgling Communist Parties. For all of the economic and political upheaval surrounding them, the Communists in America lived in a world of make believe. The unreal seemed real because after such a climactious success as the Russian Revolution nothing any more could be dismissed as unreal. To the average American Communist in 1919 their war against Capitalism, yellow dog trade unions and democratic elections was just as real and had just about as much chance of success as the Bolshevik Revolution.
>
> Possibly more important was that, after World War I, thirty-seven of the States had laws against "criminal anarchy," "sedition" and "criminal syndicalism" and prosecution broke out against Communists; many proceedings resulting in deportations. A footnote of history of interest here, is that an old Army Transport ship from the Spanish-American War, officially named the "Buford" was called back into service to take the deported 249 Russians away from the United States. It was nicknamed "the Soviet Ark" for performing this service.

There were never more than 75,000 Americans who gave their unflinching loyalty to the communist movement even though Attorney General Mitchell Palmer estimated officially that there were 5,000,000 fervent communists in the United States that were plotting to overthrow the government and establish a communist reign of terror.[281]

The communists had a simple answer to the raids, arrests, deportations and, most importantly, for the infiltrations by the Bureau of

Investigations' informants into their ranks. The American communists did exactly what Trotsky would have done, they went underground. Both the Communist Party and the Communist Labor Party therefore became clandestine, highly organized conspiratorial organizations.

They also adopted the tactic of "boring from within" when taking over a socialist organization. This meant entering an organization and reforming it or capturing it from the inside, instead of torpedoing it from the outside as in the past.[282] They would then use the organization taken over as a cover for their communist activities.

Pamphlets were distributed to the rank and file with instructions on how to work illegally; all membership cards were destroyed; meeting places were constantly changed; all-clear signals and danger signs were used for street meetings; and alias and pseudonyms were to be used in all writings and telephone calls.

A printed list of instructions for clandestine conspiratorial operations was issued to the membership:

1. Don't betray party work or workers under any circumstances.

2. Don't carry or keep on you names and addresses, except in a good code.

3. Don't keep openly in your rooms any incriminating documents or literature.

4. Don't take unnecessary risks in party work.

5. Don't shirk party work because of the risks connected with it.

6. Don't boast of what you have to do or have done for the party.

7. Don't divulge your membership in the party without necessity.

8. Don't let any spies follow you to appointments or meetings.

9. Don't lose your nerve in danger.

10. Don't answer any questions if arrested either at preliminary hearings or in court.

In May of 1920 the two communist party factions decided to cooperate and present a united front against the prosecutions. This new defense tactic was forced upon them by the vigilant and persistent investigations by the United States authorities.

A turning point of sorts occurred in an open space surrounded by woods on a farm near Bridgeman, Michigan. There, 58 delegates from the two American communist factions met with several agents of the Executive Committee of the Comintern. The purpose of this meeting was to reconcile the differences between the two contending communist groups with the hope they would emerge combined as a United Communist Party. Foremost on the agenda of this meeting was what methods could be employed to thwart the increasingly successful infiltrations of the Bureau of Investigation spies into the communists' secret meetings.

Government agents and informants had penetrated into every strata of the newly formed communist parties and the entire leftist movement in the United States. The Department of Justice had ordered extra men into regional anti-Red squads. These agents quickly and quietly infiltrated into radical ranks everywhere and went to work as party functionaries to cover their task of reporting valid, unbiased intelligence information to the government.

They worked the lowliest jobs in the leftist organizations in order to solidify their cover. They gladly worked as slop kitchen drudges in laborers' mess halls, as miners, foresters, mill hands and whatever opportunities they found that allowed them to ingratiate themselves into a group suspected of communist affiliations. They were always careful not to induce or instigate seditious or traitorous activities but only to report on the unlawful actions of the real instigators.

Several of these undercover agents achieved positions of eminent leadership in the anti-government leftist and communist groups. Many became leaders in the top echelons of the communist movement in the United States. Some of these more subtle penetrations remained in positions of leadership for many years, almost till the present, reporting to the FBI on the activities of the moribund CPUSA.

In 1920, a courier from the Soviet Union was captured in Latvia with two more letters from the Comintern to the American communists.

In these letters the Comintern ordered a combined organization to be created in the United States. The two communist factions were ordered to stop bickering and act as one united group. The Comintern letters urged a combination of legal and illegal working groups. On a separate piece of cloth there was an agreement calling for the unification of the Communist Labor Party and the Communist Party.

In May of 1920 therefore, on orders from Moscow, the United Communist Party came into existence. The United Communist Party was to be subservient to the Communist Party of the Soviet Union.

Consequently a resolution was adopted at the Second Convention of the American United Communist Party in 1920:

> The Communist Parties of the various countries of the world [including the American Communist Party] are the direct representatives of the Communist International and thus indirectly [are the representative] of the aims and policies of the Soviet Union.

The plenipotentiary Comintern representative in the United States at the time this protocol was signed by the American United Communist Party was Max Horwitz. Horwitz used the alias of Professor H. Valetski and the underground pseudonyms of Brooks and Michaelson.

Valetski's first order of business was to end the factionalism between the warring communist splinter groups. He organized that meeting of reconciliation, previously noted, that came to be known as the Bridgeman Convention - named for the quiet village, about fifteen miles from St. Joseph, Michigan.

Unknown to Valetski, the Department of Justice's Bureau of Investigations, under the direction of William J. Flynn, was also preparing for the Bridgeman Convention. One of Flynn's agents, Francis A. Morrow, who had been a Bureau of Investigation's agent penetrating radical groups since 1919, was also ready. Morrow was better known in communist circles as Comrade Day.

It is an interesting comment on CPUSA security that Comrade Day developed his entre' into communist activity by literally forming his own communist party cell. It consisted of himself, his wife, his son-in-law and two friends in Camden, New Jersey. He sought accreditation for his group from the CPUSA Central Committee and was accepted as a section organizer. He rose rapidly in the communist party hierarchy, achieving one of the highest posts in his region. All of the time he was rising rapidly into communist echelons of power, he was sending detailed reports to the Bureau of Investigations which he signed with his code designator as agent K-97.

Among other things, Comrade Day supplied the Bureau of Investigations with the most secret code of the communist underground. One night a drunken district organizer permitted Comrade Dag to help him decipher a secret message. He showed Day how the code was derived from the printed instructions on the back of a United States postal money order. It was a simple code of word and letter substitution but the code had baffled the cryptoanalysts of the Bureau of Investigations for a long time.[283]

Cracking this code was a major breakthrough for the Department of Justice. They were able to read all the messages between units of the CPUSA that they had been collecting (by one means or another) over the years. They were also able to break the current messages obtained from penetrations of the communist party and the coded messages from the telephone taps and trans-Atlantic cable intercepts.

By the summer of 1922, K-97 (Comrade Day) had become a highly placed functionary of the communist underground. He was chosen by the CPUSA to attend the Bridgeman convention. He was told that the meeting place was so secret that he would not be told exactly where the meeting was going to take place until he got there.

The communist organizers told him he would be sent on a series of routings with instructions for the next part of the journey relayed to him at the end of each day's travel. By a series of circuitous short trips over constantly changing routes from one city to another, and never knowing where he was going, he traveled generally west from New Jersey for several days. Finally, the group with Comrade Day, still not certain where they were supposed to go, arrived in Bridgeman, Michigan.

All of this mis-direction by the communist organizers was for naught. K-97 had managed to send notes to the Bureau of Investigations at every stop along the way. Each time his group changed direction, the trailing Federal agents changed direction with them.

As the Bridgeman convention got ready to open (it was scheduled to start on the 17th of August), Director Flynn of the Bureau of Investigations, sent a message to his special agent in charge of the

Chicago officer, Jacob Spolansky. The message told Spolansky that there was going to be as secret meeting of the communist parties somewhere in the vicinity of St. Joseph, Michigan. Spolansky was to locate the meeting site and keep it under surveillance. Through a series of fortunate incidents, and some very good investigative work, Spolansky located the Wulfskeel farm and placed it under twenty-four hour observation.

On the morning of 22 August 1922, Spolansky and four government agents, the local sheriff, and about twenty hastily mobilized local towns people sworn in as acting deputies, raided the convention. Unfortunately, Spolansky's presence in the area had become known to the delegates and most of the communist leaders had fled before the raid.

Only seventeen of the delegates were captured in the early morning raid but unfortunately Comrade Day, the Bureau's own K-97, was one of those captured. Comrade Day, thinking quickly, gave the raiding agents his name as Francis Ashworth. Spolansky immediately recognized Ashworth as K-97. He quickly telephoned Director Flynn for instructions on how to handle this valuable penetration agent so as not to blow his cover in front of his fellow communist prisoners. If Spolansky treated K-97 more leniently than the other communist prisoners it would be suspicious. Conversely, if Spolansky treated K-97 too harshly in order to make his arrest look genuine, K-97 might decide that being a penetration agent for the Bureau of Investigations, was just too hard on his health, and quit.

Flynn replied to Spolansky's query with the order that K-97 should be locked up and treated the same as the other prisoners. One of Spolansky's men, who was a medical doctor, swathed K-97 in bandages to make it appear that he had been brutally beaten. It would thus appear to his communist comrades that he had received the same severe treatment as they had and maybe a little bit more.

When K-97 was allowed to leave custody he was well bandaged, limping badly and barely able to talk through a supposedly beaten throat. The other comrades asked him if he was going to sue the agents of the Bureau of Investigations for his beatings. K-97 told them that he had been forced to sign an agreement not to prosecute any of the agents or they would not have released him.

Spolansky eventually caught up with the communist leaders who had escaped his raid at Bridgeman. Sixteen communist leaders, including such luminaries as William Z. Foster and Earl Browder, were apprehended and brought to trial. In December of 1922 when the list of witnesses scheduled to testify against them was published, the communist leadership was astounded to see Comrade Day on the list of Bureau witnesses against them.

By testifying Comrade Day was blown and could no longer be used as a penetration of the communist movement. Fortunately there were many more compartmented Bureau agents already in place acting as penetration agents of the communist underground.

In any good compartmented operation most, if not all, of the Executive Committee of the CPUSA could have been Bureau plants. None of these compartmented agents would have known about any of the others. In this manner the Bureau could use each agent's report as a check on the veracity of the other agents' reports. The Bureau could thus be assured that their agents were fundamentally loyal to them and not playing the Bureau against the CPUSA by withholding information. This cross-checking and constant verification is a most valuable tool for a case officer in the continual monitoring and assessing of his double agents.

Draper in his excellent book, *The Roots of American Communism* concludes with a truism:

> It [the start of the American communist movement] was a difficult birth and unhappy childhood. Like most people with unpleasant memories, the older communists would rather forget them; they prefer to give the impression that the real history of the movement started much later. It was transformed from a new expression of American radicalism, to the American appendage of a Russian revolutionary power. Nothing else so important ever happened to it again.

If Trotsky had not felt compelled to return to Russia from New York to be a leader of the Bolshevik revolution, he might have become the General Secretary of the CPUSA. If this had happened, would not Trotsky's version of Marxist philosophy have led the CPUSA away from Soviet domination and into conflict with the Comintern? Would Stalin, sitting in the Russian Kremlin, have had the courage to murder a rival such as Trotsky, sitting in the CPUSA headquarters in New York City?

Even without Trotsky as a leader, there were anti-Soviet groups formed in the United States. They provided a balance to the pro-Soviet propaganda of the CPUSA. Some had other agendas but they all were vociferously anti-Soviet, anti-Stalin and a few were pro-Trotsky. In 1922 the White Russian Peter Afanassieff, alias Peter Armstrong, arrived in San Francisco. In collaboration with a former Czarist officer, Victor de Kayville, he published the *American Gentile*, which featured anti-Soviet propaganda and an expose of the infamous "The Protocols of Zion."

In 1924 Sydney Reilly and his White Russian associates formed the International League Against Bolshevism with chapters in the United States. About the same time Nicolai Rybakoff, a former Colonel in the White Russian army, arrived in America. He founded a anti-Soviet newspaper in 1933 called *Rossiya*. Associated with Rybakoff was a former Okhrana agent, Boris Brasol, who was dedicated to the return of the Czar. He campaigned tirelessly throughout America on an anti-Soviet platform. He became a legal consultant to the law firm of Coudert Brothers in New York City and as an expert of the Soviet Union for the United States government.

Jacob Nosovitsky, disgruntled GPU agent and sometime informant for the American BI was always on the lookout for a live sucker and found a real beaut in the person of the wealthy, self-appointed communist hunter, Henry W. Marsh. Nosovitsky glibly persuaded Marsh, and his newly organized American Defense Society, of the increasing threat of communist subversion in the United States. He reaffirmed March's already firm opinion that the Soviet army was literally just a step away from their planned invasion of the United States from Mexico.

Nosovitsky convinced Marsh and the American Defense Society that for the paltry sum of $25,000 (very big money in those days) he would gather evidence of the Soviet Army's presence in Mexico. It was plannend that after Nosovitsky had the evidence that Marsh would present it as a fait accompli to the Department of Justice for prosecution.

There was not the remotest possibility that the Soviet army would invade the United States from Mexico but Nosovitsky knew he had a good thing going. He stayed in Mexico, enjoying the good life, while continuing to send back increasingly vague but titillating reports to Marsh about the impending invasion.

The American Defense Society and Marsh continued to feel sufficiently threatened by the ominousness of the forthcoming Soviet invasion that they kept the money flowing to Nosovitsly in Mexico. Nosovitsky discovered a wealthy Mexican widow with a large estate who took him in and helped him spread the money over her holdings as an additional security blanket for her retirement. Nosovitksuy was essentiallty out-foxed by a Mexican vixen!

There were other patriotic endeavors in America against the Soviet Union's attempts to subvert the United States through the CPUSA. On 13 August 1931, President Herbert Hoover stated in an interview with the *San Francisco News* that, "To tell the truth, the ambition of my life is to stamp out the Soviet Union."

In early 1927 another organization fighting the CPUSA was the National Civic Federation (NCF) founded by Ralph M. Easley. The avowed prupose to the NCF was to counter the Soviet Unmin's attempts at subverting America. A prominent member of the NCF was the Congressman from New York, Hamilton Fish. The NCF was a no-holds barred outfit that referred to the Soviet Union as the "the Red monster" and to Stalin as "that Madman."[284]

One of the first known Comintern agents sent to the United States was Trotsky's protege, Michael Borodin, whose real name was Michael Gruzenberg [285]. Born in Russia, Borodin came to the United States in 1906. He studied at Valparaiso University in Indiana and married a Russian girl who was a student there. The two left Valparaiso to establish a private school in Chicago where they taught English and other subjects to immigrants needed to qualify for citizenship. Thc Russian revolution lured Borodin back to Russia in 1917 and be became one of the Bolsheviks' better GPU/Comintern agents. He returned to the United States to work for the GPU where he arranged money transfers to the incipient communust party.

On another occasion, that reads more like a espionage thriller than a true story, Borodin was sent to the United States by Trotsky on a secret mission. The purpose of the trip was to finance the Soviet Russian Trade Delegation in Washington with some of the Czar's confiscated jewels, estimated to be worth more than one million rubles.

Thinking to avoid the discovery of the jewels by the American Customs authorities when he arrived in New York, Borodin hid the jewels in the false bottoms of two heavy leather suitcases. A little additional thinking on his part convinced him that, if the Aemrican Customs agents were looking for one man traveling with the jewels, two men traveling together might serve to confuse them. Accordingly, Borodin picked up an unwitting young Austrian ex-military officer to be his traveling companion. The young Austrian was to remain by Borodin's side on the trip from Europe to America.

In Haiti Borodin and his traveling companion were both removed from the ship by the American Customs officials. These American officials were acting on an informant's tip that the jewels might be found in the baggage of these two. No jewels were found but Borodin was frightened and decided to go to New York alone to reconnoiter the situation. He left the two suitcases with the jewels still intact in the false bottoms in Haiti in the custody of his new found traveling companion.

Jacob Spolansky, the renowned communist investigator for the Bureau of Investigation, was waiting for Borodin in New York. Spolansky, acting on the informant's tip, thoroughly searched Borodin's baggage for the jewels. He did not find any jewels and had to release Borodin.

As soon as Spolansky released Borodin, he contacted Dr. D.H. Dubrowsky, an official of Martens's Soviet Information Bureau. Borodin told Dubrowsku to expect the jewels at a certain time via a specified courier. One story is that eventually Dubrowsky received the jewels from Haiti and turned them over to Martens. There is another story, to be discussed later, that when Borodin returned to Haiti the jewels had disappeared.

The Soviet Information Bureau was operating at a loss and badly needed cash. Martens hocked the Czar's jewels (however they eventually came to him) for a ready cash loan of $20,000 to a young Irish revolutionary then living in New York by the name of Eamon De Valera. In 1938, many years later, De Valera was the Prime Minister of Ireland. He suggested to Maxim Litvinov, then the Soviet representative to the League of Nations, that if the Soviets would repay to him the loan of $20,000 he would return the Czar's jewels. There was no answer from the Soviet government. By 1948, during a hotly contested election, the presence of the Czar's jewels in Ireland had become an embarassment to the Irish government.

Finally, after twenty-eight years, the Czar's jewels were returned to the Soviet government. The tight-fisted Soviet government repayed the exact amount of the loan to the Irish government without paying any interest or even a thank you note. The Czar's jewels were finally returned to Mother Russia where they belonged. The Soviet government was only "symbolically" receiving then, is what they said!

Financing for the CPUSA and GPU activities in America has been a constant problem for the Soviet Union. Two brothers Morris and Jack Childs were high-ranking members of the CPUSA; Morris was a member of the Central Committee and editor of *The Daily Worker*. Both were intricately involved in Party financial matters and both were informants for the FBI. A New York attorney, Stanley Levinson, whom they knew was working for the Soviets, laundered Soviet funds for use by the CPUSA. Later Levinson was a close advisor to Martin Luther King which was what led the FBI to monitor and wire tap King.[286]

Count Anastase A. Vonsiasky, a former Czarist officer came to the United States and married Mrs. Marion Buckingham Ream Stephens, an American millionaress. He became an American citizen and, with his wife's money, founded the Russian Fascist National Revolutionary Party in the United States. He was able to establish branch offices in many American cities. His followers drilled with real weapons on his wife's estate in Thompson, Connecticut and actually carried out espionage and sabotage operations inside the Soviet Union. In June of 1940, Vonsiasky told a reporter from *The Horn* that he and Trotsky had parallel objectives in the struggle against the Soviet regime.[287]

In 1927, when Trotsky's fate had been sealed in Moscow, the factional fighting of the CPUSA had not yet caught up with the official Stalinist line, They were still extolling Trotsky as Lenin's comrade in arms and heir apparent.

Charles Ruthenberg, the titular leader of the CPUSA died on 2 March 1927. His death threw the aspirants for this position into frenzied squabbling for recognition from Stalin to be the new leader of the CPUSA. Everyone who thought they might have a chance to the title, hurried to Moscow to plead their case with the all powerful Stalin. These worthy aspirants included such highly placed CPUSA functionaries as Jay Lovestone, William Z. Foster, Benjamin Gitlow, Joseph Cannon, William Weinstone and others.

The Lovestone group got to Moscow first and pre-empted the argument for recognition by demanding the ouster of Trotsky from the Communist International. Lovestone and Gitlow came back from Moscow as the official Stalin-designated heirs to Ruthenberg. They had also received permission from Stalin to move the CPUSA headquarters from Chicago to New York which they considered a better place from which to promote the communist line.

Foster and Cannon lost their bid to take over the CPUSA but agreed to abide by Stalin's decision. When they returned to the United States they were pushed by the inter-party wrangling to leave the CPUSA and assume a position closer to the Trotsky camp of Marxian philosophy. Foster was shrewd enough to divorce himself from Trotskyism. Cannon, not as adroit, was subtly but persistently pushed into a position of opposition to the CPUSA. Eventually Cannon was forced to openly proclaims himself a Trotskyite, no longer taking orders from Stalin in Moscow.

Cannon's expulsion from the CPUSA in 1928 together with Max Schactman and about fifty other Trotskyite renegades was inevitable. In the winter of 1920 Solntsev, a young Russian who worked for the Amtorg Corporation in New York and Ludwig Lore, the editor of the influential Jewish newspaper, *New York Volkszeitung*, brought together a group to form the Trotskyite party of America. Beside Solntsev and Lore, there were Max Eastman and his wife, Eliena, and Dr. Antoinette Konikov of Boston.[288]

The American Trotskyites began their excommunication from the CPUSA as the Communist League of America. About six years later, in 1934, they accepted the fact that change had to come and became a section of Trotsky's Fourth International completely divorcing themselves from the CPUSA.

This schism caused many old line American Marxists who had originally hewed to the ideals of Bolshevism to wonder what they should do? If they remained with the CPUSA, they would have to renounce their interpretation of Marxist communism of continuing world revolution. This was Trotsky's philosophy and rejected by Stalin who preached revolution first in only one country. If, on the other hand, these old timers joined the Fourth International of Trotsky they would have to severe their ties with the Motherland of Soviet Russia that had spawned the only communist revolution.

This enforced separation between the CPUSA and the Trotskyites of the Fourth International had, in some cases, personal family ramifications. Individual communists could never had foreseen the heartbreaking rifts that resulted when they pledged themselves to one or the other faction.

Dr. Phillip Rosenbliett was a long serving, deep cover, GRU (Soviet Military Intelligence) agent stationed in New York City.[289] His sister was Mrs. Joseph Cannon, wife of James Cannon, the leader of the Fourth International in New York and a leading Trotskyite. In 1936 Rosenbliett made a secret visit to his sister on purely family matters. Previously the GRU had warned Rosenbliett not to have any contact with his sister or other Trotskyites.

Apparently the GRU coverage of Cannon and his wife was so thorough that Rosenbliett's visit was immediately reported. Rosenbliett was recalled to Moscow and executed. The GRU had no one watching Cannon or his wife. What they did have was Cannon's personal secretary and several other close confidants of Cannon reporting to them. Some were GPU penetrations of Trotsky's Fourth International but there was enough cooperation between the GRU and GPU to report such blatant misconduct as Rosenblietts' visit.

As it turned out, the GPU agents who were penetrating Cannon's entourage reported directly to Gaik Ovakimian. Ovakimian was one of the principal masterminds in the United States during the final implementation of the planning for Trotsky's assassination in Mexico.

On 28 January 1939, Rudy Baker, a veteran American communist wrote a brief on the work of the "CPUSA Secret Apparatus" for transmittal to the Comintern.[290]

> Compile comprehensive list of Trotskyites including names, addresses, biographies, connections, the organization to which they belong, who their friends are and so on. Our party [CPUSA] has exposed and weeded out the spies and traitors.[291]

> Formerly, Trotskyites would be called forward at a meeting and expelled from the party on the spot, and that would be the end of the matter; now he is exposed and *his connections discovered* [this means that he forced in the strictist sense to confess his associates]... The work [interrogation and coercion] of uncovering Trotskyites requires considerable expertise as well as trained and specialized cadres. This is a message from the CPUSA to the Comintern in Moscow.[292]

Caridad Mercader

To a mother, a son is never a fully grown man; and a son is never a fully grown man until he understands and accepts this about his mother.

Sydney Harris

Each event produces its own radiating shock waves. These consequential, rippling effects are conditioned by the same manipulative factors that produced the initial shock. Thus, the resulting after shocks retain the characteristics of the good and the bad of the original event.

The assassination of Leon Trotsky was a product of inexorable historical events that pitted the implacable forces of Stalin against the immoveable will of Trotsky. The moving force of the assassination, the person that ultimately brought into being the assassin himself, was a self-centered, iron-willed, completely immoral, diabolical mother.

Of course, one may wonder if it had not been for this particular mother producing this particular son, would the crime still have been committed? The answer would be a resounding affirmative. Stalin would have persisted with an infinite number of other killers until he eventually succeeded in murdering Trotsky. But, this particular mother did produce this particular killer and what kind of a woman was she?

Caridad Mercader, christened Eustacia Maria Caridad del Rio Hernandez, was born into a respectable, modestly wealthy, family in Sanitago de Cuba on 31 March 1892. She attended a Roman Catholic primary school in France operated by the nuns of the Order of the Sacred Heart of Jesus. There she learned French, English and the other usual graces that a proper young lady of that period would acquire in a convent school.

At the age of fourteen, in the year 1906, she transferred to another convent school in Barcelona, Spain. After graduation from that school she returned to Cuba to rejoin her family but they stayed together in Cuba only a short time. On 23 October 1910 the entire family went to Spain to settle there permanently.

Caridad, as many a young girl in Catholic Spain, toyed with the idea of joining a religious order. For a short time she served a novitiate in the Carmelite Order of Descalcez, which roughly translated means Order of the Shoeless. However, her wild spirit prevailed over any religious inclinations and she bade farewell to the nuns of the Shoeless order after only a few months in their convent.

Caridad at that age is remembered as quick tempered, highly nervous, unpredictable, depressingly moody at times, introspective, sometimes given to mysticism, but above all else, a very beautiful woman. In an attempt to tame her wild spirit, her family decided that marriage might bring her down to an acceptable level of propriety and lady-like decorum. On 7 January 1911 they married Caridad off to Pablo Mercader Marina. She was not quite nineteen years-of-age. Poor Pablo Mercader, having gotten the worse of the bargain, did not know it but he had already seen the best years of his life, before his marriage to Caridad, at the age of twenty-six.

Pablo Mercader was the son of a not too prosperous Catalonian family. He had been born on 26 August 1884 and was essentially what the neighbors would call "a good man and husband." He was somewhat shy, retiring from social life by nature and a political monarchial conservative. Pablo and Caridad started married life in a small fourth floor apartment at 8 Sarria street in Barcelona and remained there until the middle of 1912.

Jaime Ramon Mercader del Rio Hernandez was Caridad's and Pablo's second child.[293] He was born on 7 February 1914 where the Mercaders, still in Barcelona, lived at Illias and Vidal streets.[294] Ramon attended school at the English Institute in Barcelona.

After primary school, Ramon attended a Roman Catholic institution run by the Escolapin Fathers on Carcegas Street. There, he was in constant rebellion against the discipline of the Fathers and their categorical religious teachings. Ramon repudiated the Catholic Church and all it principles of right, justice and morality. Abetted by his mother, he denounced religion and embraced agnosticism.

Caridad, who spoke fluent French and English, soon added the Catalonian accent to her native Spanish. This idiomatic fluency, together with her already Bohemian propensities, gained her acceptance into the gypsy way of life of the liberal left cafe set. Always a bit of a coquette, she embraced all of the vices of Bohemianism that suited her. Spurning hearth and home she all but deserted poor Pablo to his pipe and slippers and left him taking care of the children. She began making the rounds of the cafes in the evenings, sleeping with whatever male partner she happened to end with and returning home to recuperate during the day for the next night's debauch.

In 1925 as she neared the age of thirty-three, she took up painting under the tutelage of a well known local Bohemian artist named Borras Abella. She soon became romantically and politically enmeshed with the pseudo-intellectuals who frequented his studio. She accepted as dogma, the communist philosophy of these hangers-on who espoused radical communist political ideas as chic.

Ramon, then and always, idealized his mother and her friends. To him their ideas were rakish and the epitome of the real world. In his mind they were interesting and doers as opposed to the image he retained of his father as a stodgy, sluggish and physically bourgeois - a dull non-personality.

The revolutionary atmosphere of Catalonia in which Ramon was raised was similar to that of the old Russia in its early revolutionary days. For example, during 1918 when Ramon was only four years old, more than two hundred bombs were exploded in Barcelona by anarchists, communists and by agent provocateurs who were in the employ of the police as penetrations of the terrorists.

Workers' strikes led by terrorists were fierce and bloody. They were intended to create chaos and class war. Each strike was introduced by bombings and augmented by bloody rioting in the streets. This was indeed heady pap for an impressionable lad with a mother who encouraged this unorthodox philosophy and his vicarious participation in the violence in the streets.

Having made life unbearable for her husband with her anarchies, her moral debaucheries and her unfettered foolishness, Caridad did the right thing and left Pablo Mercader in 1925. She took the children with her and settled in France to live life the way she always wanted. She settled first in Toulouse, then tried Dax and finally moved to Bordeaux, flitting from one gay, mad fling to the next until she found the ambient that she believed suited her way of life.

No efforts at reconciliation came close to succeeding when they were tried by either Pablo or the relatives from both sides of the marriage. There was much lamenting and breast beating from Pablo's and Caridad's relatives about trying to save the marriage and "where did we go wrong?" However, Caridad spurned all efforts to be reunited with Pablo. And, if the truth be known, Pablo was probably not too eager to resume connubial life with Caridad either!

Regardless, Caridad seemed bent on her own destruction and appeared to be driven by a force stronger than domestic convention. Her drama, like all good theater is supposed to be, once started was destined to play itself out.

The prelude to Caridad's second act was her discovery of what she believed to be the panacea of communism. A love affair with a prominent French aviator who was known for his strong communist convictions brought Caridad into the Marxist fold. There followed for Caridad a period of personal agony as she shed the last vestiges of personal moral restraint and embraced communism.

Caridad swung from unequivocal acceptance of the dialectic communist doctrine and dogma to philosophic questioning of some of the more outlandish Marxist-Leninist mish-mash. Then her early Roman Catholic training must have instinctively instilled doubts in her iron resolve of acceptance of this foreign theory; she would be tormented by both betrayals.

During this emotional upheaval of values and conscience she twice tried to commit suicide. On one attempt she took a strong dose of poison which only resulted in her convalescence in a Bordeaux hospital for several months. Her profligate life, her flagrant disregard of propriety in living openly with a man not her husband, the suicide attempts and her general moral debauchery finally became too much for Pablo. In 1929 he engineered the final break from Caridad.

Ramon, his brothers and sister had been shuttled between their father's home and their mother's houses, which was wherever the man she was currently living with happened to be. The children did however have the territorial constant that whenever their mother's houses were they were always someplace in France and their father's only home was always in Barcelona.

When Ramon reached the age of fourteen, he attended a school for apprentice chefs and learned the rudiments of carving, cooking and the preparation of exotic foods. This school was located in Lyons and also taught Ramon some of the professional aspects of operating a hotel by using his charismatic charm on the guests. This personality trait that he acquired was later to be useful when he was wheedling his way into Trotsky's group.

Ramon returned to his father's house in Barcelona for a short time after he finished chef's school, while Caridad continued to lead the "good life," as she saw it, in France. Ramon took a part-time job as a messenger at the Goya Centennial Exhibition in Barcelona. This menial work fell far short of his ambition of pursuing the career of a hotel manager, or of a master chef. At least, he rationalized, he was meeting the public and learning how to be convivial with all kinds of people.

In 1929 Ramon was employed for a short time following his chosen career as an assistant manager at the Plaza de Espana Hotel in Barcelona. Later he worked at the Ritz Hotel as an assistant chef, carving roasts and fowls and preparing ornamental food layouts. The Ritz Hotel was Barcelona's finest and the fashionable meeting place for the town's social elite.

Ramon was inordinately proud of his meat carving skills. It was this knife wielding experience at these hotels which accounted for most of his skill with carving knives. He delighted to display this surgical deftness at dinner parties when there was a roast or a fowl to be properly cut up and served.

Some commentators of the Trotsky case have erroneously alluded to the possible connection between Ramon Mercader, as the knife wielding, dismembering assassin, and the mutilated, headless, surgically carved body of Rudolf Klement.[295]

Rudolf Klement had been the translator into German of Trotsky's writing under the pen name of Walter Steen. Before that he had been Trotsky's secretary during his exile on Prinkipo Island. At the time of his murder he was the Secretary of the International Bureau of Trotsky's Fourth International.

It was on 14 July 1938 that Klement disappeared. Two days later a letter was received by one of his companions addressed to Trotsky. The letter was obviously a fabrication, supposedly written by Klement the letter excoriated Trotsky.

The letter cited Trotsky's intellectual dishonesty as his alleged reason for breaking with Trotsky and the Fourth International. The style of writing was clearly not that of Klement but was probably, in light of subsequent happenings, that of the GPU.[296]

Caridad, for the most part rid of any maternal or conjugal obligations that had held her attached to Barcelona, and freed emotionally from any vestiges of domesticity with Pablo, moved permanently to Paris. There she displayed, even more flagrantly, her dedication to wanton, riotous living of the good life. The life of a dedicated communist party worker apparently filled the spiritual void formerly satisfied by hearth, home and church.[297]

Caridad wanted to live life to the fullest and took up residence on the Left Bank of Paris which was then in its hey-day as the gathering place of renegades and cosmopolitan debauchers. Despite this pleasing ambient of moral freedom, Caridad increasingly did not care about any part of the world except the Soviet Union and her zeal to help international communism succeed accordingly. She acted as an underground courier for the Comintern between its Paris headquarters and the communists in Spain and between Paris and the Belgium comrades. By dint of completing arduous tasks she was increasingly trusted with more demanding and sensitive assignments. Starting as a novice agent she rapidly assumed more responsibilities until she was accepted as a fully vetted Comintern operative.[298]

Caridad's many sexual encounters with French communist leaders became a topic with which she regaled her less promiscuous, or more discriminating, comrades at every opportunity. Caridad's sexual appetite was also the talk of the French Quarter, the Left Bank of Paris and wherever communists gathered.

Later, much later, during her bleak years in Moscow when her active GPU career was over, and she was virtually a prisoner in the Soviet Union, she still could produce a laugh from her fellow exiles. She would describe, how one of the members of the French communist hierarchy performed in bed, how they looked in their long underwear with the seat flap askew and the confidences they breached to her under the influence of her sexual diversities.

In July of 1932, when he was eighteen years-of-age, Ramon Mercader, volunteered for military duty with the Jaen Infantry Regiment of the Spanish army. His timing was fortuitous since it was just before he would normally have been called up for obligatory military service.

Ramon took to soldiering. He enjoyed the drilling, the regimentation and the total lack of personal responsibility. In retrospect, the dictatorial maternalism of the military may have contributed to his becoming the semi-automaton who was so easily controlled by the NKVD when he murdered Trotsky. He rose rapidly in the Jaen regiment, soon made corporal and was on his way to promotion to sergeant.

Ramon remained in the army until 1934, eventually becoming a brevet lieutenant. He thoroughly enjoyed the company of the bold men and the strong discipline of institutionalization. He had always taken orders from his strong willed mother and enjoyed being subservient to the materialistic military authority.[299]

Friends and relatives noticed a drastic change in Ramon's personality and general demeanor during this period of military service. Although Ramon obviously enjoyed the military life he became cranky and irritable about little things. He allowed his appearance to droop to a slovenly state and he began to shun his former non-military friends. It turned out that Ramon was only suffering from Caridad's absence since he was used to routinely confiding in her and accepting her advice on every day matters.

Ramon was also active in the organization of the Cervantes Artistic Recreational Circle in Barcelona. A lofty name for what in reality was a communist subversive group dedicated to the overthrow of the Spanish government. He enjoyed work in this clandestine group since it required performing sabotage under the noses of the Spanish secret police. Much like a little boy, he enjoyed masquerading under cover of being someone other than himself.

The Spanish anti-terror police were fascist trained and efficient at digging out underground communist subversives. The police knew that the Cervantes Circle was a front for an underground sabotage cell of the Barcelona communist youth of Ramon's age. At the time Ramon undertook this organizational work he was living at a boarding house above a shoe repair store at 7 Calle Ancha. Although the Cervantes Circle met occasionally in Ramon's room they preferred to meet at the bar at 11 Calle Wilfredo. The group's secretary worked there as a bartender and the group could always be assured of a free drink during an evening's plotting.

On 12 June 1935 the police, tired of having this nettlesome subversive activity going on under their noses, raided the bar. Jaime Ramon Mercader was unceremoniously arrested with seventeen other members of the Circle. Ramon is described in the police records at the time of the raid as a writer, a bachelor, a native of Barcelona and as a member of the executive committee of the Circle.

With the exalted rank of executive committee member attached to his name, Ramon's arrest attracted increasing attention from the authorities in Madrid. He was soon transferred to a jail there after only three months detention in the Barcelona prison. The leaders of the secret police found permanent lodging for Ramon in a prison called Miguel de Los Reyes in Valencia which was about as far away from Barcelona as they could put him.

In February of 1936 a popular front government came into power in Madrid and the imprisoned anarchists and communists, including Ramon, were pardoned. They were thereby given free rein to resume their subversive activities. Shortly after his release from prison, the GPU directed Ramon to acquire more experience in labor agitation and he helped organize a communist school teachers' strike.

On 18 July 1936 Caridad was ordered to return to Barcelona for a publicity stunt that was needed to bolster support for the flagging loyalties of most of the population for the communists. The GPU propaganda section felt they needed a symbol of a valiant female fighter to help galvanize the complacent women of Spain to rally to the Republican cause. Accordingly they asked Caridad to help stage a mock rally of "Peasant Women Against the Loyalists" with a daring display of a make-believe battle with feminine leadership.

The stage was set! Caridad supposedly led a group of these rehearsed valiant peasants against the mock-up machine gun emplacements of Franco's make believe troops. The actors on both sides of this drama played their roles superbly and Caridad carried the day to victory when she stormed the Franco fortifications for a Republican victory.

Caridad became something of a folk hero after the press release of her daring deed hit the streets. The GPU publicity department got their story of a valiant fighting woman leading Spanish peasants into a battle against the evil of Generalismo Franco's fascist forces. Everyone gained something from this charade except maybe the gullible peasants in the country who believed the story!

Caridad emerged unscathed from this staged battle to become a popular heroine of the liberal faction of the Spanish people and the adored darling of the communist Republicans. The people whom she had impressed elected her to take charge of the Union of Communist Women with a palatial office at the party headquarters at the Hotel Colon. Caridad loved this adulation, she always had been a compulsive actress. She enjoyed the limelight and felt this instant popularity was justly due her. This inclination to accept unearned acclaim was similar to the vicarious adoration she basked in years later with the reflected glory of her son while she was in the Soviet Union after he had murdered Trotsky.[300]

The Republican government in Madrid called for volunteers to form a peoples' militia to fight Franco. The clarion call to arms was emotionally patriotic and undeniable. Caridad and her adult sons immediately joined the Republican forces. Caridad was not an intellectual but she knew that Franco represented the forces that were against communism. Since she was a good communist she therefore knew she was against Franco. She would fight on the side that was fighting Franco.

While she was on the way to the front to join the militia forces, Caridad was wounded in the shoulder by a bullet or piece of shrapnel from a German straffing plane. Although the wound was not life threatening Caridad had to be evacuated to Lerida for treatment. From there she was moved to Montserrat Clinic on Calle 22 Septimana in Barcelona.

As soon as she was sufficiently recovered from the wound, the GPU decided to give her a rest as part of her convalescence. Early in October of 1936 they chose her to head a delegation to Mexico to reconnoiter the area where neither they nor she were yet certain that her son would be required to play out his life's destiny.

Caridad's ostensible mission to Mexico was one of charity and mercy supposedly on behalf of the impoverished children orphaned by the Spanish Civil War. In reality she went to Mexico to lay the foundation for the support structure the GPU might need in many countries where there was a chance of Trotsky eventually settling. These elaborate, by clandestine or any other standards, support structures were created both in the anticipation of the possible need in the Trotsky operation but also, in general, as part of the GPU's own administrative expansion of covert operations in the anticipated war against Nazi Germany.

Meanwhile Ramon remained in Spain, still in training for his future assignment wherever it might eventually lead. He had risen to the rank of lieutenant in the Republican army and had become the political commissar of the 27th Division on the Aragon front.

PART II

THE KILLERS GATHER

> You have to be a bit of a villain for that kind of work [espionage].
>
> William Skardon, British MI-5

Stalin had vowed to liquidate Trotsky regardless of where he eventually was brought to bay. Stalin's minions stalked the prey over most of Europe without discovering a propitious opportunity to strike. The hunt would take the wolves to many places in the world. Stalin had to have his bureaucrats develop many flexible contingency plans to cover all conceivable locales and situations for Trotsky's assassination.

About the time the Civil War began in Spain in the summer of 1936, the Political Bureau of the Liga Comunista Internacionalista, a Trotskyite political organizations in Mexico, received a letter from the International Secretary of the Fourth International in New York. The letter inquired what the ambient political attitude would be in Mexico if Trotsky tried to settle there.

Of course, the GPU contingency plans had to be reviewed to include this new area. As soon as the possibility was broached to the headquarters of Trotsky's own Fourth International, the information was known to the GPU. The GPU penetration agents working in the office of the director of the Fourth International, James Cannon, probably Sylvia Caldwell (aka Franklin) immediately relayed the information to her GPU case officer in the CPUSA. Thus the GPU planning for Trotsky's murder in Mexico began even before Trotsky himself knew that resettlement there was a possibility being considered by his own organization.

The Mexican Liga immediately started a propaganda campaign to curry popular opinion in Mexico in favor of Trotsky's asylum. By their own admission, the Liga did not believe there was any real possibility that Trotsky would consider coming to Mexico. There was an even less likely chance that the President of Mexico, Lazaro Cardenas, would tolerate Trotsky in Mexico.

The idea of Trotsky seeking political asylum in Mexico was first suggested in correspondence between Manuel Rodriguez, of the Liga, and the Fourth International headquarters in New York. At the time Rodriguez was working for General Francisco Mujica, a member of President Cardenas's cabinet. Rodriguez speculated in his letter, that if the occasion should arise, that the Cardenas administration might be willing to favorably consider the idea of granting Trotsky political asylum.

Subsequently, on 21 November 1936, Diego Rivera, a pseudo-Trotskyite and a world renowned Mexican muralist, received a cable from the New York Trotskyites stating that there was a life and death urgency to determine if President Cardenas would, in fact, grant Trotsky asylum in Mexico. There was imminent danger, the cable stressed, that unless Trotsky, then precariously residing in Norway, received a Mexican visa, he might be handed over to the Soviet government by the Stalin lackey, the Norwegian Minister of Justice, Trgvie Lie.

Before the Trotskys themselves were certain that they might be accorded political asylum in Mexico, an advance party of Spanish communists sailed from Barcelona bound for Mexico, Their mission was to survey the area for the assassination of Trotsky should he ultimately decide to settle there. Thus, the information gleaned from the GPU penetrations of the Fourth International brought an immediate reaction.

The principal member of the advance scouting party traveling to Mexico aboard the ship S.S. Manuel Arnuz was a beautiful Spanish woman of commanding demeanor. This strikingly beautiful women, with raven black hair, streaked with wisps of grey and patrician Castillian features, was Caridad Mercader.

Among the passengers traveling with Caridad aboard the Manuel Arnuz were several other Spaniards. There was the nineteen year old Lena Imbert, who was rumored aboard ship as the fiance of Caridad's son, Ramon.[301] Lena was also described as a near perfect Castillian beauty; dark and vivacious. There were also two male Spanish communist teachers traveling as part of Caridad's group.[302]

Caridad's party all traveled on forged Mexican passports. The ostensible purpose of their trip was to generate Mexican public opinion in support of the Spanish Republican cause. They were also to ask the Mexican people to donate funds for the care of Spanish children orphaned as a result of the Civil War.[303]

Caridad's official group, that is Lena and the two school teachers, were identified as soon as they arrived at Vera Cruz, Mexico on 6 November 1936, as Spanish citizens traveling on bogus Mexican passports. The group was formally detained for investigation by Mexican immigration authorities. (The GPU planners should not have exposed their agents to such a possibility of being discovered using false Mexican passports as they tried to enter Mexico. There was not any operational reason the group could not have traveled on Spanish passports using an alias which were easily obtainable.)

Caridad, in real trouble and not knowing what to do about the Mexican immigration authorities, called a Spanish friend living in Mexico. The friend tried to arrange her status and that of her group as political refugees so they could enter Mexico legally. Eventually, a sympathetic Mexican Minister of the Interior sent word from the capitol that the group could remain in Mexico but only on visitors' visas.

On 10 November 1936 the Mexican Foreign Office queried its Consul in Barcelona with the question of how could these obvious Spaniards ever have been documented as Mexican nationals. There was no valid answer and the solution probably could be attributed to the theft of valid Mexican passports from the Mexican Consulate in Barcelona or Madrid by person or persons unknown.

As soon as Caridad and her group were released from the custody of the Mexican immigration authorities in Vera Cruz, they immediately traveled by train to Mexico City. There they went directly to the headquarters of the Mexican Communist Party.

Caridad's covert mission on this trip was probably to spot and make initial assessments of potential agents for a cadre of willing workers who later would follow her orders, or those of her designee, without question. A firmly recruited agent needs to be fully assessed only once although that one time may actually be a long process covering many months and many probational tests. Once fully assessed and motivationally recruited, the agent will usually be amenable to working for any designated case officer.

During this trip Caridad was spotting and assessing raw agent material. The actual recruiting might come later for those stalwarts whom she selected after their dossiers had been evaluated in Mexico and later by the GPU in Moscow. The final approval for recruitment would probably be given to someone who would handle them as a case officer in Mexico.

The GPU was constantly canvassing everyone they came in contact with for any indication of agent potential. Each comrade was queried, be he or she an official in the Soviet embassy, a worker in Amtorg, the Soviet trading company, or an indigenous member of the host country's communist party. All were directed to bring to the attention of the next higher echelon in the party apparat the names of everyone with whom they came in contact. It was for the GPU alone to decide which potentials they would seek additional information on and whom they might select for recruitment.

Higher up in the GPU hierarchy a decision would be made if a spotted candidate was of interest for operational use. The person spotted might be considered either as an support agent (safe house keeper, surveillance team member, etc.) or as an action agent who would be the source of intelligence information. Naturally, anyone working on a highly technical matter, in the military or in the government of the host country, would immediately be selected for further, more intensive, development.

The principle criterion for selection of either kind of agent is ability and access. Does he or she have the physical ability to be a good surveillance team member? Does the prospective information agent have access to the target the GPU is trying to penetrate or control? If he does not have direct access, does the prospective agent have indirect access, i.e., does he have secondary access through a relative that might be coerced or a lover who is vulnerable?

Further, and this is the critical step, even if the prospective agent does have access to the target area, can he get the information out of the target to the GPU? Failing that, could he be in a position to sabotage the target if needed during war time? For example, he may go to work every day in an experimental atomic energy plant. If he works in the finance section or in the motor pool, his access to the secrets of fissionable material usage in the laboratory is severely limited. Conversely, working in the finance section or in the motor pool may make him an ideal agent for recruitment as a prospective sabotage agent for political terrorism.

As the first stage of the spotting and assessing process, the person who made the initial evaluation that there might be operational potential in the candidate's use, submits an extensive biographic history on the subject. The person's background, dependability, political astuteness, relatives and their political orientation, financial status, education, friends, physical characteristic and much more must be written down for evaluation by higher authority.

After the spotter has submitted the basic biographic information cited above, a more operationally-attuned series of questions are posed and answered. How should the recruitment be handled? What are the prospective agent's psychological motivations for accepting recruitment? Is he more susceptible to a straight out money approach instead of an ideological one? Can he be blackmailed? How should he be treated as an agent? Should he be recruited under a false flag recruitment? How should he be paid? How shall he transmit his information to his case officer or headquarters? What are the bona fides to be used in re-contacting the agent if contact is lost? What would be appropriate danger, safety signals considering the agent's position in the community? What meeting recognition signals and dead drop procedures might be used considering his employment and time he can take away from his job? What telephone code or message code could be properly used with this agent? Can the spotter's name be used in the initial approach for recruitment? What emergency and alternative contact procedures can be used? What termination procedures could be securely used? (It is axiomatic that an agent is never even considered for recruitment until a secure termination plan has been developed!) There are literally thousands of operational questions about the agent that must be answered before the ideal recruitment takes place. (There are some situations where having all these questions answered before recruitment is attempted is a nicety that cannot always be had.)

For the GPU the single most important question might be, how to best handle the agent's termination (or elimination), if he becomes an embarrassment or a threat to the GPU or the Soviet government. Also, how to counter the propaganda if he defects and turns hostile?

In the meantime the overt work of Caridad's group, of raising funds to aid the orphans of the Civil War, was quickly being accomplished. Another of Caridad's overt tasks that was designed to give real diversionary substance to the group's cover work was the mobilization of the Mexican Teachers' Union in support of the Spanish Republican cause.

Potentially the country-wide Mexican Teachers' Union could give a broad national base of support to the Republican efforts of fund raising in Mexico. This Union was already communist controlled, and the Mexican Ministry of Education had several communist sympathizers in high-level positions.

The pro-communist ambient within the Teachers' Union galvanized by Caridad's efforts was graphically illustrated by an incident that took place while the group was traveling though Yucatan. There, the Mexican teachers in anticipation of Caridad's visit, had taken down the Mexican flag from atop the school house and hoisted the Soviet Union's hammer and sickle flag. These rural teachers, carried away by the euphoria of seeing a real life Spanish communist, had put up a picture of Stalin in place of the government supplied obligatory picture of the Mexican president, Lazaro Cardenas.

On 18 November 1936, dressed in the blue uniform of the Spanish Republican Women's Militia, Caridad addressed the Mexican Chamber of Deputies. Her speech was well received. On 21 November she was allowed by the Mexican government to speak to a large crowd of enthusiastic Mexicans assembled in the Zocalo of Mexico City.

The Zocalo is an enormous town square in the center of the old section of the city and the site of the most important departments of the Mexican government. The Zocalo probably has more historical, religious and Mexican Indian cultural reverence associated with it than any other area of Mexico. For Caridad, a Spaniard, to be allowed to speak in the Zocalo, venerated almost as holy ground by the Mexican people since Aztec times, revealed the amount of political clout she and the communists were able to generate.

Caridad was a good speaker and swayed the crowd that day with her eloquence. She told them that the world had reached a cross roads. They must chose between the good path to communism or the bad path to fascism. The right way, she said, was with the benign patriots of communism who had already mobilized to help the Spanish workers and peasants throw off the shackles of oppression and capitalist domination and exploitation.

For the next two months Caridad and her group toured the outlying areas of Mexico accompanied by representatives of the communist Mexican Teachers' Union. They made friends everywhere they went since the Mexicans were predisposed to like most Spaniards. (Some Mexican commentators challenge this assertion. They claim that many Mexicans still vehemently dislike Spain and things Spanish, including Spanish people. The reason being Spain's former conquest and colonial exploitation of Mexico.)

In February of 1937 the Mexican Department of the Interior told the Mexican police to round up Caridad and her group for over staying the period of time permitted them on the temporary visas. The Department told the police that Caridad and her people were to be sent back to Spain.

Caridad and her group tried every bit of clout and conniving to remain in Mexico since Trotsky and his wife had arrived in Tampico, Mexico on 9 January 1937, but they were refused permission to remain. The Mexican government did not know why Caridad was trying to so desperately to remain in Mexico, but it would not have made any difference; the group had over stayed the legitimate period of their temporary visas and they had to leave the country.

This is another example of poor GPU planning. If Caridad was sufficiently important at that time to the Trotsky elimination operation, the GPU could have contrived some way to have her remain but they were not in a position to do so.

CAST OF CHARACTERS

Agents must be in the intelligentsia; they must not shrink from the last sacrifice at the critical moment.

Order of the GPU #185,796

Jose David Alfaro Siqueiros

Another of the daring-do killers to be trained and brought up to GPU murdering standards in the Spanish Civil War was the Mexican muralist, Jose David Alfaro Siqueiros.[304] It might be noted that spiritually Siqueiros never left Mexico when he went to Spain. He was born and raised in Mexico but spent his adult life wrapped in the cocoon of self-deluding fantasy and the warm spit of communist ideology.[305] Thus, although he may have been away from his native land in body, be it Moscow or Spain, he was still living the life he had chosen, as a dedicated Mexican Communist wherever he was. His ambient was therefore always familiarly similar; he made certain there were no terrors of the unknown for Siqueiros.[306]

When the assassin, Ramon Mercader, arrived in Mexico in the latter part of 1939, he was welcomed into the assault team already in place by Siqueiros whom he had known in Spain. Siqueiros had known his mother, Caridad Mercader, in Spain where they were reported to have enjoyed an intimate friendship as lovers, and were politically dedicated communists wholly committed to Trotsky's elimination. Caridad had also introduced Siqueiros while he was in Spain to her son and the men became close personal friends.

Siqueiros claimed that part of the justification for his attack on Trotsky was the shame the Trotskyite party in Spain brought upon the Mexican contingent fighting there with the Republican forces. Supposedly the Trotskyite, Partido Obrero de Unificacion Marxista (POUM) had been disloyal to the Republican cause and unreliable in fighting against the Franco army. There is no discernible truth to this charge. Further, the POUM did no fighting as a unit since its members had been massacred by the GPU as Trotskyites. This pap from Siqueiros is a palpably thin excuse without foundation in fact.

The POUM, Siqueiros believed (or at least pretended to believe to bolster his justification for his subsequent criminal actions), was disloyal and undermined the Republican's allegiance to Soviet communism. As a consequence the other Republican fighting men blamed the Mexican contingent for the POUM's disloyalty solely because President Cardenas had granted Trotksy political sanctuary. Siqueiros, his delicate sensibilities supposedly shattered by this criticism, vowed, together with several other Mexican volunteers of equally wounded sensibilities, that Trotsky's base in Mexico must be abolished in order to regain the respect of the other Republican soldiers.

According to the cover story concocted by the GPU, Siqueiros and Vitorio Vidali were supposed to have attended a meeting together in Spain where La Pasionaria (Dolores Ibarruri) severly chastized the Mexicans for permitting Trotsky to be granted asylum in their country.

This trumped up scenario had supposedly so insulted their revolutionary manhood that they allegedly vowed to oust Trotsky from Mexico.[307] (This version clearly omits the crucial element of the operation that Stalin had ordered the GPU to eliminate Trotsky and the GPU was ordering its agents what to do, how to do it and when to do it. There would be no tolerance by the GPU of two of its agents going off on a frolic of their own to eliminate Trotsky.)

Reports of this nature, attributing the raid on Trotsky's home on 24 May 1940 as a personal whim of Siqueiros acting without orders from the GPU displays a naivete that unfortunately taints the otherwise acknowledged veracity of these sources. The presence of Leonid Eitingon in Mexico directing the 24 May raid is ample evidence of the GPU's operational direction.[308]

More than a year passed after Siqueiros returned to Mexico from Spain before the wrath of his wounded pride compelled him to raid Trotsky's home. Siqueiros's reasons for the raid were, he claimed, three fold: to cause "commotion" by randomly, but without lethal intent, firing their guns inside Trotsky's home; to demonstrate that the Mexican government could not guarantee Trotsky's safety, and to steal some of Trotsky's documents.[309]

In a post-raid attempt to mitigate responsibility for his criminal acts, Siqueiros claimed his intentions were "from the beginning it was agreed to avoid all bloodshed" and the "violent death of Trotsky" would have been counterproductive. He claimed that all he really intended to do was create an "uproar that would force [President] Cardenas to expel Trotsky from Mexico."[310]

When Siqueiros was fourteen he became a sergeant in the "battalion Mama," an army group in the Mexican Revolutionary army made up of teenage boys. By the time he was fifteen he was a battle proven veteran of many skirmishes and was commissioned a lieutenant. Later, when he was seventeen, he was appointed a staff officer to Venustiano Carranza, a leading Mexican revolutionary and future president of Mexico.

Siqueiros was soon given the opportunity to learn more about painting and the outside world when he was assigned as a nominal military attache to the Mexican Embassy in Paris. There he experimented with various forms of painting and the use of plastics as a technique. He did not acquire much new expertise in painting but he did live the life of a Bohemian expatriate to the fullest.[311]

In 1920 Siqueiros went to Moscow and, it is reported, that he was trained there in the arts of espionage and subversion with a emphasis on guerrilla tactics, explosives and assassination techniques. There was a rumor at the time Siqueiros was in Moscow that the Soviet Union was planning to wage a guerrilla war against the government of Mexico for the purpose of establishing a base there for future subversive operations against the United States. The idea for a guerrilla war against the United States from Mexico was abandoned, for the time being at least, when the GPU had to concentrate its full attention on the Trotsky assassination.[312]

Around 1928 Siqueiros was reputedly inducted into a quasi-staff capacity of the GPU, that is he was accepted into staff ranks only as far as any non-Soviet citizen can ever be. At this time he would have taken the oath, as every staff GPU officer must take, and swear that he would do everything in his power, even lay down his life, in the service of Stalin and the Soviet Union. (It should be noted that it was the man, Stalin, during that epoch that always came first in the GPU oath, and the country, the Soviet Union, was always secondary.)

In Spain Siqueiros joined the Fifth Regiment in February 1937. This regiment was commanded by Vitorio Vidali, who used the name Carlos Contreras or Commandante Carlos.[313] Siqueiros, for a very short time, was put in charge of a tank unit in the Fifth Regiment. Angelica, his wife, joined Siqueiros in Spain, ostensibly as a reporter for the Mexican newspaper *El Nacional* in March of 1937.[314] They both returned to Mexico in November of 1937.

Siqueiros returned alone to Spain in December of 1937. He was then sent to Italy by the Republican Servicio Inteligencia Militar (SIM) to obtain a copy of a magazine that had the entire order of battle of the Italian detachment that Mussolini sent to Spain. Even though the issue had been withdrawn, Siqueiros, by a lucky fluke, found a copy in a deserted tavern in Rome and brought it back to the SIM representative in Paris. In September 1938 when the war was over, Siqueiros gathered the remaining 52 Mexican volunteers out of the initial 533 and led then back to Mexico.

The Siqueiros group of Mexicans were not allowed to disembark in New York until President Cardenas intervened. They were allowed to take buses from New York to the Mexican border escorted by the United States Marshals. The returning veterans and Siqueiros arrived back in Mexico City in January of 1939. Angelica and Siqueiros were the organizers of the movement to arrange Spanish Republican refugees and their families to settle in Mexico. For this purpose they formed the Sociedad Francisco Javier Mina.[315]

In Spain Siqueiros was the leader of the Mexican volunteers. He attained the rank of colonel, largely by intimidating the Mexicans in the rank and file of his unit, in what was considered to have been a rigged election, Siqueiros nevertheless wore the dubious distinction of being a brevet colonel, as a badge of honor. (For the remainder of his life he insisted upon being addressed as "colonel.")

Siqueiros was a blow hard and a braggart but there was no doubt as to his bravery whenever his courage was put to a test. He spent little time, however, in the mud and filth of the front line trenches in the Spanish Civil War. Most of his time was wisely and more safely spent behind the lines at the Union of Revolutionary Artists in Barcelona.

Ramon Mercader, the future assassin of Trotsky, was sponsored by Siqueiros for membership in the Union of Revolutionary Artists even though Mercader had no talent for painting and could barely wield a paint brush. The future assassin later became Siqueiros's close companion in Spain and something of a protege of his during Mercader's initial period in Mexico.

There was a breakdown in GPU compartmentation when Ramon Mercader arrived in Mexico City. In Spain, Siqueiros and Mercader had what must be acknowledged as an overt relationship. There had been no attempt at concealment or the establishment of any basis for subsequent plausible denial if their association was ever questioned. Therefore, when Mercader arrived in Mexico, he and Siqueiros were openly seen together by many of Siqueiros's friends and no effort was made to protect the relationship of the two agents from compromise.

Furthermore, Mercader's common law wife, Sylvia Ageloff, once asked Mercader where he worked in Mexico City. Mercader told her he worked in an office in room 820 in the Ermita building. A few days later Sylvia's sister, Hilda, innocently and with no attempt to check up on Mercader's story, was in the neighborhood and went to that address to pay a social visit to Mercader. She discovered that there was no such room as 820 in the Ermita building.

When confronted with this apparent discrepancy, Mercader offered the explanation that he had been mistaken or that Sylvia had misunderstood him. He corrected this misunderstanding and told Sylvia that his office was in fact in the Ermita building but in room 620 instead of 820. Room 620 in the Ermita building was the office of Jose David Alfaro Siqueiros. This was another compromising detail built into the cover story concocted by the GPU for Mercader.

If it was not compromising enough for the GPU to taint one agent by allowing him to openly associate with another agent, it was even worse to have that agent use as his cover address the actual business address of the other agent. What could have motivated the GPU planners when they developed this cover story for the assassin? Was there some obtuse operational justification that escapes rational explanation? Normally much thought and backstopping, sometimes years of work, go into the preparation of a proper cover legend that will usually defy compromise.

The GPU planners may have assumed Mercader's ultimate role to be suicidal and the fact that he might be compromised might not have made any difference to them. Did this mean they were planning to also throw Siqueiros to the wolves once Trotsky's murder had been committed? On the other hand, why permit Siqueiros to be blown needlessly? He might be good for years to come as a powerful agent of influence in the Mexican political arena. Further, Stalin's command that no Soviet participation be shown in Trotsky's assassination should have made the GPU very cautious not to compromise agents during the operation. Lastly, at the time the cover story was created there was every expectation that Mercader would never be used as the assassin.

The elements of the operational plan to assassinate Trotsky were predicated upon the supposition that a future raid led by Siqueiros would succeed in eliminating him. If Mercader represented the fall back plan should Siqueiros fail, available records fail to reflect this possibility. Mercader's role was to be that of a penetration and reporting agent, gathering information upon which the final murder plan for Siqueiros would be based.

Vittorio Codovila

Another prime mover of the GPU assassination team was Vittorio Codovila, who came from the Spanish Civil War to Mexico. He was born an Italian but spent most of his life in Latin America and was the founder of the Argentine Communist Party. He had been a GPU agent extraordinaire for many years before he joined the International Brigade in Spain; and he was a militant disciplined GPU agent for many years by the time he reached Mexico in 1938.

Codovilla went by the name of Medina in Spain and was noted as speaking Spanish with a strong South American accent. He was the chief Comintern representative to the Partido Comunista Espana (PCE) in 1935-36 and, as such, exercised ultimate power within its ranks. He played a major role in the Party's popular front policy in Spain prior to the outbreak of the civil war. He was to remain a powerful figure behind the scenes throughout the Spanish Civil War.[316]

Codovila, by his successful Comintern work over the years, had earned the support of most of the communist rank and file. He had gained their respect the hard way. Communist party hacks throughout Latin America regarded him as the titular leader of the Comintern in the southern hemisphere. (At the time of the Russian revolution the Bolsheviks knew little about South America and, conversely, Latin American socialists and liberals knew little of Bolshevism or communism. Lenin considered all of Latin America as neo-colonies of Britain. Codovilla overcame this apathy of Lenin's by forming a viable communist party in Argentina.)[317]

Together with Siqueiros, Codovila was a natural for the assassination team. They quickly recruited a small guerrilla force made up of the miners from the state of Jalisco and former members of the Spanish Republican army chosen by the GPU to be part of the murder team.

In Spain Codovila had acquired the reputation of being a ruthless but efficient murderer of human beings. He worked tutorially under the infamous General Kotov who surpassed every other exterminator for viciousness, brutality and awesome efficiency. Kotov's true name was Leonid Eitingon. This is, of course, was the same Leonid Eitingon who was Caridad Mercader's lover and Ramon Mercader's mentor.

Codovila was linked with one of the most revolting crimes of Stalin's atrocities in Spain, the brutal torture and murder of Andres Nin, the leader of the Trotskyite, POUM.[318]

Vittorio Vidali

Most students of the GPU would consider Vittorio Vidali, (alias Enea Sormenti, Carlos Contreras, Commandante Carlos, General Carlos) as one of its foremost and deadliest executioners. Vidali's proficiency in this art might be considered second only to that of the notorious George Mink.

Vidali, as Commandante Carlos, was the organizer of the infamous Fifth Regiment of the International Brigade in Spain, The Fifth Regiment was partially responsible for some of the most heroic battle actions of the Brigade's history. When the Regiment was first committed to combat it halted Franco's seemingly unstoppable advance. Commandante Carlos commanded this battle-hardened regiment by the systematic use of methodical terror.

> Vidali and his common law wife, Maria (sic) Modotti were old Comintern functionaries who, like others of their kind, had been assigned to work in Spain even prior to the outbreak of hostilities. Vidali was untouchable even by members of the Politburo of the PCE. One time he countermanded an order of Vicente Uribe, a member of the Politburo of the PCE. Vidalio came to Spain as a representative of the "Socorro Rojo" (Red Aid).[319] (The Socorro Rojo was a cover for the GPU under the guise of a charitable organization similar to the American Red Cross.)

The executions of Vidali's men for trivial breaches of discipline were common knowledge throughout the Republican army. Vidali was reputed to be the most feared of all GPU enforcers in Spain.

Vidali's background presents a typical Comintern maturation. The Anti-Fascist Alliance of North America was a front organization that was eventually captured by the communist. It was organized in 1923 to counteract the influence of Mussolini's pro-fascist sympathizers in the United States. It had welcomed socialists, communists, anarchists and others into its ranks. Its secretary was the radical poet, Arturo Giovannitti, who had enjoyed the confidence of all factions. Three years later the communists seized control; most of the membership left and formed the Anti-Fascist League for the Freedom of Italy.

The new secretary of the captured Anti-Fascist Alliance was Enea Sormenti (later known as Vittorio Vidali). Sormenti was a native of Trieste who had been forced into exile by Mussolini's regime because of his communist activity. Soon after arriving in the United States, he was appointed secretary of the Workers' Party Italian Bureau and editor of its official newspaper *Il Lavoratore*.[320]

Vidali first came to America in the early 1920s using the name Enea Sormenti. His principal duties for the Comintern at that time were for the fledgling CPUSA to organize strong arm goon squads. These thugs were used by the CPUSA to convince democratic workers, by systematic beatings and terror, to vote either pro-communist in the labor elections or face almost certain savage consequences. Vidali was deported from the United States for these activities in 1927 and a short time later surfaced in Mexico.

Isaac Don Levine, one of the foremost historians on the GPU commented that one of the most formidable Stalinist agents to reach Mexico for the Trotsky assassination was Vittorio Vidali. When he arrived in Mexico, Vidali purged the PCM and reconstituted it, on Stalin's orders, in preparation for the Trotsky assassination.

The leader of the PCM, Hernan Laborde, and his deputy, Campa, had become complacent while adhering to the popular front ideology. They had directed the PCM to strongly support the Cardenas presidency. They favored complete cooperation with the leftist revolutionary government of Cardenas as part of the popular front campaign that had been ordered by the Comintern. They were therefore, according to this Comintern policy, putting the PCM on the side favoring Mexican sanctuary for Trotsky.

The preparation for the assassination of Leon Trotsky called for a dramatic reversal of the popular front policy of the PCM. Vidali convened a purge commission with himself, Vittorio Codovilla and James Ford, the CPUSA perennial candidate for president of the United States and the token black in the party heirarchy. The purge commission voted to oust Laborde and Campa, thus setting the stage for a PCM reversal of policy and the outright condemnation of Trotsky and his refugee status in Mexico.

Vidali established a Comintern Committee to cleanse the PCM of its anti-assassination leadership. When the dust settled in March of 1940, the former leaders of the PCM had been ousted in disgrace on false charges of personal dishonesty and deviation from strict party dogma. An "honorary" presidium of Comintern leaders assumed the stewardship of the PCM. Among the notables on the presidium was Vidali "whose record as a master in the art of assassination was well know", according to Levine.

Trotsky himself wrote that Vidali's appointment to the temporary leadership of the PCM was part of the moral preparation for Stalin's executioneers getting ready for the final task of murdering him.

In Mexico Vidali became the lover of one of the most beautiful Italian/American women to ever cross the scene of GPU activity in Mexico. Her name was Tina Modotti. She had come to Mexico from California as the mistress, model and help mate of the world famous American photographer, Edward Weston. She was already well known in international art circles from the nude photographs of her that Weston had taken in California and her own acknowledged ability as a photographic artist.

> There was a cafe rather like the Dome where they all hung out along with heavily armed politicians, bull fighters, criminals, prostitutes and burlesque girls. The most spectacular person of all was a photographer, artist model, high-class courtesan and Mata Hari for the Comintern, Tina Modotti. She was the heroine of a lurid political assassination and was what I guess is called an international beauty..... She was exterminated in the Great Purge. A wiry little tart, who later became an actress, used to dance naked on the table, clad only in a rhinestone glued to her navel, while people banged guitars and everybody shouted. I think at the time she was Siqueiros's girl; at least she was always at his

> table, and years alter when she was rich and famous, killed herself because she was about to have an illegitimate baby."[321]

Jose Vasconcelos, the Mexican author, was one of those enamored with Modotti and describes her as the character Perlotti in his memoirs saying,

> Her legend was a dark one. One husband she had liquidated in California, kept there in a lunatic asylum because of venereal excess. At the time of which I am writing, she keeps two strong men pallid and gentle rivals: the famed photographer [Weston] and Rondio [the pen name of the Mexican writer Gomez Robelo]. Rondio would pour out tears of sensual tenderness. Slowly the unwholesome passion wasted his body, drugged his will and jealousy produced a fever in him. She. for her part, maintained herself apart.

This over-blown passage is included in a book about the Trotsky assassination in order to better understand how Modotti might have had a calming effect on Vidali with her siren like seductive beauty and vivacious personality. Despite Modotti's calming effect on Vidali, however, he would still drag her through the torments of hell. He would persist that he must remain as a GPU executioner, instead of succumbing to Modotti's entreaties to leave the GPU and Comintern.

Modotti, in addition to being Vidali's mistress, was at various times also the mistress of the communist Mexican artists, Javier Guerrero Saucedo and Diego Rivera. Modotti was a communist ideologue before she left California to come to Mexico with Weston. No doubt her association with the communist artists in Mexico strengthened her previous incipient tendency toward militant communism.

Vidali was one of the principals involved in the assassination of Trotsky. In this context his long ago relationship with Modotti is as pertinent as the then current one. In order to better understand Modotti's background, we go back to 10 January 1929.

Modotti and her current lover were walking along the street of Abraham Gonzalez, just off Avenida Juarez, enjoying a late evening stroll. They were returning to Modotti's apartment nearby when the man with Modotti, Julio Antonio Mella, was shot. Mella was a communist agitator from Havana, Cuba who had organized an illegal student strike against the Cuban government of President Machado.

The assassin was a Cuban gunman hired by the secret police of President Machado to kill Mella. The gunman was arrested by the Mexican police who had an air-tight case against him. To everyone's surprise the Cuban gunman managed to escape from the Mexican jail and flee to Cuba. The Mexican police attempted to conceal their complicity in the escape and the fact that the Cuban government had obviously bribed them to allow the killer to escape.

In a complete about face, the Mexican police then claimed that Mella's murder had not been politically motivated at all, but had been a crime of passion engineered by none other than Modotti herself. The Mexican police issued a statement that the weight of the evidence (newly discovered they alleged but never produced nor seen by anyone) was that the real killer could only be the woman that was with Mella. (In fact Modotti was close by Mella's side when he was shot and the forensic experts agreed that the shot that killed Mella had been fired from some distance away.) Despite overwhelming evidence exonerating her, Modotti was arrested for Mella's murder![322]

Diego Rivera, one of Modotti's lovers, ever one to seek publicity and finagle a way to hog the limelight, came to Modotti's defense. Rivera had the Mexican Communist Party (PCM) institute an investigation into Mella's murder for the purpose of ascertaining who had been the actual killer. The Mexican press delighted in wallowing in this sleazy story; Rivera's well known proclivity for women, the famously beautiful "nude American model" plus the Cuban political details of the Mella/Modotti affair, all made for sensational yellow journalism. The press even hinted that Rivera was the true love of the gorgeous Modotti and he had hired the gunman to get rid of his rival so he might have her affections all to himself.

The case became such a cause celebre with daily rehashings of the sordid details, that the Mexican government wanted to be rid of the entire squalid mess. The Mexican police raided Modotti's apartment and confiscated her papers. Among them they found letters which, they claimed, showed that Modotti was a lewd person who had come to Mexico for immoral purposes.

It is difficult to imagine how a more hypocritical piece of contrived evidence could have originated from a more cynical and jaded police force. Apparently they had not been paid a "mordida" (bribe but literally bite) of sufficient size by Rivera to lay off Modotti. Modotti did not have any money to pay them and Rivera was going through one of his lean periods when his paintings were not selling. Not being able to extort any money, the Mexican police ordered Modotti's deportation.[323]

In the ensuing years Modotti went with Vidali to Moscow where he conferred with GPU officials and then on to Paris where they both worked for the Comintern under cover as working for Socorro Rojo or Red Aid. She later accompanied him to Spain where he commanded the Fifth Regiment.[324] In Spain Modotti worked tirelessly as a nurse with Vidali for the Republicans in the International Brigades. The Brigaders loved her and she became famous among them as a revered mother figure whom they lovingly called Maria.[325]

After the Spanish Civil War Modotti and Vidali escaped with little more than their meager belongings. They flew back to Moscow for a brief visit where there is every likelihood that Vidali was briefed on the plans for the Trotsky assassination. Using their cover as Red Aid volunteers, the couple returned to Mexico where President Cardenas dropped the phoney charges against Modotti that were still pending from the Mella murder.

Vidali immediately become involved in the planning for the Trotsky assassination. Valentin Gonzelez (El Campesino) former Spanish communist general in the civil war, who was then in Moscow, reported that in August of 1939 he sent to Vidali in Mexico a squad of three GPU agents led by Pedro Martinez Carton, a former colonel in the Spanish Republican army. Gonzalez described Martinez Carton as a "cold and unscrupulous man, fanatical and dedicated, [and] instinctively cruel."[326]

Martinez Carton and his German wife, Carmen Salot, were close friends of Modotti and Vidali.[327] Modotti and Vidali were already deeply committed to an highly sensitive operation in Mexico when Martinez and Salot arrived.

Margaret Hooks who has written the definitive work on Tina Modotti summarizes:

> While there is no evidence to link Tina [Modotti] to the attack [24 May raid on Trotsky], it is possible that she knew about it beforehand. The flamboyant nature of the assault

> would certainly have appealed to Vidali's style of action, although when Mexican police later interrogated him about his participation they could not find sufficient evidence to press charges. Nevertheless, the rabidly anti-Stalinist, former Spanish communist leader Jesus Hernandez later accused him of having instigated [the raid] and handled a Thompson machine-gun in the assault. Hernandez claimed to have been in Moscow after the news arrived of Vidali's involvement and quoted Comintern chief Palmiro Togliatti as having shouted angrily: "He was ordered to have nothing to with it![328]

It is interesting to contrast the moral vacuum of Vidali with what is known about Modotti. She was genuinely liked by most women, and, in western cultures at least, it is axiomatic that women are a better judge of moral turpitude in another woman than are men. Despite her unorthodox behavior at times and her adherence to a moribund political philosophy, she was in the strict sense of the term, a lady.

Modotti's final association with Vidali is epic and tragic. According to the official Mexican coroner's report, and what is written on the death certificate, Modotti died from a heart attack. Although Mexican official documents have been known to be less than reliable there appears to be some basis of fact in this certificate. However, the GPU would have been keenly interested in making her death appear to be from natural causes if they had poisoned her.

Modotti had expressed a desire to be rid of Vidali and she had no doubt assured him, and her GPU case officer, that she would not reveal any of the skeletons from Vidali's past life. In fact she just wanted to be rid of Vidali; free from his womanizing and from his way of life. Modotti said on more than one occasion, "I hate him [Vidali] with all my soul and I have to follow him till death!"[329]

On 5 January 1942 Vidali and Modotti attended a dinner party at the apartment of Hannes and Lena Mayer's on Ramon Guzman street in Colonia Roma. Toward the end of the evening, after most of the other guests had departed, there were three couples remaining, Modotti and Vidali, Patricio Azearate and Cruz Diz, and the Meyers. Vidali left about midnight to go to the newsroom of *El Popular* where he claimed he had a meeting. About one o'clock Modotti said she felt ill and Ignacio

Aguirre, who had joined the party, accompanied her downstairs where she was able to hail a cab.

Federico Trejo, the driver of the cab took her to the General Hospital as she had requested. When they arrived Modotti was refused admittance because the hospital did not treat emergency cases. By the time Trejo took her to the Cruz Verde hospital on Victoria street she was dead in the rear seat of the taxi.

The coroner's report claimed that Modotti had died from a heart attack but there was enough evidence in the autopsy report to raise the suspicion that she had been poisoned.[330]

After Trotsky was assassinated on 20 August 1940 Vidali left Mexico and surreptitiously returned to the United States.

In order to complete the character profile of Vidali we must look ahead to 11 January 1943. On this date Vidali is alleged to have been involved in another assassination operation that he supposedly also master minded. Carlo Tresca, an anti-communist radical was shot to death by a gangster thug during a dim-out as he walked along Fifth Avenue and Fiftieth street in New York City. (The gangster had been hired by persons unknown but very possibly Vidali.)[331] Tresca was attempting to obtain the cooperation of the United States Attorney for the New York District to indict several members of the CPUSA for the murder of Juliet Stewart Poyntz (aka Juliet *Stuart* Poyntz). Poyntz had been kidnapped and murdered by Vidali's henchman, George Mink.[332]

Later, according to Isaac Don Levine, in March of 1945, Vidali was still in the good graces of the GPU and Stalin and was put in charge of attempting to arrange the escape from the Mexican prison of Ramon Mercader, the assassin. The fact that Vidali was involved in this proposed escape attempt caused Mercader to refuse to cooperate with the attempt to free him. Mercader was well aware that Vidali was a killer and he was convinced that Vidali would murder him while he was trying to escape.

Years later the GPU retired Vidali to the sunny part of Europe. He was appointed the Communist Party chief in Trieste, Italy.

George Mink

George Mink was one of the masterminds behind the Trotsky assassination. He was more than a principal agent, he was one of the sadistic catalysts that drove the other killers.

Mink had little formal education. He managed to exist, and to do rather well, with the meager amount of native intelligence for self-survival he developed in the alleys of his youth.[333] He was extremely wary and cocky, quintessentially ruthless with a consummate conceit. He was a "nifty" dresser, "with slightly Jewish features, full of watchful, cynical arrogance, short but strongly built. His mouth was small and cruel, his teeth were irregular, and his eyes were of a greenish-brown color with a faint animal glint."[334]

Mink was born in Russia but came to the United States at an early age. He joined the pre-CPUSA Marxist group in Philadelphia and was known as a communist since open membership was the official party line at the time. He worked for a while as a taxi driver in Philadelphia. He had other odd jobs while building his cover legend in the United States that would later prove useful on GPU assignments.

Mink started on the docks as a marauder, a pirate breaking into shipments and stealing. His cronies called him "Mink the harbor pirate." He moved to New York in 1927, wrote confidential reports to Solomon Lozovsky, the chief of the Soviet Profitern, and was accepted as a secret Profitern agent.[335]

Mink was eventually assigned to penetrate the Marine Workers Trade Union (MWTU). Aided by generous donations from the Profitern, a cadre of paid grunt workers and strong arm goons from the CPUSA, plus a hard sell propaganda campaign from the Party agitprop, Mink was elected leader of the MWTU for the eastern seaboard.

At that time, in the mid-1920s, all maritime labor organizations throughout the world were being given special attention by the Comintern. Mink's assignment, after ascendancy to the leadership of the MWTU, was to use his position to further Comintern operations. The sailors who belonged to the MWTU were a fertile field to be recruited as couriers during peace time and as saboteurs during a war.

Mink was called to Moscow in 1928 for special training and instructions. He then returned to America with an alias passport and considerable money. He toured the United States, Mexico and Latin America on secret Profitern business. From 1930 on he spent more time in Berlin and Moscow.

The one-time editor of *Soviet Russia Today*, Lister Oakes, testified in 1947 about Mink's years in the CPUSA. He told the HUAC that "...

George Mink transferred industrial information and patents [from the United States] to the Soviet trade agency, Amtorg."[336]

One of Mink's better known murders was the assassination of Hans Wissinger, a GPU agent who had refused to obey an order to return to Moscow after he botched an assignment. Jan Valtin was a GPU maritime courier working out of Hamburg, Germany when Mink murdered Wissinger.[337]

Shortly after Wissinger's death, Valtin accosted Mink in Hamburg at the Fuerher Verleg International Seamen's Congress. Mink as usual was drunk. He was singing and carousing with his ever present bevy of promiscuous secretaries from the local communist party headquarters. Valtin knew that Mink had murdered Wissinger but decided to tease him. He asked Mink if he knew Wissinger to which Mink replied yes, with a so-what attitude. Valtin pressed on to ask if, just maybe, Wissinger had been innocent and it had been a mistake to kill him? Mink replied with the classic GPU answer, as he shrugged his shoulders in finality and contempt for weakness, "We never make mistakes! We never strike at innocent men!"

Talent such as this was soon rewarded, and, since he was Russian born, Mink was eligible for promotion to staff in the GPU. In 1934 the GPU apparat in Germany, that had been so effective in collecting intelligence and conducting covert operations, was destroyed by the Nazi Gestapo. The hierarchy of the former GPU group in Germany was transferred to Denmark. Mink was placed in charge of this third country operation based in Denmark which was operating back into Germany.

In May of 1935 Mink was arrested in Copenhagen for the attempted rape and sodomy of a chambermaid at the Nordland hotel. The girl's screams were heard throughout the hotel and the guests who rushed to her assistance were able to restrain Mink. Codes, accommodation addresses, false passports, other spy paraphernalia and three thousand American dollars of suspicious origin were found scattered on the bed in Mink's hotel room by the Copenhagen police.

One of the leading members of the CPUSA, and a principal GPU operative in the United States, Leon Josephson, was conferring with Mink in his hotel room at the time of Mink's subsequent arrest.[338] Mink and Josephson were accused of espionage, plotting the assassination of Adolph Hitler and directing a secret GPU underground intelligence network into Nazi Germany from Denmark. On 30 July 1935 Mink was convicted of these charges and sentenced to a prison term of eighteen months. He served only a small part of the sentence due, no doubt, to communist influence in the Denmark government. Upon his release from the Dutch prison, Mink went directly to Moscow.

Leon Josephson and Nicholas Sherman, other communists who had been with Mink, were imprisoned for four months. At the trial Josephson used his legal talents to obtain his own acquittal while Mink and Sherman went to jail. Josephson was subsequently lavishly entertained by the American Embassy in Copenhagen. (Later Josephson was convicted of Contempt of Congress for refusing to tell them about the Soviet espionage labyrinth he had created.)[339]

Mink probably should have been executed when he returned to Moscow for the serious breach of security in compromising the codes, identities in the false passports and the accommodation addresses. Apparently he could make a case for leniency, or his purported relative, Solomon Lozovsky, had sufficient clout to save him. It has been opined that maybe Stalin recognized a kindred spirit and decided to give him another chance.

In 1936 the GPU supplied Mink with a false passport in the name of Adolph Hertz and sent him to Barcelona to help maintain discipline among the volunteers of the International Brigades. One American who met Mink in a Barcelona hotel in 1937 reported that he bore striking similarities to the stereotype of a gangland "torpedo." "He got drunk and boosted of his NKVD work... He showed me a roll of bills and was wearing a London tailored suit which cost him $150 [a sizeable sum in those days]. He had very expensive leather luggage and said that was the way to get a good reception in a hotel."[340]

Mink did not fight in the front lines in the Spanish Civil war but chose to operate from his quarters at the Continental hotel.[341] Mink gained considerable notoriety in Spain as the NKVD executioner and was involved [there] in arranging for Trotsky's assassination in Mexico.[342]

In Spain Mink was to enforce discipline particularly in the American Abraham Lincoln Battalion of the International Brigade. Lister Oakes, a communist editor and member of the International Brigade stated,

> I met George Mink who boosted about his part in organizing the Spanish GPU and offered me a job ... to put the finger

> on 'untrustworthy' volunteers entering Spain to fight Fascism." (Oakes claims he refused Mink's offer.) "William McQuiston, a former friend of Mink's testified that he had been present when Mink and Tony De Maio, killed two members of the International Brigade.[343]

In Mexico, on April 1938, Trotsky announced that the Killer, George Mink, an American communist, was on his way to Mexico to liquidate him.[344]

When he fled Spain after the defeat of the Republican forces, circa 1939, Mink appeared in Mexico as part of the Trotsky assassination team. The GPU sent its assassins to Mexico through Paris where Narcisco Bassols was the Mexican Ambassador and could grant them Mexican visas. Among those entering Mexico through this route were George Mink of the CPUSA and Vittorio Vidali (Contreras) of Trieste.[345]

Due to his bragging, the rumors spread about the preparations for the assassination. A former communist official, Maurice L. Malkin, later testified,

> A Mexican Trotskyite recognized him [Mink] and he vanished. I do not know what happened. I know that plenty of American loyalists [Spanish Republicans] who came back [from Spain] would like to get their hands on George Mink. He was responsible for shooting many Americans in the back over there.[346]

In 1938 Stalin replaced Nikolai Ivanovich Yezhov, with Lavrenti Pavolovich Beria as chief of the NKVD. Beria was a kindred soul with Mink. Beria is remembered for saying about one of his subjects who was about to be interrogated, "Let me have him for one night, and I'll have him confessing he is the king of England."

The presence of a known GPU executioner in Mexico, obvious from all other indications that he was there waiting for an appropriate moment to strike Trotsky, should have alerted Trotsky's guards. They could have moved Trotsky to his friend's home in Cuernavaca or to another house in sprawling Mexico City where the GPU could not find him.[347]

If no other action, why did not Joseph Hansen, the American chief of Trotsky's security guard force, at least report these rumors to his personal contact in the Legal Attache's office in the American Embassy. (The Legal Attache was the cover for the Federal Bureau of Investigation's office in the American Embassy in Mexico City.) Hansen is known to have been an informant of the Bureau's and, at that time, was a fairly frequent visitor to their office in the American Embassy.

Joseph Hansen bragged to American Consul George P. Snow in Mexico City that he and an associate had "liquidated Mink" shortly after the 24 May raid on Trotsky's house. Hansen claimed they had tied up Mink up and thrown his body down a crater thirty miles from Mexico City.[348]

Preparation in America for Trotsky's Assassination

Jacob Golos

Concurrently with the preparations for the Trotsky operation in Spain and Mexico, there was intensified activity by the GPU in the United States to provide a support mechanism. This activity was not exclusively in anticipation for the Trotsky assassination since the GPU also had a priority target to establish an intelligence reporting network in America. This base was needed to keep the Soviet Union apprised as to the military capability and strategic intentions of their World War Two ally whom they already recognized would be their post-war rival super power.

We must go back to 1933 when the Roosevelt-Litvinov Agreements were signed between the United States and the Soviet Union. Although there were mutual benefits as a result, the Soviets gained considerably more than did the United States. The net result of these Agreements was to permit the GPU to maintain a secure base from which they could quasi-overtly conduct espionage operations against the United States.

Alexander Orlov testified before the Senate Internal Security Subcommittee in 1938:

> There was here [in the United States] a chief director resident of the NKVD, by the name of Gusev, a man who had been in former years my assistant. Gusev had six assistants, from the CPUSA usually who were the contact

> men with the spies in the United States. Each of the Russian assistants took care of at least three rings. So you multiply three rings by six assistants and that makes eighteen rings, eighteen spy rings.[349]

There were two strategic tentacles of the Comintern/GPU intelligence operations existing in the United States at the time the Agreements were signed. One division was specifically aimed at the collection of political and military intelligence and was centered in Washington, D.C. under the direction of J. Peters.[350] Peters recruited as the principal agent for his network Harold Ware, who was the son of the grand old dame of American communism, Ella Reeve "Mother" Bloor.

In 1928 under Peters's direction, Ware obtained employment in the Department of Agriculture. Early during his first term, in 1933, President Roosevelt appointed Henry Agard Wallace as his Secretary of Agriculture. Wallace was pro-communist and a dedicated Marxist-Leninist in his political orientation. Immediately after his appointment Wallace created the so-called "Triple A" (Agriculture Adjustment Administration) a extreme socialistic venture.

The Triple A attracted Marxist ideologues who knew Ware and others who were directed there by Peters. Among those who formed the nucleus of this political-military Comintern/GPU spy network within the Department of Agriculture were John Abt, Nathan Witt and Lee Pressman. All of these were later identified as spies of the GPU by the defected GPU agent, Elizabeth Bentley. Also, tucked away in the legal section of the Department of Agriculture at that time was another Marxist ideologue who would later gain wide notoriety as a convicted perjurer and alleged Soviet agent, Alger Hiss.

The Peters group could have constituted a shadow government of the United States as these comrades left Agriculture for senior positions in many other departments of the government.

The other division of strategic Comintern/GPU intelligence collection apparat concentrated on industrial espionage against the United States. This section was headed by Jacob N. Golos who had a cover job as the director of the GPU controlled World Tourist, Inc. with offices at 41 Union Square in New York City. Golos was described as short, heavy set, red-haired and a man whose clothes appeared to have been pre-disheveled.

Golos was an espionage professional. He was born in Russia in 1890 as Jacob Raisin in the Pole of the Settlement, a ghetto where the Jews were confined under Czarist law. In 1908 Golos came to the United States with his parents and in 1916 became a naturalized citizen. At the beginning of World War One he joined Louis Fraina's wing of the fledgling CPUSA. In order to avoid being drafted for service in the American military he dodged the draft call-up by using the alias of Jacob N. Davis.

On 10 June 1929, operating as Jacob N. Golos, Golos chartered World Tourist, Inc. under New York incorporation laws with a capitol of $15,000 and himself as secretary. The ostensible purpose of World Tourist was, as stated in the incorporation papers, to be the official Soviet travel agency. It was also to arrange tours, individual accommodations on ocean liners and trains, and the mailing of packages between America and the Soviet Union. The geographical area to be covered by the business activities of World Tourist, within which it would service customers, was the entire United States, all of Canada and Mexico.

In 1934, no doubt due to the excellent work he had been doing for the Comintern/GPU, Golos was promoted to be the head of the Central Control Commission of the CPUSA. This was not only a great honor for past meritorious service but also gave Golos great power over all CPUSA members and activities.

Also in 1934, Golos and Earl Browder, the Secretary General of the CPUSA, founded the United States Shipping Corporation with offices at 212 Fifth Avenue in New York City with a capitalization of $15,000. The $15,000 seed money for the corporation came from Anatole B. Gromov, the Second Secretary of the new Soviet Embassy in Washington, D.C..

Both of these Soviet cover companies were financially successful. They were able to covertly pass large sums of money from profits to the CPUSA to keep it from going broke. The success of these two companies, with Golos's intense participation in operating both of them, became too much of a burden for him. He therefore had to select some trustworthy and dedicated comrade to assist him.

The new assistant was not needed to help him with the legitimate business of the two companies but with the covert intelligence collection and agent handling problems. Golos already had lawyers, accountants and managers running the legitimate businesses but he needed help with the day-to-day legwork of paying agents, collecting reports, handling agents' personal problems, changing collection requirements and the thousand and one problems that daily beset a case officer in handling his agents.

After much searching and vetting Golos hired a Vassar graduate and CPUSA member named Elizabeth Terrill Bentley. She was to work as his street alter-ego, secret courier, confidential administrative assistant and, as it later transpired, his stand in for making policy decisions and determining the operational direction of all his GPU cases.

At the time of which we write, Golos's wife, Celia, had gone back to Moscow with their son, Milton. Golos, was alone in New York, with a Bohemian Greenwich Village mentality, and only his new assistant to keep him company. Bentley was a good looking young lady who wanted to do everything she could to please her new boss. The inevitable happened and Golos and Bentley took an apartment together on West 54th Street in New York City.

Even though Bentley had been originally recruited to act as a courier and general factotum for the industrial group, Golos soon assigned her similar duties with the political group. Thus, Golos broke one of the cardinal rules of clandestine security when he breached the golden rule of compartmentation.

Bentley was used to service a mail drop for Sam Carr, Fred Rose and Tim Buck; all were Canadian communists involved in the theft of atomic secrets. She also serviced another mail drop used by the Arenal brothers who were involved in Trotsky's assassination operation in Mexico City.[351]

There is a continuing connection between the Trotsky case and Canada — more as a coincidence rather than any sinister connection. When Igor Gouzenko, the GRU (Soviet military intelligence) code clerk defected from the Soviet Embassy in Ottawa, he exposed the deeply entrenched Soviet atomic espionage network that included Buck, Carr and Rose.

Gouzenko's defection unmasked twenty-six Soviet spies in the first debriefing. In the first 48 hours after his debriefing, thirteen members of the Canadian government were arrested. Gouzenko was a GRU (Soviet military intelligence) code clerk and his superior was Colonel Zarubin. The NKVD resident in the Soviet Embassy at the time was Vitali Pavlov.[352]

The Soviet Ambassador in Canada, at the peak of this Soviet atomic espionage activity, was Kudryavtsev, who was also a colonel in the GRU.

Kudryavtsev was replaced by Alekseyev (whose real name was Shitov) who had been an officer in the GPU for twenty-five years. Later Alekseyee was the Soviet Ambassador who welcomed Ramon Mercader, to Cuba when he was released from the Mexican prison for Trotsky's murder and fled to Cuba.[353]

When Bentley defected to the FBI, after Golos's death, she was able to give them twice as much information about the two groups than if her activities had been properly compartmented.

Bentley, acted as the courier between Golos and the Canadian, Herbert Norman. In his own right, Norman was famous for having been a Cambridge associate of Anthony Blunt, the notorious British Soviet spy and comrade of Philby, Maclean, et.al. Norman was a Soviet agent and homosexual who had given North American defense secrets to a Japanese Marxist before Pearl Harbor. Norman was sent to Japan and Egypt as the Canadian Ambassador after World War Two. He was revealed as having been a Soviet espionage agent the entire time and was recalled to Ottawa for investigation. He committed suicide by jumping from his apartment in Cairo before he could be interviewed by the Canadian investigators.[354]

Gregory Rabinovitz

Louis Budenz, editor of *The Daily Worker*, the newspaper of the CPUSA, and a Comintern/GPU agent, was a principal agent in the United States who coordinated many of the key support assets that made the Trotsky operation a success. He accepted this role because his GPU case officer convinced him it was necessary to monitor the Trotskyites in order to counter their plots to murder Stalin and their plans to overthrow the Soviet Union by subversion.

Budenz collaborated with the GPU because, as he says,

> ...this was the period of the great purges... I collected and took them [the GPU] all the available information... of secret Trotskyites... and their relations to left wing socialists. At that time I had a number of agents... planted in the Trotskyite groups."[355]

Sylvia Franklin (aka Sylvia Caldwell. Her last married name was Sylvia Doxsee) was the principal penetration Budenz had of the Trotskyite headquartters in New York. Franklin was the private secretary to James Cannon, the head of the Fourth Internaitonal, Trotsky's world organization that was fighting Stalin and the Soviet Union. Franklin made it possible for the GPU/NKVD to steam open and read Trotsky's mail from Mexico to his New York followers. Franklin had complete access to all files and communications of the Fourth International.[356]

Budenz was turned over by his original GPU handler to another case officer named "Roberts" (he used the street aliases of "Rich" and "John") and was told he must concentrate on the Trotsky operation. Roberts was an intellectual, an expert on Trotsky and fanatically persistent in eliminating Trotskyism. Budnez discovered that Roberts was really Dr. Gregory Rabinovitz, working under cover as the official representative of the Soviet Red Cross in the United States.

Although supposed to be a humanitarian agency, the Russian Red Cross was believed by the Soviets to be the last place the American counterintelligence authorities would be inclined to look for an espionage organization. The only legitimate head of the Russian Red Cross from its inception in 1921 had been Dr. David Dubrovsky. In 1933 Jacob Sternglass, a former GPU station chief in Afghanistan, came to the United States as Dr. Dubrovsky's assistant and devoted all his time to GPU operations. Dubrovsky complained about Sternglass to Moscow and the Soviet Embassy in Washington but nothing was done about him. Dubrovsky later told his story to a congressional committee investigating Soviet espionage in the United States, all about the use of the Russian Red Cross by Soviet intelligence.[357]

When the Nazis came to power in Germany in 1933 they were able to eliminate almost all of the Soviet intelligence operations there. Moscow Center dispatched one of its senior case officers, Dr. Gregory Rabinowitz (variously spelled Rabinovitch and Rabinovich), a Jewish doctor, to Germany to rebuild the Soviet espionage apparatus.

Rabinowitz would later impress, at least one of his agents, who described him as having "a sadness but also intelligence in his deep brown eyes... His excellent tailored but conservative clothes enhanced his appearance of stability and solidity." Rabinowitz cut back the Soviet networks operating in Germany until they were functioning only as an underground political opposition to the Nazi government. In the early 1930s Rabinowitz left Germany for the United States.[358] "He might easily be some recently arrived European businessman — and that was supposedly his role."[359]

Rabinowitz recruited a Stalinist member of the CPUSA, Thomas L. Black, to go to Mexico and help prepare the assassination of Trotsky. Black later testified before the Senate Judiciary Committee that Rabinowitz told him "there are other GPU agents in the Trotsky household."[360]

Dr. Rabinowitz had apparently checked on Ruby Weil before he suggested Budenz contact her. (More on Ruby Weil in subsequent chapters.) He was confidant that she had the requisite ambition and loyalty to the Party for secret work. Weil was chosen from the party's rank and file and convinced that she should help Rabinowitz solely because she was a friend of the three Ageloff sisters, Ruth, Hilda and Sylvia.[361]

Jacob Golos was the intermediary between Budenz and Rabinowitz. Golos told Budenz to meet Rabinovitz for the last time before Rabibowitz left to return to the Soviet Union. At that last meeting Rabinowitz told Budenz that Ruby Weil had done an excellent piece of work for the GPU in Paris. For the last meeting Rabinowitz met Budenz in the apartment of Irving Franklin, the husband of Sylvia Franklin.[362]

Gaik Badalovich Ovakimian[3]

Ovakimian arrived in the United States in 1933 to head the Soviet espionage network in the United States. He was short, stocky, with a lively disposition, blue eyes and brown hair. According to his passport he had been born in Russian on 11 August 1898 of Armenian parents. He had been educated as an engineer but proved to be an excellent intelligence officer. He was married to a woman named Vera and they had a daughter named Egina.

From the time of his arrival in the United States in 1933 Ovakimian was involved mostly in industrial espionage using an American named Armand Feldman as his agent. A large number of agents also

[3] The FBI referred to Ovakimian as "the wily Armenian." (Lamphere, Robert J. and Tom Schachtman, *The FBI-KGB War*, Berkeley Books, New York, (p.24.) (Ovakimian was an Armenian pseudonym for a Russian born Soviet intelligence officer whose true identity remains unknown. (Dallin, David J., *Soviet Espionage*, Yale University Press, New Haven, (1955), p.405.)

worked under him: among them were Robert Haberman, who operated in Mexico City and the United States;[363] Eda Wallace and Fred Rose, who worked in Canada;[364] Aaron Markovich and Adolph Stark, who worked on obtaining passports; Simon Rosenberg, an industrial spy [in the United States] from 1932 to 1938; Jacob Golos, [noted elsewhere] and an attorney in the Department of Justice who furnished reports from the FBI to Ovakimian.[365]

After the defection of Whittaker Chambers in 1938, Soviet espionage in Washington, D.C. was taken over by the NKVD New York resident, Gaik Ovakimian. Henceforth, the United States became *a* major target (by the end of World War Two *the* major target) for Soviet espionage rather than a base for intelligence operations directed elsewhere, such as the Trotsky operation in Mexico.[366]

On 12 June 1940, after the fiasco of the Siqueiros raid of 24 May on Trotsky's house, a change was obviously called for in the Trotsky execution plans. When Stalin was told of Siqueiros's blundering failure to assassinate Trotsky, he asked only one question, "Will this [compromise] Ovakimian's operations in the United States and Mexico?" Pavel Sudoplatov, chief of Soviet Special Tasks, assured him that the Siqueiros debacle did not jeopardize Ovakimian's operations.[367]

Vladimir Petrov, an NKVD officer who defected in Australia with his wife, reported that the KGB (NKVD) file on Trotsky that he had examined in Moscow Center, clearly showed that the Trotsky assassination was an operation planned and directed by NKVD headquarters. Its organization on the American continent was in the hands of the NKVD Resident in the Soviet Consulate-General in New York, Gaik Ovakimian.[368]

After the Siqueiros disastrous raid on Trotsky's house, Ramon Mercader was ordered by the GPU/NKVD to report to his superiors in New York. He made a personal call on Trotsky before he left to bid him goodbye and left his Buick sedan for Trotsky's guards to use while he was away.

According to his mother, who told this information to a Spanish communist friend, Ramon went to New York to confer with Leonid Eitingon who had fled Mexico to New York immediately after the botched Siqueiros raid. He also met with the permanent NKVD resident who was attached to the Soviet Consulate General and the director of the Trotsky operation in the United States, Gaik Ovakimian.[369]

Ovakimian operated in America for over a decade with complete freedom. After the Trotsky operation was over, he was associated with the Soviet atom spy ring before it was discovered by the FBI.

Semyon Semyonov was an NKVD agent in the United States working for Amtorg. Semyonov had attended MIT and knew a large number of the American scientists working on the Manhattan atom bomb project. Semyonov (code name Twain and street alais of Sam) was the cut out who ran Harry Gold, the courier in the Rosenberg atomic bomb case, who was, in turn, directly working for Ovakimian.[370] In 1950 when Gold told the FBI of how his friend Tom Black had recruited him, the FBI did not know who had induced Black to spy. When Black finally talked to the FBI, he identified Ovakimian as his recruiter.[371]

On express orders from Ovakimian, Black had been ordered to join the Socialist Party and had infiltrated the ranks of the Trotskyites. He had been given money and told to go to Mexico to spy on Trotsky's household. He told the FBI he had simply not bothered to go. A few months later Trotsky was murdered and his Soviet superiors never mentioned the matter again.[372] Black was to have monitored Trotsky's telephone calls from a clandestine base close to Trotsky's home in Coyoacan.

It was Ovakimian who recruited Julius and Ethel Rosenberg and turned them over to Semyonov for day-to-day handling. It was Anatoli Yatskov, the NKVD New York control officer on atmoic espionage, that sent Semyonov's agent, Harry Gold, to Los Alamos to contact David Greenglass. Before that Gold had been the sole cut out/courier for Klaus Fuchs, a British atom scientist working at Los Alamos. Greenglass was Ethel Rosenberg's brother and an Army machinist at Los Alamos. Yatskov broke the fundamental rule of espionage, to never allow an agent from one cell to have contact with a member of another group, i.e., always maintain compartmentation.

Yatskov had received permission from the Center in Moscow to substitute Gold in the operaton to meet with Greenglass. When the FBI was able to roll-up the Soviet operation because of this breach in compartmentation, the Center blamed Semyonov and Ovakimian for the derelection. Semyonov was discharged and Ovakimian was demoted.

On 5 May 1941, Ovakimian was arrested in New York on a charge of not registering as a foreign agent. At that time he was officially listed as an employee of the Soviet Amtorg Trading Corporation which was a well known cover company for Soviet espionage.

Amtorg Trading Corporation used a very secure code for its radio messages back to GPU headquarters in Moscow. In 1930 Congressman Hamiltaon Fish subpoenaed over 3000 coded Amtorg telegrams for intensive investigation by American code breakers. Neither the Navy's nor the War Department's code breakers could read the mnessages. These agencies tried for two years without success and could not break the Amtorg code.[373]

Ovakimian tried to claim diplomatic immunity when he was arrested but that did not apply to him as a non-diplomatic employee of Amtorg. He claimed that he was a "purchasing agent" of "defense goods" for the Soviet government but had been unable "for political reasons to negotiate any substantial transactions."[374]

The Soviet Consulate General put up a bond of $25,000 for his release. He was permitted to leave the United States for the Soviet Union without being prosecuted in exchange for a Soviet promise to release six Americans unlawfully detained in the Soviet Union.[375]

Ovakimian was Golos's superior (although Golos never knew Ovakimian's name). After Ovakimian's arrest, Golos and Bentley were tailed for a time by the FBI and had to lie low. Once the State Department let Ovakimian return to the Soviet Union, the two resumed their espionage activities.[376]

Ramon Mercader evidently received considerable money during a meeting with Ovakimian in New York. He strapped envelopes full of money, sealed with red wax, to his forearms under his shirt sleeves to take back to Mexico. He told Sylvia Ageloff, who was staying with him in the New York hotel, the packets contained Mexican money for his use in Mexico. Mercader left the New York meeting with Ovakimian about 1 July to return to Mexico.

Ovakimian became the director of the American Department at the Center of the NKVD in Moscow in 1945, after his expulsion from New York. He reported that he was certain the FBI had successfully infiltrated his espionage networks. The FBI conversely, always maintained that they uncovered the Soviet espionage networks through cracking the codes the Soviets used to communicate from their Embassy in Washington and their Consulate General in New York with the Center in Moscow.[377]

Years later, Nikita Khrushchev, the leader of the Soviet Union, ordered about one hundred retired generals and colonels who had served in the security services stripped of military rank and purged from the party. Major General Ovakimian's name was on the list.[378]

Gerhardt Eisler[4]

Gerhardt Eisler, Comintern/GPU agent extraordinaire, arrived in the United States in July of 1933. He had been working with Soviet intelligence in Shanghai, China, since 1929, with the GPU agent Agnes Smedley as part of the Far Eastern Bureau of the Comintern.[379] Eisler's cover in the Far East was that of a salt merchant. Also working with Eisler were a husband and wife team of GPU agents, Arthur Ewert and his wife Elsie — called Mrs. Szabo because of her maiden name of Sabrowski.[380]

Eisler was to assume command of all Comintern/GPU operations in the Western Hemisphere. This extensive jurisdictional mandate was supposed to include Golos's domain. However, Golos was so deeply entrenched with his two cover firms and very productive intelligence networks that Eisler wisely had little to do with overseeing Golos's operations.

In the United States Eisler operated under the name of Hans Berger. In October 1946 the former GPU agent, Louis Budenz, charged in a radio address from Detroit that "the equivalent of a representative of the Communist International" still functions in America. The *New York World Telegram* identified the representative Budenz referred to as Hans Berger-Eisler. Gerhardt Eisler was then forced to publicly confess that he was Berger.[381]

Eisler was born in Leipzig in 1887 into an intellectual, upper-middle-class family. All the Eisler siblings developed into thinkers and academics. Gerhardt's brother, Hans, became a script writer in Hollywood and Elfriede, his sister, better known to her comrades by her party name of Ruth Fischer, became head of the German Communist

---

[4] Gerhardt Eisler was married at one time to Hede Massig. See her book, *This Deception*, Duell, Sloan and Pearce, New York (1951).

Party. Whether through parental persuasion, or personal conviction, at least two of the Eisler children became ardent adherents of communism.

Unfortunately for family and party unity, Elfriede had sided with the Trotsky faction in a bitter ideological dispute between Stalin and Trotsky when she was chief of the German Communist Party. For her deviation Stalin expelled Elfriede from the German Communist Party and the Comintern. Again, unfortunately for Gerhardt, this expulsion had an adverse effect on Elfriede and she developed a deep hatred for her brother for remaining in the Stalinist communist movement after she had been expelled.

At first Elfriede began to publicly denounce Gerhardt to other comrades. Later she went farther and denounced him as a Comintern/GPU spy to the police of whatever country Gerhardt happened to be in at the time. When Gerhardt arrived in the United States he feared his sister's denunciations would reach the long ears of the FBI. But by one means or another Gerhardt was able to keep his sister from reaching America for over seven years while he continued to operate for the GPU.

Eisler's assignments in the United States were simple. In addition to being in charge of all Comintern/GPU operations in the Western Hemisphere, he was supposed to bring about real cohesion and a working partnership among all factions of the various communist party splinter groups. A secondary objective of Eisler's assignment was to be prepared to lead an insurrection against the United States government. Stalin was certain that economic chaos would follow the disastrous leadership of President Franklin Delano Roosevelt's New Deal economic policies.

Stalin prophesied that American business and culture would collapse as a result of Roosevelt's National Recovery Act (NRA). Fortunately, the New Deal, as the NRA was called, was a success. Eisler never got the chance to lead the revolutionary forces of communism against the capitalist United States.

Eisler's accomplishments in the United States also included ordering and directing the murders of Carlo Tresca and Juliet Stuart Poyntz. This aspect of Eisler's leadership role leads directly back to George Mink.

Steve Nelson, veteran NKVD agent in the United States was attempting to recruit scientists working on the atomic bomb Manhattan project. Nelson was involved with the Joint Anti-Fascist Refugee Committee (JAFRC). The JAFRC was a communist front and Eisler worked in the New York offices of the JAFRC under the name of Julius

Eisman. He was paid $150 per month for this work. This is a situation where a communist front organization is paying the salary of a GPU agent.[382]

On 30 August 1934 Eisler and Leon Josephson were co-owners of Cafe Society and Downtown, two restaurant hangouts for society and fellow travelers.[383] They went to the passport office where they swore out a passport application for Eisler under the name of Samuel Liptzen. Eisler used his picture and signed the application as Liptzen while Jospehson swore to the truth of the application — both were federal offenses.[384]

Juliet Stuart Poyntz and George Mink Revisited

The story of Juliet Stuart Poyntz is included here to demonstrate another facet of Mink's modus operandi.[385] Since he was such an important participant in the Trotsky assassination, but so ephemeral a personality, every example of his operations is pertinent to his conduct in Mexico.

Juliet Stuart Poyntz was an intellectual by nature and a schoolteacher by profession. She was a beautiful woman with a few unfortunate mannish characteristics. She was, at one time, in charge of the illegal section of the CPUSA and knew many secrets of underground Comintern operations. Due to her underground activities she had been a close conspiratorial comrade of Jacob Golos and Elizabeth Bentley. In 1936 Poyntz went to Moscow at Stalin's personal invitation. In Moscow, at the same time, was her fellow American communist and her former live-in lover, George Mink. Many comrades wondered what a lovely lady like her could see in a person like Mink.

Poyntz and Mink had been lovers for some time but had apparently been more discreet in New York, because few people knew of the romance. In Moscow they flaunted their relationship in front of the other American communists who were attending party functions in the Soviet capitol. The lovers would have had a idyllic vacation in Moscow but it was marred for Poyntz by the brutal purges of party members that Stalin was conducting at the time.

When she returned from Moscow in 1936 Poyntz confided to a few close friends that she had been sickened by the purges. She told them she was disillusioned with communism, especially Soviet communism, where such horrors were not only permitted but encouraged.

Poyntz also told one or two of her trusted friends that she was leaving the CPUSA to join Trotsky. She considered Trotsky to be more ideologically suited to her concept of what the benefits should be to a person under Marxist/Leninism. She also confided to these trusted friends that she intended to write a book exposing Stalinism and the espionage networks of the Comintern/GPU in the United States.

These trusted friends sought to dissuade Poyntz from these rash and potentially suicidal intentions. Poyntz persisted and categorically vowed that she was going to go through with her plans. The leaders of the non-GPU section of the CPUSA to whom these infractions were duly reported by Poyntz's trusted friends assumed that she was engaging solely in bluster and would soon see the error of her ways. Others in the non-GPU section of the CPUSA who knew of Poyntz's declarations believed she was going through the ritual purging of her overt party trappings in preparation for an undercover assignment.

Poyntz had been through that type of ritualistic recantation in 1934 when she really did leave the overt party for underground work. Therefore, those in the know, such as Golos and the members of the Central Control Commission, were concerned that maybe this time Poyntz was really going to do what she threatened to do. They definitely knew she was not going through the stylized motions of purging herself for an underground assignment.

On 3 June 1937 Poyntz received a telephone call in her room at the American Women's Association Club at 353 West 57th Street in New York City. She picked up her purse, put on her coat, left the Club without leaving a note and was never seen alive again. She had not packed a bag for an overnight trip. There was nothing missing from her room that she would ordinarily have taken with her on a trip. There was no sign of a struggle and there was no clue to explain her abrupt departure. Further, she had not mentioned to friends or the doorman that she planned to be away.

Six months passed without any clue to her disappearance. No one had any inkling as to her whereabouts until February 1938. A leading New York anarchist, Carlo Tresca, told the Assistant United States Attorney for New York, Lester C. Dunigan, that Poyntz had been kidnapped by her former lover, Schachno Epstein. He also categorically stated that she had been murdered by George Mink.[386] Epstein was a member of the CPUSA and a GPU agent. Assistant District Attorney

Dunigan did all he could to prepare this meager information for prosecution but there simply were not enough hard facts to proceed with an indictment. Dunigan had to drop the prosecution.

Tresca persisted for five years in his investigation of Poyntz's disappearance. He tried to collect additional hard evidence of Mink's guilt that would stand up in court. Before he could succeed, however, Tresca was murdered by two shots from an assassin's gun on Fifth Avenue in the heart of New York City. Evidence was introduced that the assassin was a small time Mafia punk. Dorothy Gallagher's well researched book *All the Right Enemies* indicate the shooter was one Carmine Galante who murdered Tresca. Other evidence implied that it was the GPU through cut-outs that ordered Galante to do the job.[387]

Shortly after Tresca's murder Benjamin Gitlow, the former Secretary General of the CPUSA and a defector to the FBI, spoke out. He reported that Tresca had told him a good portion of the evidence he had discovered concerning Poyntz's murder and what he was about to reveal to the United States Attorney. Gitlow put it all down in his best selling book *The Whole of Their Lives*.[388]

Gitlow said that several members of the Central Control Commission of the CPUSA and Jacob Golos had a meeting with the GPU representatives in the United States where it was decided that Poyntz was too dangerous and must be eliminated. She must, at all costs, be prevented from writing her book about the Comintern/GPU activities in America. As was usual with such GPU assignments, more than one killer was assigned to carry out the task. Three murderers were assigned to the Poyntz assassination and one of them was Poyntz's former lover, George Mink.

Mink intimidated Schachno Epstein, Poyntz current lover with the threat of severe bodily mayhem, to act as the Judas goat to lead her into the slaughter pen. Mink had Epstein telephone Poyntz at the American Women's Association as previously described. Poyntz then left her room and met Epstein. The two walked in the park as if they were sweethearts reconciling after a petty quarrel.

Mink later bragged that Epstein was a sniveling coward who cringed at the threat of the slightest physical hurt. Epstein was more afraid of Mink than he was of some unforeseeable punishment from the American authorities for being an accessory to murder. He decided the wiser course of cowardice was to cooperate with Mink.

Taking Poyntz by the arm, Epstein led her up a side path in the park where a black limousine was waiting by the curb. As they drew abreast of the car, two men jumped out, grabbed Poyntz, knocked Epstein to the ground in a mock attack, threw Pouyntz into the back seat and sped away, all within seconds.

Mink and the other man in the back of the car (Mink never revealed the identity of the other man but other factors strongly indicate that it probably was Vittorio Vidali) drove the struggling Poyntz to a wooded area in Duchess Country in New York. This was a secluded spot on the bluffs overlooking the Hudson River. Mink shot Poyntz in the head and they buried her body in a deep gully and camouflaged the grave. They piled rocks and forest debris over the grave to better conceal it.[389]

In a Party cell meeting the man who killed Poyntz, probably Mink, reported that he almost felt sorry for Poyntz. She had pleaded for her life and was crying hysterically when he shot her. Ever a cynic, he summed up his feelings about Poyntz's murder. He said she deserved to be shot. She had behaved like a common peasant woman, crying and carrying on as if her life was more important than the Party's security. She had not acted at all like a good communist, he philosophized; she was afraid to die for communism when she knew there was no other way![390]

Poyntz's death, it is believed, was also tied in with the Robinson-Reubens case. Poyntz no doubt included a reference in her intended expose of GPU activities in America to the passport mill in New York City that had issued passports to Robinson and Reubens.

In December of 1937 a Mr. and Mrs. Donald Robinson, both traveling on United States passports, were taken from the posh National Hotel in Moscow by the NKVD. After six weeks the Soviet authorities said that they had been arrested for espionage. Their passports had been obtained from Ossip Garber, a Fifth Avenue photographer in New York City. Garber, a well known fellow traveler of the CPUSA ran a fraudulent passport operation for CPUSA members and GPU agents. Robinson was in fact a Lett by the name of Adolph Arnold, who was also known in the United States as Ewald. His American born "wife", Ruth Marie Reubens, was also known as Ruth Boerger. Both were well known members of the CPUSA. Mrs. Reubens refused all assistance offered by the United States authorities in Moscow. She was eventually released but stayed in the Soviet Union. "Her husband" was nver seen or heard from again.[391]

The Robinson-Reubens case figured peripherally in the Alger Hiss case. One of the pieces of information turned over to Whittaker Chambers by Hiss dealt with this affair. The Soviets wanted to know just how far the American government was prepared to go in protesting the treatment accorded Mrs. Reubens in Moscow.[392]

Leonid Eitingon[5]

Pivotal to any discussion of Trotsky's assassination is the GPU case officer working at the scene of the crime in Mexico. In December of 1938, Lavrenti Beria assumed command of the NKVD. He encouraged even more terror than had his predecessor Yezhov. Beria purged ruthlessly on Stalin's order and thus became one of his favorites.[393]

Even though he was still working under Stalin's whim, the guidance from his GPU superiors in America and the direct orders from GPU headquarters in far away Moscow, Leonid Eitingon still had to call the tactics on the spot in Mexico.[394]

There were other GPU senior case officers directing other aspects of the Trotsky operation, such as Rabinovitch, Ovakimian and Golos in the United States. In the Soviet Union there was General Sudoplatov, director of the Department of Special Tasks which specialized in assassinations and kidnappings. However, there was still only one person charged with carrying out Stalin's orders in Mexico and that was General Leonid Eitingon.

At the age of nineteen, in 1918, Eitingon joined the Red Army and was soon transferred to the Cheka. He joined the Bolsheviks in 1920 and earned glory in putting down an uprising of White Army officers in Gomel. He joined the GPU and went to China where he worked with Richard Sorge, the renowned Soviet master spy. After other field assignments he returned to Moscow to be deputy director of Special Tasks. At one time in his career Eitingon was arrested in one of Stalin's purges because he was a Jew.[395]

[5] Leonid Eitingon was born Naum Isakovich Eitingon on 1 December 1899. He changed his name in the 1920s to Leonid Alexandrovich Eitingon because it was politic for Jewish Cheka officers to adopt Russian names at that time so as not to attract attention to their Jewish origins.

Although Russian born and a Soviet citizen, Eitingon came to the Trotsky assassination operation from the Spanish Civil War. He had commanded the Spanish branch of the GPU's Division of Special Tasks. This division was a highly efficient murder and enforcement unit created by the GPU to maintain discipline among the International Brigades.

Eitingon had considerable experience with successful murders before assuming field command of the Trotsky case. One of Trotsky's secretaries, an intelligent, likeable young man, named Erwin Wolf, had been expelled from Norway in December of 1937 before Trotsky left that country for Mexico. Wolf literally had no place to go until his mentor, Trotsky, was settled in some place other than Norway. Wolf made the mistake of taking refuge in Spain during this interim away from Trotsky. Eitingon (as General Kotov as he was known in Spain) discovered that Wolf was in Spain and had him murdered as part of the on-going anti-Trotsky campaign.

From the beginning of his professional career Eitingon specialized in murder and kidnapping. His first overseas assignment was to Paris in the early 1930s, where he organized the kidnapping of General Alexander Kutepov in broad daylight from a heavily traveled downtown street. This sensational kidnapping galvanized all of Europe against the GPU's high-handed techniques as had no other previously known GPU outrage. Despite its adverse effect on Soviet prestige, the kidnapping and subsequent murder was still considered a success by the GPU and Eitingon was promoted. Mark Zborowski, as previously noted, was an agent in Eitingon's network at the time.[396]

Kutepov, the victim, had been the leader of a well organized anti-Soviet, White Russian emigre group known as the Federation of Imperial Army Veterans. This group was active in denouncing Stalin and the policies of the Soviet Union which automatically placed their leaders on the GPU list of future victims. At the time of Kutepov's abduction, Eitingon was a young, relatively inexperienced GPU case officer, under diplomatic cover in the Soviet Embassy in Paris. He was operating under the cover name of Sakhov and using the street alias of Valery and Lyova. (He also occasionally used the alias of Tom and Pierre.)

Kutepov was lured into a taxi by a GPU agent who was penetrating the Federation of Imperial Army Veterans. The bait in this kidnapping was General Skoblin, a White Russian refugee in the pay of the GPU. Skoblin and Eitingon were also later involved in the GPU kidnapping of

General Miller who was Kutepov's successor as the leader of the Federation.

Skoblin and his wife, otherwise known as Nadine Plevitskaya, a very popular cabaret singer of Russian folk songs and reputed to be the one of the most beautiful woman in Paris, enticed Kutepov into the taxi. They drugged him and drove to a road on the northern fringes of Paris where they met Eitingon. There, on a side road concealed by woods, Kutepov was transferred to Eitingon's private car and taken to a villa near the seaside resort of Malo-les-Baines, which was close to Dunkirk. The following day Kutepov, still heavily drugged, was put aboard a Soviet trawler bound for Russia and his execution. Eitingon and the Skoblines returned to Paris to get on with the business of luring more emigres back to the Soviet Union.

Skoblin fled Paris once step ahead of the police but Pleviskaya was captured and sentenced to twenty years by a French court for complicity in the General Kutepov's disappearance. She died in 1949 still in a French prison.[397]

Later in Spain Eitingon adopted the nom de guerre of General Kotov and it was in Spain that he met Caridad Mercader. Caridad was the mother of Trotsky's future assassin, Ramon Mercader. Shortly after Eitingon and Caridad met, Caridad became his mistress. Eitingon, because of his down to earth machismo with his Spanish comrades was also known as Comrade Pablo.

Caridad Mercader at that time was a strikingly beautiful woman. In addition to her physical Castillian beauty (although Cuban born) she had a pleasingly vivacious personality, cute and pert. She was an all-around charming woman when she was with a man and wanted to be pleasant and enticing. This attractiveness was her charm despite the fact that she was a rotten mother, extremely cruel to underlings and a sadistic seductress to many of her lovers. All in all, she was Eitingon's kind of woman.

Caridad had a continental sophistication that appealed to a Soviet peasant like Eitingon. To top it off, she also had a well developed provocative figure that brought the Spanish machos out on the street in the midday sun. (This may have been the first and last time a Soviet GPU agent gave flowers and silk underclothing to another GPU agent!)

Eitingon was short, fat with deep set eyes and heavy eyebrows. He had a small scar on his chin.[398] His association with the Trotsky assassination case began in 1936. Alexander Orlov, the chief of the GPU in Europe, received orders from Moscow to proceed to Spain to help Eitingon set up assassination training schools. Ostensibly these schools were to train the Spanish communists in guerrilla warfare, sabotage and political assassination. Eitingon was to chose the men who would attend these six schools and then dispatch them where they could create the most havoc and mayhem behind enemy lines. Ramon Mercader attended one of these schools.

On one occasion Ramon Mercader was brought back from a mission with a slight wound in his right elbow.[399] He was placed in the special hospital for communist cadre in the Pension Bank Building in Montejuich. While he was in this hospital he contracted jaundice.

Ramon was not too incapacitated in the hospital to ignore an opportunity for relaxation. He had an affair with a tall blond Dutch girl Fani Castedo who was in the room next to his. Castedo was a leader of the communist movement of Holland.[400]

On one occasion Ramon displayed the youthful trait of braggadocio when he confided a secret to a nurse at the hospital with whom he had been having an affair. He told her that he had to leave Spain to go on a very secret and dangerous mission. He further dramatized this venture of self-starring melodrama by telling the nurse that it was a spy mission so secret that he could not even tell her about it.

In December of 1937 Eitingon and Caridad Mercader left Spain for Moscow. Ramon had been ordered to Moscow for additional training and probably a briefing on his future with the GPU. Caridad soon discovered that Eitingon, despite his protestations of abiding love for her, had wives and families in Moscow that he had no intention of leaving.[401]

Ramon arrived in Moscow for this training to find his mother established in a comfortable apartment, happy and prepared to pamper him as always. His mentor, Eitingon was in constant attendance at his mother's side, despite his basic loyalty of remaining faithful to his wives and families. (This domestic tranquility in Moscow during the time of the Spanish Civil War was different than the ostracism Caridad experienced from Eitingon when they both returned to Moscow from Mexico after the Trotsky assassination.)

When Eitingon fled Spain in 1939, after the Fascists won the Spanish Civil War, he went to France where he revamped the Guy Burgess espionage operation. (Burgess was part of the notorious British spy ring that included the Maclean, Philby, Blunt et.al. previously referred to.) Eitingon suggested that despite the defection of Alexander Orlov, who had directed Burgess and other members of the British spy ring, that the operation could be salvaged and be continued. He did not believe that Orlov would immediately disclose all the secrets he knew about Soviet intelligence and thereby immediately compromise Burgess and the others.

When Eitingon left Spain for Mexico he took with him Vittorio Codovilla, a founder of the Argentine Communist Party and a trusted Stalinist, and Vittorio Vidali. Codovilla had been deeply involved in the vicious murder of Andres Nin in Spain and Vidali was known as "another of the most formidable killers from Spain."[402]

Eitingon's direction of the Trotsky assassination reveals how precisely the entire operation was planned, with the exception of a few monumental mistakes such as the Siqueiros raid. The general GPU method of operation reveals that they did not shirk from assassination, provided it was properly planned. The Standard Operating Procedure (SOP) of a GPU for assassinations was:

1. The assassin must never be a person who could be traced as a direct agent of the GPU.

2. Wherever possible the assassin should pose as an adherent or associate of his victim.[403]

In 1940 when Eitingon and Caridad Mercader returned from Mexico, they, together with Sudoplatov, were decorated by Soviet President Kalainin in the Kremlin for their participation in Trotsky's assassination.

Eitingon was named Sudoplatov's deputy in the Department of Special Tasks on 5 July 1941 by order of Lavrentia Beria, the chief of all Soviet intelligence. In 1951 Eitingon was sent to prison and not released until 1953. He was arrested again and not released until 1964.[404] When he was released the second time he was immediately hospitalized for general exhaustion and an ulcer.[405] He then rejoined his second wife, Olga Naumova, and went to work as a senior editor for the Foreign Languages Publishing House.

Eitingon died in 1981. His former wife, Yevgenia Puzirova, was initially denied the privileges of a war widow to which she was entitled in her own right. When Eitingon's memory was rehabilitated in the Soviet hierarchy in April 1992 she was awarded these privileges. Eitingon was cremated and his ashes interred in the wall of Donskoi Monastery cemetery.

VIVA MEXICO!

Stalin seeks to strike, not at the ideas
of his opponent, but at his skull!

Leon Trotsky[6]

On 19 December 1936, after nine years of exile, Trotsky and Natalia boarded the oil tanker S.S. Ruth in Oslo, Norway. They were headed for what they thought would be a safe haven in hospitable Mexico. The Norwegian government had apologetically explained their decision to expel Trotsky, claiming that they were being intimidated by the Soviet government who threatened to curtail its imports of Norwegian herring.

The real reason, of course, was that Trygiv Lie, the Minister of Justice in Norway was a Soviet toady who eagerly agreed to expel Trotsky to a country where the GPU hoped to have a better chance to quietly murder him. Lie was also ecstatic to be rid of the "Trotsky problem" and welcomed Mexico's offer to accept Trotsky.[406]

Trotsky's requests for asylum away from Norway had either been turned down politely, flatly rejected or abjectly ignored by almost all western nations. Mexico's unexpected offer of sanctuary had only the one condition attached, that Trotsky not interfere with the internal political process of Mexico.

Mexico's largess in granting Trotsky asylum was because the Mexican President, Lazaro Cardenas, had been convinced by Diego Rivera that it was the humane thing to do for a world leader of Trotsky's stature.

The voyage took about three weeks. Trotsky used this enforced leisure to read books about Mexico and to write a journal on the Zinoviev-Kamenov trial which became part of his book *Stalin's Crimes*.[407]

---

[6] Comment by Trotsky while en route to Mexico after his expulsion from Norway aboard S.S. Ruth.

Diego Rivera

Diego Rivera was a well known Mexican muralist. He had then, and would have until his death, the mentality of a spoiled adolescent in a great oaf's body. Rivera enjoyed dabbling in real world politics about which he knew very few of the facts and realities. His leftist political peers indulged him by seeming to accept him but, in a large measure, were able to ignore his amateurish meddling.

More often than not Rivera produced a Don Quixote aftermath with his interference and someone else had to clean up after him.

> His talk, theories, anecdotes, adventures, and his successive retelling of them were an endless labyrinth of fables." "If Diego had...merely talked and had his talk set down, Baron Munchausen would have to look to his reputation." " If Diego rarely told the simple truth, he did not tell simple lies either.[408]

The leftist intelligentsia and communist party cadre encouraged Rivera to think he was helping their cause. His name was a drawing card and his presence at rallies increased the crowds who came out of curiosity to see the fabled artist.

Rivera had gone full circle in his political convictions. In the space of weeks he would go from neo-conservative to radical leftist. He was never stable in his political orientation but always unpredictable. His liberal political thinking would gyrate from socialist left to orthodox Marxism to Trotskyism. Initially he espoused strict Stalinist communism. Later, without any apparent philosophical reason, he recanted Stalinism and embraced Trotskyism.

The first Mexican Trotskyite Party (PMT) was formed in the late 1920s by Russell Blackwell, an American communist who came to Mexico to organize the "Pioneers," the communist children's organization.[409] Blackwell used the name Rosalio Negrete in Mexico. Blackwell's first convert to the PMT was Manuel Rodriguez. He then recruited Rodriguez's wife, then Jose Revueltos (who became a famous novelist) and later Pedro Maria Anaya who became the Secretary of Education and editor of the government newspaper, *El Nacional*[410]

Communism held a theoretical, intellectual and emotional appeal for Rivera, probably stimulated by his happy association with young Russians and other revolutionaries in Paris — he liked to think of himself as a radical.[411]

Subsequently, bored with Trotskyism, Rivera drifted into the quagmire of Mexican leftist socialism. Still later he engaged in a bitter self-castigating dialectic expiation for his deviation into Trotskyism and begged, literally on his knees, to be allowed to crawl back into the Mexican Communist Party. When it appeared he might not be readmitted, he pleaded for just a chance to prove his sincerity if only on a probationary basis.

Diego Rivera did not ever jeopardize his ability to get a United States visa that would permit him to flee north of the border to sanctuary in the United States when he perceived any threat to his personal well being in Mexico. Rivera had been an active informant for the FBI who were stationed in the American Embassy in Mexico City. He reported everything to the Bureau about PCM and Trotskyite activities and his fellow communists in order to continue to be able to get visas to the United States where he also received large cash fees for his painting. (Revista *Reforma*, "Los Rugidos de un Leon," (The Roars of a Lion), 25 August 1996, Mexico City, Mexico.)

Rivera was characterized politically as "blowing with the wind." His politics and love life had a lot in common. Rivera was a notorious and inveterate womanizer. He was in a constant state of emotional flux whether to remain even partially loyal to his lovely wife, Frida Kahlo, or to continue to accede to his adolescent urges to philander by trying to bed every woman who gave him the slightest encouragement.

His self-image, fantasizing that he was a "muy macho hombre" (very tough man), required that he carry a gun. He hoped this would make him feared as a tough "pistolero" (gun fighter).[412] Unfortunately Rivera never was a tough guy or real fighter. He was essentially a poseur! He did, on the other hand, impress people with his marvelous paintings.

In 1934, with the election of Lazaro Cardenas to the presidency of Mexico, the political climate in that country went far to the left. In 1936, Cardenas nationalized the Mexican oil industry, expropriating most of the foreign oil holdings, capital assets and other foreign investments. At that time, Rivera was on the outs with the Mexican Communist Pary (PCM) and was being attacked by them for his Trotskyite sentiments.

Before that, in 1933, Trotsky had created the Fourth International. He had become convinced it was impossible to remain in the same international with Stalin whose narrow views of revolutionary philosophy were incompatible with his. Rivera, grandstanding as usual, publicly declared his support of Trotsky's philosophy of simultaneous world revolution in more than one country. This decision put Rivera solidly aligned with Trotsky but he did not officially join the Fourth International until 1936.

Rivera claimed that he had always admired Trotsky but his political likes and dislikes were nothing but childish whims without substance or thought.[413] He enjoyed creating dissension in leftist Mexican circles with his bold, controversial remarks. He would be cantankerous and contrary merely to provoke controversy, many times for the sole purpose of simply promoting his own publicity; after all, publicity, good or bad, leftist of conservative, helped sell his paintings. Ramon Beteta, Mexican Under-Secretary for Foreign Affairs, said in 1940, "I know Diego very well. He would do anything for publicity!"[414]

Rivera's pendulum admiration for Trotsky was on the high side at the time Trotsky desperately needed help in finding a place of asylum away from Norway. Trotsky literally had no place to go except back to Stalin's Soviet Union where he would have been immediately executed.

Rivera had included a portrait of Trotsky in his mural at the Trotskyite headquarters of the Fourth International in New York City. He had also added a portrait of Trotsky to the second version of the so-called Rockefeller Center mural in the Mexican Palace of Fine Arts. In this latter mural, Rivera had Trotsky holding aloft a pennant inscribed with the statement, "Workers of the World, Unite in the IV International." From all this it would seem that Rivera, at that particular time, was genuinely pre-disposed to helping Trotsky.

His reason, Rivera claimed, for breaking with Stalin was the dictator's revolutionary philosophy to attempt to bring about socialism first in one country only. Stalin wanted to delay the internationalization of socialism in other countries until the socialist/communist revolution was complete and solidified first in Russia. Rivera embraced Trotsky's philosophy that the socialist revolution must go forward simultaneously in all countries. Rivera claimed, to anyone who would listen to his prattle, that Trotsky's philosophy was the true interpretation of Marxist doctrine.

The PCM, hewing to the accepted Stalinist dogma, became increasingly critical of political deviates like Rivera. Their criticism of Rivera's support of Trotskyism included, for good measure, some virulent attacks on Rivera's professional competence as an artist and painter.

There is a story about Rivera that illustrates very well his capricious nature and his defense of his so-called artistic integrity.[415] One time Rivera defended his professional position as an artist with a lecture entitled "The Arts and Their Revolutionary Role in Culture." He gave this talk at the Palace of Fine Arts in Mexico City on 26 August 1935. He was fairly well received, at least his speech did not provoke any controversy that afternoon.

The following day, David Alfaro Siqueiros, a totally committed Stalinist communist, and a Mexican muralist of considerable stature, decided to rebut Rivera's contentions with a speech of his own.

The next afternoon when Siqueiros tried to speak, Rivera shouted his objections from the audience. Not achieving any recognition, Rivera pulled a pistol from his belt and demanded equal time to answer Siqueiros's attack. The chairman of the meeting, rather than risk being shot by a stray bullet, conceded Rivera's right to reply. He hastily scheduled a verbal debate, instead of a bloody duel with pistols, to take place the following day.

Siqueiros and Rivera traded verbal jibes the next day across a balcony overlooking the crowded hall. The debate rapidly degenerated into pettiness. The most argued issue became which artist had compromised his immortal Mexican artistic soul by selling the most of his paintings to "gringo" tourists. This mundane debate was a far cry from the lofty theme of revolutionary artistic philosophy that was supposed to have been the core of the dispute.

Frida Kahlo, a Rivera booster, angered at the absurdity of the debate, took her frustration out vicariously on the crowd in general. She screamed at the audience that they should know that "the gringo bastards come down [to Mexico] from their fucking country for nothing else than to make fun of us!"

Joining in the general turmoil that marked the end of this childish oratorical contest, Rivera threw a punch at a spectator who he thought was bothering Frida. For good measure he called the poor spectator, "a son-of-a-bitch Stalinist!"

Frida Kahlo

Frida, at the beginning of her association with Trotsky genuinely admired him as a person. She never completely accepted his political philosophy nor did she join the Fourth International.

Frida was notorious for her flamboyant bisexualism. She often declared her preference for lesbianism to one and all who would listen.[416] Some detractors claimed she was an alcoholic. There is no doubt that she enjoyed drinking liquor. In later years she carried a little flask of liquor disguised as a perfume bottle hanging on a small chain around her neck. She would whip out the flask, as if to freshen the perfume around her face, but really to take a short swig.[417]

In later life, toward the end, the numbing effect of the liquor helped smother the pain in her legs.[418] A streetcar accident in her childhood had caused a severe injury to her leg and body. The hurt never really left and the ever present pain was an increasing torture for her in later life.

Frida occasionally used drugs. She was not addicted to drugs as her enemies have charged but, similar to the liquor, she enjoyed the calming effects of some mild drugs. Frida's reported over-use of drugs came much later in her life when she was very ill and physically tormented by the excruciating pain. The drugs at this time in her life were only those prescribed by her physician specifically for her symptoms and could never be described as Frida's catering to any drug addiction.

In early 1928 , as a young woman, Frida was introduced into a group of friends surrounding Julio Antonio Mella and Tina Modotti. Frida and Modotti became close personal friends and it was through Modotti that Frida joined the PCM and met Diego Rivera.[419] Frida wore a leather jacket, blue jeans and black or red shirts with a pin of the hammer and sickle. This manly, devil-may-care, macho garb may have been one of the reasons Rivera was attracted to Frida.

Frida and Rivera were married on 21 August 1929. Her parents, although they consented to the wedding, said it was the union between an "elephant and a dove." Rivera, they claimed, looked like a "fat, fat, fat, Brueghel." Rivera was forty-three and Frida was nineteen.[420] The party after the civil ceremony took place on Tina Modotti's roof.[421] A month after their wedding, on 30 October 1929 Rivera was expelled from the PCM and Frida resigned in sympathy with him.

Frida sometimes dressed as a Tehuana, as the women from the Isthmus of Tehuantepec are known. These women are famous for being stately, beautiful, sensuous, intelligent, brave and strong. It is a matriarchal society, where the women run the communities, handle the money and dominate the men. The costume is an embroidered, blouse with a low cut neckline and a long skirt of purple or red, with lots of jewelry made from gold coins which may constitute the girl's dowry.[422]

Despite having a loving, sensuous wife at home, Rivera had many affairs, one notorious one with Cristina Kahlo, Frida's sister. Frida pretended not to be hurt by Rivera's infidelities but a desire for revenge may have made her susceptible to Trotsky's blandishments when he made them.[423]

The Trotskyite Party in Mexico, which Frida refused to join, never had many members. It was mostly a few non-combatant, ineffectual intellectuals, abetted by some misguided fanatical socialists and a few maverick trade union people. Frida saw no camaraderie in formally aligning herself with, what she considered to be, a motley group of dissident have-nots.

Frida's formal political affiliation remained relatively dormant until 18 July 1936 when the outbreak of the Spanish Civil War galvanized her political conscience.[424] She and Rivera, together with a group of other Republican sympathizers, formed an activist unit in Mexico. They hoped to raise money for the Spanish Republican cause and send relief supplies to Spain. Frida's caring nature, to always side with the underdog and the weak, played a large part in determining her political orientation in this instance, and throughout her life.

Rivera was at the Blue House, Frida's ancestral home, in Coyoacan on 21 November 1936 when he received a telegram from Anita Brenner in New York. The message simply stated that it was a matter of life or death that the Mexican government grant Trotsky's request for a safe haven and political asylum. Rivera and Octavio Fernandez, a leader of the Mexican Trotskyite Party,[425] went immediately to request an interview with the newly elected president of Mexico, Lazaro Cardenas.[426]

Cardenas was in Torreon, a northern state of Mexico, supervising his newly promulgated land distribution program. Rivera and Fernandez hurried there to present their petition. (They were driven by two chauffeurs in relays, stopping only for meals.) Rivera eloquently explained to Cardenas the need for urgency in granting Trotsky asylum.

Rivera appealed, as a Mexican citizen and world renowned muralist, that Cardenas would be benefitting mankind if he would compassionately grant asylum in Mexico to a great revolutionary humanitarian like Trotsky. After such a plea there was little Cardenas could do but grant the petition subject to the condition that Trotsky would not interfere in the internal politics of Mexico.

On 16 December 1936 Mexican Foreign Minister Eduardo Hay, sent a coded telegram to the Mexican Embassy in Stockholm:

> Department informs you quote it has been agreed Leon Trotsky be permitted to enter Mexico and admitted with his secretaries and members of his family with the status of political exiles stop. They will not be asked to fill in any forms as this agreement is to be taken as an exemption from presenting regulation papers stop. They will be required only to undertake to respect our laws and not propagandize their political and social doctrines on national territory unquote. Please communicate to Trotsky concepts outlined in final paragraph. Respects. Transmitted in code. Eduardo Hay.[427]

Trotsky and Natalia would have accepted any safe haven, with or without preconditions. They knew they must escape the imminent danger of assassination and adverse world opinion that was being created against them by remaining in Norway.

It was a strange ambient Trotsky and Natalia were entering into in Mexico. The two people who were leading them into it, Rivera and Frida, were not the most stable. Rivera, like a reed in the storm, blew any way the fashionable political wind was blowing. Possibly even more foreboding, and dangerous for them, was the rumor that Rivera might be involved in a plot to lure Trotsky to Mexico where the GPU could more easily murder him.

For all of Rivera's political immaturity, and lack of moral and ethical convictions, he was still lucid with an ability to reason. For him to have been a part of such a complicated entrapment scheme with the GPU is illogical. The GPU, under orders from Stalin for immediate, but secret and efficient action, would not have trusted a political neophyte such as Rivera with any part of such a sensitive assassination operation.

Although the specter that Rivera, in order to curry favor with the GPU, and because of his infantile political posturing, might have had a part in such a scheme, all credible evidence militates strongly against it. On balance, Rivera could not and did not lure Trotsky to Mexico so that the GPU could murder him.

At this period in Rivera's political gad-flying he sincerely considered himself on the world's stage as a bona fide Trotskyite. He was looked upon with favor as a true disciple by Trotsky himself. This close personal friendship with one of the leading political figures of the day was too compelling a political magnet for Rivera to risk losing.

Having disposed of the possibility that Rivera might have been the witting or unwitting dupe of the GPU in luring Trotsky to Mexico, it is accepted that Rivera's actions in arranging the asylum were genuine. It must also be recognized that Rivera desperately desired to bask in the reflected glory of having been the saviour of this great revolutionary hero. It was the kind of hammy, histrionic ploy that would appeal to Rivera's juvenile conceit. The anticipated surge of publicity for Rivera's art work and for his position as the great revolutionary painter were also greatly enhanced by his participation in saving Trotsky.

Frida, as equally devoid of political chicanery as her husband was addicted to it, was a spirited, brazen woman, at times irrational but always with a genuine compassionate loving and tender nature. She was fiercely proud of her Mexican Indian heritage.[428] She was as much an artist in her own right or, according to some eminent critics, a much better artist than her husband. (Picasso said she had no equal in the world in painting heads.) She probably had as much, or more, of the unstable, volatile, artistic temperament than did Rivera.

However the mix was applied and analyzed, Rivera and Frida were an odd combination. On the first meeting they probably made life in Mexico seem unpredictable and chaotic to Trotsky and Natalia. At the same time, after years of living in exile, sometimes barely existing, life in Mexico was welcome. Regardless of how erratic and whimsical it seemed, it represented for them a bountiful, friendly and, most importantly, a secure home in an otherwise hostile world.

On that bright and sunny morning of 9 January 1937 when the steamship Ruth arrived in Tampico harbor, Natalia was wary about the disembarking arrangements. When she was asked to leave the relative safety of the ship, she adamantly refused unless she was given some proof of a friendly welcome.

Trotsky suggested to the ship's captain that his wife would be convinced it was safe to disembark if they could see a face they recognized in the welcoming party. The ship's captain had just decided to forcibly evict the Trotskys if they further delayed his ship's departure, when Natalia looked through the porthole and saw a government launch approaching.

The launch carried the welcoming party of several carefully selected Mexican journalists and Frida Kahlo who had come in Rivera's stead because he was hospitalized with a minor ailment. Also, in the launch was Max Schachtman, a founder of the American Trotskyite party who was well known to both Trotskys, and George Novak, who was the secretary of the American Committee for the Defense of Leon Trotsky. Both of these men were old friends of the Trotskys and they waved and shouted a welcome to Natalia.

There were also several Mexican officials from the Federal government plus some local Tampican leaders in the launch which completed the party of official greeters. Everyone else was unimportant as far as Natalia was concerned for; as soon as she saw Schachtman and Novak, she knew it was safe to land.

Trotsky, nattily dressed in a tweed knicker suit, with his cap at a jaunty angle, carrying a brief case, walked down the gangplank with Natalia on his arm. Trotsky commented informally to the Mexican officials who had received him with a little formal ceremony that, after many months of exile, the meeting in Tampico with friends was especially welcome. He added that he felt that he would always owe a debt of gratitude to Rivera for having arranged his asylum in Mexico.

President Cardenas had sent his personal train, El Hidalgo (The Nobleman) to transport Trotsky and his party from Tampico to the capital of Mexico City. The train secretly left Tampico after ten o'clock at night in order to thwart any GPU agents who might be monitoring Trotsky's arrival. The entire group of welcomers and the arrivals stayed in a state of constant readiness to repulse any attack by the GPU against Trotsky. Fortunately for Natalia's frayed nerves, the journey was uneventful.

On 11 January, before daybreak, the special train arrived at a small milk depot just outside of Mexico City called La Lecheria. Rivera, who had been given dispensation from the hospital to meet the train, was in the welcoming reception. The leaders of the Mexican Trotskyite Party, some Mexican officials and a group of special police were there to guard against any GPU action against Trotsky. To confuse the GPU watchers, if in fact there were any such observers in the pre-dawn darkness, the Mexican Trotskyite Party had sent a group to the main railroad station to fake a formal welcome for Trotsky.

Reporters and photographers however saw through the diversions the Mexican Trotskyites had thrown up as a decoy and were at La Lecheria when the train arrived.

Trotsky was the first to descend from the train and warmly embraced Rivera. Natalia and Frida, already apparently having become close friends on the trip from Tampico (prophetically, sisters under the skin), linked arm-in-arm, were next off the train. The entire party went immediately to several waiting taxis and were swiftly driven to Frida's Blue House in Coyoacan.

Coyoacan had been the area where Cortez had established his headquarters circa 1522 when he was attacking Montezuma in Mexico City. At that time the city was built on islands surrounded by water. Coyoacan was on high ground at the edge of the lake. In 1940 it was a quiet and slightly derelict area, too far removed from the center of town to be incorporated into Mexico City but close enough to be almost be considered fashionable for artists and Bohemians.[429]

The Blue House was (and is today) a lovely Spanish colonial one-story house with a high surrounding wall, painted a bright but muted blue. The Blue House had been Frida Kahlo's girlhood home and had been occupied by the Riveras as a married couple. The Trotsky party arrived about noon at the Blue House where they were to live rent free for the next two years.[430]

The house was ideally suited for the protection of Trotsky, surrounded by a high wall of sturdy masonry construction which, in turn, was surrounded by a compliment of Mexican police and soldiers thoughtfully provided by President Cardenas.

Two or three of the Mexican Trotskyite group would come to the Blue House every evening to stand guard through the night so that Jean van Heijenoort and Trotsky's other personal guards could get some sleep. Joseph Hansen, an American who would become Trotsky's chief guard and be with him until the end, joined the group in the summer of 1937. Hansen drove a new Dodge automobile down to Mexico from Ohio accompanied by Max Sterling, the husband of Rae Spiegel, Trotsky's Russian secretary at the time.[431]

President Cardenas had appointed Antonio Hidalgo to be the liaison contact between the Mexican government and the Trotsky group.

A journalist who interviewed Trotsky for a *Time/Life* picture story at this time reported:

> Walking into the room where Leon Trotsky stood was like entering the presence of a death ray. I hardly crossed the threshold before I felt the angry magnetism of the man, and by the time I had reached his desk and shaken hands with him I had the feeling of a personality so dynamic and destructive that it was actually terrifying.[432]

Neither Trotsky nor Natalia spoke any Spanish and Rivera spoke no Russian. The house guests found a common language with their hosts in pidgin English. Both Frida and Trotsky could speak English with considerable fluency while Rivera spoke only a little English and Natalia practically none. Therefore, English by necessity became the lingua franca for the household. It was to be a disguised blessing for Frida and Trotsky and an obscure curtain for Natalia and Rivera.

Frida supplied the Trotskys with servants from her own household staff whom she knew to be trustworthy. Heijenoort hired a Spanish-Russian typist, Rita Jakovlevna, who started work on 16 January 1937. Frida had provisioned the house with the special foods and liquors she knew the Trotskys enjoyed. She also arranged that either she or her sister, Cristina, would be available to serve as translators, companions and guides for the Trotskys. Frida and Rivera were living in their two houses in San Angel about two miles away from the Blue House.

Trotsky's guards who joined him in Mexico soon had the Blue House secured as an impregnable fortress. Jan Frankel, a guard/secretary, arrived from Europe on 19 February and Bernard Wolfe, a member of the American Trotskyite group came to live at the house and work as an English-language secretary.[433] They blocked all windows facing the street with adobe bricks and mortar. Their schedules were arranged so that some of them would always be with the Mexican police on the outside perimeter. The house settled into a routine of domestic harmony and tight security with visitors admitted only after screening.

Trotsky was convinced that in early January 1937, the first probe in preparation for an assassination attempt in Mexico was made by an unknown man. The man presented himself at the door of the Blue House with a forged letter of introduction form a Mexican political figure.

A similar incident occurred in October of 1939 when one Manuel Fernandez Barrechena told the guards he had an urgent message for Trotsky. He claimed he was a refugee from Spain, a former pilot in the Republican Air Force and a member of the PCE. He said he lived at the Hotel Moderno in Mexico City. The following day Trotsky received a telephone call from a man who purported to be General Jose Miaja, who had been in charge of the Republican forces defending Madrid during the Spanish Civil War. According to the General, Manuel Fernandez was a person of sterling qualities and of considerable wealth whose greatest desire in life was to present Trotsky with an automobile.

Trotsky thanked the General and telephoned the real General Miaja who had never heard to Manuel Fernandez or the proposed gift of an automobile. One of Trotsky's guards went to the Hotel Moderno and interviewed Manuel Fernandez. He was nervous, had no documentation, no explanation for his actions and appeared to be from South America not Spain.[434]

As a result of incidents like these an alarm system was installed in the Blue House and a twenty-four hour a day guard schedule was instituted.[435]

Security at the Blue House was enhanced when Rivera purchased the land and house immediately adjacent. There had been a rumor that the GPU might try to launch an attack on Trotsky from the convenient next door site. There was also speculation that the GPU might buy this property through a strawman and use it as a base of operations against Trotsky. Rivera prudently bought this property and connected the two properties to form one large spacious and charming garden for the Trotskys to enjoy.

Ecstatic with a new sense of freedom, Natalia began to relax. Being in an ostensibly friendly country, under the protection of the President and living with two of the country's leading citizens, who were also good companions, Natalia felt peace for the first time in nearly a decade. She often spoke of the Riveras eagerness to do everything they could to assure the Trotskys' comfort. Surrounded by staunch personal guards and uniformed Mexican police put both Trotskys at ease and into high spirits.

The Trotsky's developed friendly relationships with some of the Mexican Trotskyite leaders. Sra. de Fernandez remembers a Christmas party at the Octavio Fernandez house which was attended by the Trotskys, to the delight of his grandson. The "Old Man," as Trotsky was called, participated in breaking a pinata, the traditional Mexican papier mache figure filled with toys and candies. Trotsky was also an active participant in pin the tail on the donkey.[436]

The Trotskys were charmed with the Blue House with its picturesque colonial architecture, Mexican furnishings and pre-Columbian art. Frida's unique collection of decorations of brightly colored Mexican handicraft added a sparkle to the effect. They enjoyed strolling in the newly enlarged garden or sitting on the flower filled patio. They treasured the quiet meditative time they could spend together among the many paintings of Frida's and Rivera's that were hung throughout the house.

Shortly after Trotsky's arrival at the Blue House, an international magazine reported that:

> At the latest reports, Host Diego Rivera had to return to the hospital with a kidney ailment; Mrs. Trotsky had gone to bed with what seemed to be a recurrence of her malaria; Guest Trotsky, respectfully watched and waited upon by the dark-eyed young Hostess Rivera, had resumed dictation to his secretaries of his monumental biography of Lenin begun nearly two years ago.[437]

Before he left Norway, Stalin's purge trials had charged Trotsky with a myriad of subversive crimes. Once comfortably settled in Mexico he had to prepare his defense against these charges. On 23 January 1937 a new trail opened in the October Hall in Moscow and Trotsky was the principal defendant.

The cynicism of the purge trials that started with the carnage of executions in 1935 is evident when it is remembered that only those defendants who were fully rehearsed and abjectly ready to confess were brought to public trial. There were literally thousands who were executed without trials because they would not publicly confess to crimes they never committed.

Trotsky's gullibility in believing the earlier purge trials were based on valid charges is incomprehensible. He accepted the validity of the early trials and never doubted that a workers' state, albeit a degenerate workers' state, could do such a thing. He even protested the leniency of the sentences given some of the alleged leaders of the non-existent subversive gangs. When the trial of old Bolsheviks whom he knew began, Trotsky then realized his original opinions had been drastically wrong.[438]

In the enormous amount of so-called evidence collected by the overly zealous Soviet prosecutors only a minuscule amount lent itself to independent corroboration. In every instance where independent verification of the evidence against Trotsky could be made, the facts turned out to have been manufactured and utterly false.

For example, there was supposed to have been a meeting between one of the defendants and Trotsky in a Copenhagen hotel. Trotsky's investigation disclosed that this particular hotel had been torn down years before the alleged rendezvous. One poor defendant was forced to confess to having flown from Berlin to Oslo to get instructions for a subversive operation from Trotsky. The Swedish police records for the dates in question revealed that Trotsky had been under continuous round-the-clock surveillance with his every visitor logged in and out during this time. Therefore, he could not have met the alleged contact. Further, there were no flights in or out at the airport cited during the time period.

One witness at the purge trials testified that he had delivered an important traitorous message to Trotsky at a rendezvous in Paris. This meeting was alleged to have occurred at a time when Trotsky was demonstrably several hundreds of miles away, and again, under round-the-clock police surveillance.

Trotsky was charged with conspiring with both Japan and Germany, in the belief that war was inevitable and that the Soviet Union would be defeated. In return for their agreement to make him master of all Russia, he was supposed to have promised Germany an "independent Ukraine" and Japan outright possession of the maritime provinces.[439] The Soviet prosecutor contended that Trotsky's campaign had already caused many mining disasters, railway accidents and factory sabotage accidents

Joseph E. Davies, the newly appointed United States Ambassador to Moscow, and a die-hard capitalist, attended the purge trials and found nothing suspicious in the eagerness of the defendants to "heap accusation upon accusation upon themselves." He wrote to Washington that most of the diplomatic corps in Moscow were of the same opinion that the Soviet prosecutors had clearly established the existence of a conspiracy to overthrow the government.[440]

Trotsky consulted with Natalia, his secretaries and the Riveras and decided that he must present himself for trial before the world in order to respond to Stalin's charges against him. Trotsky would ask world socialism to form an international tribunal of inquiry to evaluate and judge whether the evidence presented against him at Stalin's purge trial was credible or not.

Trotsky himself had given the Mexican communists grist for plotting against him and to try to have him expelled from Mexico for violating the conditions of his asylum. The grist for this plotting came not only from the charges in Stalin's trials but inadvertently from Trotsky himself when he agreed to testify before the Dies Committee of the United States House of Representatives. The PCM charged him with engaging in political activity in contravention of the terms of his political asylum as originally demanded by President Cardenas.

The PCM and Stalinist press vilifications against Trotsky became so intense that President Cardenas had to intervene. The official periodical of the PCM, *El Machete*, was the most vicious in attacking Trotsky and his followers.[441] In an interview published in *La Prensa*, President Cardenas declared that Trotsky was a man of honor who had scrupulously kept his promise not to interfere in Mexican politics.[442]

In 1937, long before the witch hunts of the McCarthy era in the U.S. Senate, the Dies Committee in the House was carrying out an investigation into un-American activities. The purpose of the inquiry was to gather evidence for legislation to suppress the CPUSA on the grounds that it was an agency of a foreign power.

Trotsky had agreed to testify before the Committee. He intended to use this platform to expose the GPU's murderous activities against him and his followers in the United States. Unfortunately, Trotsky was never allowed to appear before the Dies Committee. The entire matter was rendered moot when the Department of State refused to grant Trotsky a visa to enter the United States.

Yet the mere fact that Trotsky had been willing to appear before the Dies Committee made it easier for the PCM to build a network of lies. They accused Trotsky of conspiring with Dies and the oil interests of the imperialist United States against the Mexican people and government.

In protest against this unholy alliance, the PCM were able to marshal some 20,000 uniformed communists to march through Mexico City. The march began at the Zocalo (the main town square) and ended at the Monument to the Revolution with lettered banners and flags of brilliant colors demanding Trotsky's ouster from Mexico.[443] Fortunately for Trotsky, President Lazaro Cardenas, although himself a staunch leftist, chose to uphold the doctrine of political asylum and not succumb to mob pressure.

The Catholic Church joined in the propaganda campaign against Trotsky. Due to Trotsky's avowed dedication to godless communism, the Catholic Church in Mexico attacked him through his close relationship with President Cardenas. The church organs constantly harped that "Cardenas never makes a move without first consulting the 'Red Demon of Coyoacan,'" meaning, of course, Trotsky.[444]

To prepare his rebuttal against Stalin's lies, the Blue House was turned into a giant editorial office while Trotsky marshaled his defenses for this self-imposed public inquisition. His archives were combed by his secretaries for evidence in the form of statements, memoranda, letters and news stories that would support his denials of Stalin's spurious allegations.

Also helping Trotsky prepare his defense was his Russian secretary, Raya Dunayevskaya, aka Rae Spiegel. She was Trotsky's secretary in Mexico from 1937 to 1938. At the outbreak of World War Two she broke with Trotsky in his defense of Russia as a "workers state though degenerate." She held instead that the Stalin-Hitler Pact gave the green light to World War Two and revealed that Russia was a capitalist state.[445]

Ruth Ageloff, a Trotskyite from New York, and the sister of Sylvia Ageloff who figures prominently in the assassination of Trotsky and Bernard Wolfe, an American novelist and a Trotskyite from New York, together with Jan Frankel and Jean van Heijenoort made up the secretariat in the Blue House.[446]

The Dewey Commission that was formed to determine Trotsky's guilt or innocence was chaired by the eminent American philosopher and educator, John Dewey. The actual hearings took place from 10 to 17 April 1937. The other members of the Commission were John Chamberlain, Carlton Beals, Alfred Rosmer, Professor Edward Alsworth Ross, Otto Ruehle, Benjamin Stolberg, Carlo Tresca, Francisco Zamora, Wendelin Thomas and Suzanne La Follette. John F. Finerty was counsel. A subcommission to gather information in Europe was headed by the exiled

Italian socialist G.E. Modigliana and included as members Mme. Cesar Chadburn, M. Mathe, Jean Galtier-Boissiere, Professor Jacques Madaule and Maurice Delepine.

What amazed many objective observers was that hundreds of American communist men and women who had prated for justice all their mature lives, now jumped with both feet on John Dewey, the octogenarian professor in this thankless undertaking.[447]

Eighty-eight American intellectuals charged that the Dewey Commission was out of order and intruding on the domestic affairs of the Soviet Union. "A country should be permitted to decided for itself when its leadership is threatened."[448] The Jewish press in New York berated Trotsky for charging the Soviet Union with being anti-Semitic. The *TAG*, a Jewish newspaper of New York, said that the Soviet Union was their only consultation and "... it was unforgivable that Trotsky should raise such a groundless charge against Stalin."[449]

The published volume of the Commission's findings was very detailed. It was hoped by Trotsky that any objective mind would find the report utterly destructive of the entire Stalinist fabrications of evidence against Trotsky. There were others, however, who had a different opinion.

When the GPU learned from their agents inside Trotsky's Fourth International, and from those inside Trotsky's personal coterie, that he planned an independent inquiry into the charges against him, they immediately began a counter-offensive. They planned a crash campaign of denunciations against the idea of an independent commission and of vilification against the qualifications and objectivity of the individual members of the Commission.

The GPU orchestrated this campaign by initially branding all of the proposed members of the Commission as being completely subservient to Trotsky. They claimed the members were therefore incapable of rendering an impartial judgment. In fact not a single member of the Commission was a Trotskyite or an adherent of Trotsky's philosophy of communism. Trotsky freely admitted Alfred Rosmer was a close personal friend but claimed he was also an objective academician and above reproach. The GPU launched a media blitz that only the loyally blind communists could understand. They claimed that any attempt to question the orthodoxy or fairness of the Moscow purge trials would only play into the hands of Hitler and fascism.

The CPUSA synchopants jumped on the smear bandwagon. They branded anyone who even indicated a glimmer of objectivity about Trotsky's guilt, and did not join in the condemnation of the Dewey Commission, as a deviant Trotskyite. To those who may have sincerely wished to remain and work within the framework of the liberal leftist political movement, being branded a Trotskyite amounted to virtual ostracism by all pro-Marxist elements.

The Commission members were subjected to a barrage of vile, abusive and threatening telephone calls made at night for maximum terror. Hundreds of scurrilous letters and telegrams to members of the Commission and their families were designed to unnerve them and try to make then resign. Delegations, often consisting of big names in the entertainment and motion picture worlds, called upon members of the Commission. These luminaries pleaded and exhorted the Commission members to find Trotsky guilty as charged in the Moscow trials.

One GPU master stroke was to have many big name fellow travelers express initial support for the objective quality of the Commission. They were then to claim that they had been mislead by originally believing that the members of the Commission were objective and impartial. After their resignations they were to smear the Commission with charges of favoritism to Trotsky; all, of course, to be accompanied by maximum fanfare in the communist press.

For there to be only one Mexican on the Commission seemed odd. The reason was that few Mexican socialists wanted to bc openly associated with Rivera. Although they could have been objective about the charges against Trotsky, they did not want to be known as ever having had anything to do with Rivera. All who were asked to be on the Commission had politely begged off.

The choice to head the Commission was a brilliant one. It was designed to stifle any protest the communist might try to make that the Commission was weighed in Trotsky's favor. John Dewey, an eminent philosopher and academician from the United States, agreed to serve as the chairman of the Commission.

The Blue House where the Commission was to meet was sandbagged outside and fortified inside to protect its structure and occupants from bombs and rock throwing counter-demonstrators. An area for about forty members of the press, and other qualified observers, was roped off next to the large room where the actual trial was to be held.

The opening of the trial was a solemn occasion. A long table provided by the Riveras was filled with Trotsky, Natalia, Trotsky's secretary and members of the Commission. The trial opened on 10 April 1937 and lasted one full week with a total of thirteen formal sessions. Frida was the belle of the trial. She was dressed in her Tehuana Indian finery and Tarascan jewelry. She sat as close to Trotsky as she could. Rivera showed up for some of the sessions, usually trying to steal the limelight sporting a ludicrous wide-brimmed sombrero adorned with a large peacock feather.[450]

Carlton Beals, subsequently identified as a GPU stooge on the Commission, asked Trotsky if he had sent Michael Borodin to Mexico in 1919 or 1920 to form the PCM. Trotsky denied the allegation and Beals claimed that Borodin had himself told Beals that this was true. This left the clear implication that Trotsky was lying. This caused considerable embarrassment for Trotsky's Mexican protectors since it appeared that he had been engaging in internal Mexican politics. Beals refused to disclose Borodin's whereabouts and resigned from the Commission claiming it was not objectively judging Trotsky.

James T. Farrell, an ardent Trotskyite was present at the Dewey Commission hearings. (Farrell was the famous American author of the *Studs Lonigan* trilogy). In April of 1977 Farrell wrote that the man who directed Carlton Beals to try to trap Trotsky was Harry Black, a Stalinist,[451] and Herbert Solow, who was an "agent" (presumably a GPU agent).[452]

Frank Kluckhohn, was the *New York Times* correspondent covering the Dewey Commission hearings. His reports to the *New York Times* distorted the atmosphere at Coyoacan and were openly biased in favor of the Kremlin and reiterated the validity of the Moscow purge trials. The editors of the *New York Times* ordered Kluckhohn to stop editorializing his reports and to stick to the facts. Kluckhohn brought Frank Jellinek, a known GPU agent, to the Commission hearings. Jellinek sat with Pavon Flores, a member of the Political Committee of the PCM. Jellinek was part of the Trotsky GPU assassination apparatus.[453]

Trotsky was at his best during the hearings and answered his interrogators with detailed cogency. He never stinted on his answers nor was he evasive or overly qualifying .

The Soviet Embassy in Washington, the CPUSA, the PCM and leading Mexican communist labor organizations had been invited to send representatives to cross-examine Trotsky during the Commission hearings. They either ignored the invitations or obviously refused the opportunity.[454] Trotsky also placed his complete archives and files at the disposal of the Commission.

Trotsky had accumulated an enormous amount of material with which to refute the charges of the Moscow purge trials. He took a final turn on the stage of the Dewey Commission with this exit line:

> The experience of my life, in which there has been no lack of either success or failure, has not only not destroyed my faith in the clear bright future of mankind but on the contrary has given it an indestructible temper. This faith in reason, in truth, in human solidarity, which at the age of eighteen I took with me into the workers' quarters of the provincial Russian town of Nikolyev — this faith I have preserved fully and completely. It has become more mature but no less ardent.

John Dewey, caught up in the emotion of the moment, summed up the feelings of the Commission by saying that anything he might add after Trotsky's speech would be anti-climatic. Five months later the Dewey Commission delivered its verdict. Trotsky had been vindicated and there was not the slightest shred of evidence to support any of the allegations made against him by Stalin in the Moscow trials.

The Dewey Commission established in extreme detail that: first, the framework of the indictments in the Moscow purge trials rested on confessions of those defendants who claimed they acted as intermediaries between Trotsky, his son Sedov and the alleged plotters in the Soviet Union; second, in every instance these claims, being subjected to investigation, were proved false; and third, therefore, these confessions were invalid.

All the purge trials were nonsensical, which Nikita Khrushchev publicly admitted at the Twentieth Congress of the Russian Communist Party. According to Khrushchev the trials were nothing but frameups.[455]

The decision of the Dewey Commission exonerating Trotsky of all charges made practically no impression on the public. In Great Britain and France it was scarcely mentioned, nor did it excite any interest at all outside of the communist, left wing and, to a lesser extent, among the liberal elements in the United States.[456]

The "trial" ended as a farce because it was apparent from the beginning that no actual trial was being held. The setting was fake and Trotskyites were using the benevolent patronage of good old Dr. Dewey and his cohorts as a front behind which they could, across half a world, defy Joseph Stalin. Trotsky, one of the greatest propagandists of the era, played his role to the hilt, declaring, gesturing, screaming and sneering at his enemies beyond the seas.

> The thing that struck one most, in view of the great vogue of parlor pinks in the United States at the time, was the hypnotism which the dynamic and ruthless Red leader exercised over the Dewey Committee. With their idealistic faces turned in admiration to him, they sat spellbound and completely magnetized by Trotsky's stronger will and unscrupulous trickery. When I have since read of United States intellectuals being exploited and led around by the nose by communist agitators under the guise of 'liberalism', I have recalled the Trotsky trial and always thought of them like lambs being led to the slaughter. They are led to the slaughter of the very principles to which their lives are dedicated, because they are the weak-willed victims of the ruthless architects of world revolution.[457]

At this time Trotsky had not been accused of complicity in the alleged projected military coup of Soviet Marshal Tukhachevsky. Supposedly Tukhachevsky was part of the Trotsky plot to take over the Soviet government by military force. This was to occur only if the terror and subversion of the Trotskyites, abetted by the Nazi and Japanese fifth columns, did not accomplish the overthrow. On the 11th of May 1937, the Soviet government defused what was a paper tiger plot by demoting Marshal Tukhachevsky to a minor command in the Volga district.

Several military commanders allegedly involved in the coup committed suicide. On 11 July 1937 Tukhachevsky and seven other Red Army generals faced a special Military Tribunal charged with conspiring with enemy powers against the Soviet Union. On 12 June they were all found guilty and were shot within twenty-four hours.[458]

Chosen by the CPUSA to answer these detailed rebuttals of the Dewey Commission verdict was Corliss Lamont, a millionaire communist fellow traveler. Lamont was a stooge of the CPUSA and was often used to draw money from the gullible donors. Lamont did not have the intellect nor the moxie to answer to the Dewey Commission's exoneration of Trotsky.

The CPUSA expected Lamont to substantially refute the Dewey Commission's findings point by point with facts. All Lamont could offer, since he had no facts, was that he was convinced that the Soviet Union had acted in a just and proper manner by their original verdict at the Moscow trials. He did not offer a single point to contest the findings of the Dewey Commission.

Summing up his refutation of the Dewey Commission findings that the Moscow trials were a fraud upon the world, Lamont weakly contended:

> The Soviet regime and its achievements are indivisible and we cannot believe that its system of justice is completely out of step with its splendid accomplishments in practically all other fields.

Although much relieved by the Commission's findings that he was innocent, Trotsky was about to face another trial much closer to home that had been brewing since the Trotsky's moved into the Bluc IIouse. Natalia was going to be the judge who presided at this new trial and she alone was to be the jury that decided whether or not Trotsky was guilty.

TROTSKY, FRIDA AND RIVERA

> I hope that the exit is joyful... and I hope I never come back.
>
> Frida Kahlo, confiding in her diary just before her death,

Frida did some of her best paintings during the time the Trotskys were living in the Blue House where she maintained her studio. Every day Rivera continued to do his painting in his own studio in San Angel, a suburb of Mexico City close to Coyoacan. Trotsky was working on his biography of Lenin in his study in the Blue House. Natalia had no work but left the Blue House almost every day for shopping or sightseeing. Consequently, Natalia and Rivera were far away from the Blue House for most of every day which left Trotsky and Frida alone together.

As it turned out, this largely unintentional pattern would nudge Frida and Trotsky into intimacy. They were together for long periods of time with only Frida's docile, unobtrusive servants and Trotsky's discreet guards. Both the servants and guards were too loyal, and too subservient, to tattle should any indiscreet conduct occur.

In addition to this daily intimacy, the two families ate together most of the time, went on picnics together and drove to the country on holiday excursions together. The Riveras enjoyed acting as guides for the tours into the environs of Mexico City. On these outings Rivera would paint and Trotsky collected cacti. The patio of the Blue House was soon filled with Trotsky's blooming acquisitions.

Ordinarily Trotsky did not respond quickly to new people but with the Riveras it was different. Rivera could visit Trotsky in his room at any time without an invitation. He was one of the few people Trotsky ever received without having a third person present. (Ramon Mercader, the assassin, was the only other person known to have been received by Trotsky without someone else being present.)

Trotsky was given the use of a house in Taxco by one of his American followers.[459] Taxco is a beautiful old silver mining town that has retained its colonial architecture by law as a national shrine. The

Trotsky's, accompanied by his ever-present entourage of guards, would go to Taxco for a visit of several days. The Riveras would join them for the weekends. Rivera and Trotsky would talk together for hours. Trotsky particularly enjoyed the unbridled enthusiasm and wit of the rabelaisian Mexican Bohemian.

Trotsky was delighted that such an international figure as Rivera would join him in the Fourth International. He lauded Rivera as the greatest artistic interpreter of the Russian October Revolution. Trotsky commented that he considered a mural by Rivera to be, "not only a painting but a living part of the class struggle!"

When the two families returned to Coyoacan the inadvertent juxtapositioning continued. Frida would take a break from her painting just as Trotsky, coincidentally, would take a rest from the toil of writing Lenin's biography. When the two exhausted workers came out of their respective cells they would naturally meet on a common ground; such as the romantic, flower filled, secluded patio.

Trotsky had a preoccupation with sex. Around women Trotsky became animated and always quick-witted with the double entendre. He was imbued with the philosophy that if a man did not ask a woman, he would never know if she was available for some extra-curricular sex.[460]

Natalia was fully aware of Trotsky's errant ways and had devoted a good part of their married life in keeping him out of harm's way. Therefore, Trotsky's opportunities were all too few as far as he was concerned. Obviously the chances he did have he made the most of since his successes were reputed to be legion.

Trotsky's bearing and demeanor were impressive. The military mein and aristocratic presence commanded awe from impressionable females. He had an engaging sense of humor that had a disarming effect on an already predisposed prey. It was his severity and unrelenting follow-through of never giving up, while sweeping aside all objections, that ultimately made him successful in his amorous conquests.

Frida's relationship with Trotsky can be likened to a game of hare and hounds, with Trotsky pursuing her around the patio until she allowed herself to be caught. Then Frida and Trotsky would joke about going to see one of Frida's paintings that was hung in her bedroom and off they would go to inspect the painting. They would use the bed that Frida had decorated with Mexican artifacts which had a large full length mirror on the underside of the canopy.

Trotsky was not above candidly, forth-rightly and without resorting to guile or subtleties, of using the crudest of four letter English equivalents, to make an indecent proposal to woman he had just met. A woman sitting next to him at a dinner, was either delighted or abashed, depending upon her point of view.

Invariably Trotsky would begin to fondle her knee under the table as soon as they were seated. If she did not object immediately to the knee fondling, and push his hand away or let him know in some other very forceful way that she was displeased with his groping, Trotsky would inch his way up her thigh. At the same time, if he got that far, he would whisper the crudest, most indecent proposals to her despite the fact that her husband may be seated at the same table.[461]

To illustrate this aspect of Trotsky's insatiability and preoccupation with sex, consider the case of Cristina Kahlo, Frida's sister. Cristina was a very attractive lady who was rumored to be no slouch in the sexual arena herself. Cristina had been Rivera's mistress while he was married to Frida and the two continued their relationship even after Frida found out about it.[462] Although Cristina was aware of how sexual games were played, she was apparently not ready for Trotsky's international class roue routine.

In order not to arouse Natalia's suspicions, Trotsky planned an elaborate fire drill that would allow him to leave the Blue House during the melee of the drill. He planned to climb over the ten-foot high wall surrounding the Blue House, run to Cristina's house several blocks away, consummate the act and run back to Natalia's bed without her noticing his absence. This Olympian obstacle course required a practice run of a fire drill for the entire household while Trotsky timed his run. Trotsky intended to run through the mock fire drill in order to have the real thing timed down to the exact second. Unfortunately for Trotsky, and probably also for Cristina, cooler heads prevailed.[463]

Trotsky's staff pointed out to him the foolhardiness of such an exercise, not to mention the idiocy of the whole thing if he were caught. It turned out that Cristina was not really too keen on the idea herself. Trotsky bowed to wiser counsel and abandoned the fire drill.

Frida by any standards was also an international class sexual participant. She was not promiscuous since promiscuity implies unrestrained selection, or no discerning choice of a sexual partner. Frida either thought she loved her sexual partners, respected them as artists or genuinely cared for them with an emotional, almost maternal, affection. In Trotsky's case Frida, at first anyway, really cared for Trotsky and respected him as a world leader of socialism.

Frida affectionately referred to Trotsky as "pinochitas," meaning Little Goat. This endearment was because of Trotsky's goatee which did resemble a billy goat's chin whiskers. Frida also greatly admired Trotsky as a revolutionary hero. His intellect, persuasive personality, force of character, all attracted her.

Rivera's admiration for Trotsky may have played a part in Frida's feelings for Trotsky. She may have felt that the perfect retaliation for Rivera's affair with her sister would be to have an affair with Rivera's best friend and political idol.

The lovers spoke to each other in English which the people around them, especially Natalia, did not understand. Trotsky and Frida would refer to each other as "lover" and constantly use amorous phrases. When they parted, even for a short while, they would bid each other lingering adieus. They would say, "All my love to you lover, until I see you later." and "My love to you until we meet again."

They deprecated Natalia's intelligence in this insulting manner. Even though Natalia could not understand their silly little speeches, she could not help but sense the underlying sexual current of the repartee, when the lovers, blind to everything but each other, made them in front of her.

Pictures of Frida taken at this time, show an unusually attractive woman. She was radiant, as a total woman whose inner compassion and ardor clothed her with a powerful attraction for men. She was the epitome of everything that makes a woman sexually attractive to a man.

Trotsky became kittenish and senile in his pursuit of Frida. He started writing her love notes which he would slip into books that he gave her to read. He would tell Frida that she should read the very interesting passage on such-and-such a page where he had hidden his note to her

It was shortly after the Dewey Commission finished its deliberations that Frida and Trotsky found themselves alone together for the first time with all of the circumstances just right. Rivera was out painting and Natalia was out shopping. The two lovers consummated their long simmering relationship. Cristina, obliging loaned her house on Aguayo street, just a few blocks from the Blue House, to the lovers for their first few assignations.

For some reason Rivera never tumbled to the fact that he was being cuckolded despite all his claimed worldliness. Natalia, with a woman's intuition, suspected that Trotsky was being unfaithful to her with Frida. Natalia soon became depressed and grew moody. She also became extremely jealous of Frida.

Trotsky's associates, those who were more politically sensitive and worldly, were concerned that the torrid Trotsky-Frida couplings could be devastating GPU propaganda used to change world opinion against Trotsky. The affair also worried Trotsky's bodyguards because an awakened Rivera might shoot Trotsky. Regardless, none of Trotsky's entourage had the courage to caution him of these dire consequences if he continued his affair with Frida.[464]

Frida finally did break off her relationship with Trotsky. Whatever the real cause there was no doubt that it had been Frida who initiated the break-up and not Trotsky. Petulant and juvenile Trotsky wrote Frida a nine page letter pleading with her not to terminate their relationship. Frida was growing tired of "the old man," the name which, with undisguised scorn and ridicule, she now used to refer to Trotsky.[465]

Passionate embers burn out slowly and Frida must have had some vestigial feelings for Trotsky. Months after she had ended the affair, on 7 November 1937, the anniversary of the Russian Revolution, and also Trotsky's self-proclaimed birthday, Frida gave Trotsky her exquisitely painted self-portrait. Frida inscribed this beautiful work of art, "For Leon Trotsky with all love I dedicate this painting on 7th of November 1937. Frida Kahlo in San Angel, Mexico."

In February 1938, in a burst of girlish enthusiasm, Frida wrote to a close childhood girlfriend that Trotsky's coming to Mexico was the best thing that had happened in her life. The Dewey Commission had concluded its deliberations in September of 1937, Frida and Trotsky became lovers soon after the Commission ended. They had apparently broken-up before the end of 1937, yet in early 1938 Frida wrote that Trotsky is the best thing that ever happened to her.

In January of 1939 Frida sailed for France. This was a time of great world tension; Hitler had just been appeased at Munich, the Spanish Civil War was drawing to a close and Europe, as any perceptive fool could have foretold (retrospectively), was on the brink of a catastrophic war. Frida keenly lamented the loss of the Spanish Civil war by the Republican forces.

Frida, with Rivera's help, arranged to have some four hundred Spanish Republican refugees resettled in Mexico. Many of these refugees were pro-Stalinist communists and soldiers from the anti-Trotsky troops in Spain. This was a factor that neither Frida nor Rivera seemed to consider, or even be aware of, when they petitioned to have these refugees admitted into Mexico.

Their concern at the moment was apparently solely for the relief of the human misery and suffering of the refugees. There was never any sinister plot against Trotsky that Rivera or Frida were involved in by bringing these Spanish communists to Mexico.

Shortly after she reached Paris, Frida heard from Rivera that he was leaving the Trotskyite Party. Frida had no qualms nor regrets about following her husband out of the political movement of her former lover and readily joined Rivera's defection from the ambient of Trotskyism.

Rivera broke with Trotsky mainly because Rivera had the mind of an adolescent and vacillated on whim.[466] Rivera was neither consistent to his political convictions nor was he intellectually profound. Basically Rivera was a shallow mental sponge that could only soak up one uncomplicated thought at a time. He claimed that he was disillusioned with Trotsky's Fourth International. He was convinced, he said, that it would never amount to anything.

Rivera parroted the PCM propaganda line that the Fourth International was made up essentially of inactive old men and that it was not doing anything specific to capture world leadership of the socialist movement. Rivera was also angry with Trotsky because he would not permit him to put down his paint brush and assume a more militant political role in the Fourth International. Trotsky had persisted that Rivera was more help to the movement with his paint brush than he could ever be with his bumbling sword of political persuasion.

On 4 January 1939 after Rivera's defection could not be reversed, Trotsky issued a press communique that he no longer felt any moral solidarity with Rivera. He could no longer accept Rivera's hospitality by remaining in the Blue House. As a balm to Rivera's hurt feelings, Trotsky offered to pay him rent for his share of quarters in the Blue House but Rivera adamantly refused.

Trotsky had given Charles Curtiss 100 pesos to deliver to Rivera as rent on his portion of the Blue House. Curtiss was an American communist in Mexico who was attempting to reunite the warring factions of the Mexican Trotskyite Party under its old name of LCI.[467]

On 5 May 1939 the Trotskys, their servants, the secretaries and guards left the Blue House. They moved into a house at 45 Viena in Coyoacan which was about three blocks from Frida's Blue House. Trotsky originally rented this house but a rumor persisted that the GPU would buy the house to harass him. Trotsky had no other recourse but to buy the house before the GPU. He was able to obtain the necessary money from his followers in the United States.[468]

The entire Coyoacan Trotsky group was short of money. Trotsky gave his permission for Joe Hansen to go to New York to the headquarters of the CPUSA to offer to sell a draft of Trotsky's newest book, *Life of Stalin*. This would supposedly stop the publication of the book since it was to be touted as the only copy in existence. Hansen and Trotsky were the only people in Coyoacan who knew about this scheme. Natalia guessed the truth of the matter later and had to be told the details. The only people in New York who knew about the scheme were Max Schachtman and James Cannon.

The CPUSA prospective purchasers had to cable Moscow for instructions. They procrastinated for as long as they could. They appeared to be more interested in learning details about the interior of the compound of the new house from Hansen than they were in purchasing the draft of the book. Hansen did his best to bring them to a firm deal but eventually he backed out and took the manuscript back to Coyoacan.

In early 1950 there were allegations that Hansen had had secret contacts with the GPU. These accusations were made by Pearl Kluger who had been a secretary in the organization the "Defense of Leon Trotsky" and was associated with Reverend A.J. Muste and Louis Budenz. One V.T. O'Brien had been at Coyoacan and knew about Hansen's authorized efforts to sell the manuscript to the CPUSA/GPU. O'Brien tried to tell the story of Hansen's contact with the GPU, that Trotsky had authorized, to Hansen's detractors in order to clear Hansen's name.

There was another episode of a similar nature that opened Hansen to attack from later day overly pious critics. The basis for the suspicions may well have occurred at the same time as the above incident when Hansen was in New York. Hansen had cleared this contact with the GPU with Trotsky and had also told the American Consular officer, Robert G. MacGregor, on 31 March 1940, all about the matter.

When Hansen was in New York in 1938 he was approached by a GPU agent who identified himself as John. (Subsequently identified as Dr. Gregory Rabinowitz, senior GPU officer in the New York area.) John asked Hansen to defect from Trotsky and the Fourth International and join the Third International (Comintern). Impliedly, joining the Comintern would mean also joining the GPU. Hansen, as noted, reported all of this to Trotsky who instructed him to "play along" with John to see exactly what he wanted. After several fruitless months, Trotsky told Hansen to drop the contact with John, which he did.[469]

The reason the Trotskys chose the house in Coyoacan on Avenida Viena was because they already knew the neighborhood. They had found a convenient house they liked in a setting they found compatible with their way of life and it was within Trotsky's financial means.[470]

In early 1940, Cass Canfield, the head of Harper Brothers publishers, and his wife, visited Trotsky. The impression they had of the guards, the guns and the siege mentality gave the impression of a Victor Herbert comic opera. Trotsky told Canfield that he would like to go to the United States if it were not for the "man in the White House" (Roosevelt). Canfield asked what Trotsky would do if he was allowed to enter the United States? Trotsky replied unequivocally, "Start a revolution!"

When Trotsky was in the United States in 1917 he was considered to be the revolutionary leader. Now, twenty-three years later he honestly believed that all that was needed to start a revolution in the United States was the push of a professional revolutionary. Although the idea appears ludicrous when his small force is compared to the might of the established order in the United States, however, giving the"devil his due" when he arrived in Petrograd in 1917 the force at his command was considerably smaller.[471]

On 27 February 1940 Trotsky wrote his will and each line seemed to be permeated by his sense of approaching death or a serious calamity. He felt that he might die a natural death or by his own hand but there is no indication he foresaw death by assassination. He wrote, "My high (and still rising) blood pressure is deceiving those near me about my actual condition. I am active and able to work. But the end is evidently near."

On 3 March he added a codicil. "I feel a surge of spiritual energy because of the high blood pressure, but this will not last." He hoped that death would come to him suddenly, through a brain hemorrhage.[472]

When Trotsky moved out of the Blue House he left behind Frida's self-portrait that she had painted especially for him and an expensive fountain pen she had given him.[473] Frida had gone to the trouble of having a facsimile made of Trotsky's signature and had this signature engraved on the fountain pen.

In 1946 when Rivera was seeking readmission to the PCM, he demanded that Frida loan him the pen so he could use it to sign his application. Frida refused to let him use Trotsky's pen. Rivera thought by using Trotsky's pen for readmission to the PCM, he would be displaying his revenge on Trotsky by this ultimate act of snideness.

If certain facts are twisted and given a false interpretation they might seem to fall into apparent perspective that might imply more complicity for the Riveras in Trotsky's assassination than the facts warrant. For example, what could be more convincing than for Frida to have had a rendezvous with Ramon Mercader, the assassin, in Paris before Mercader came to Mexico? Frida wrote to her friend,

> In Paris [this would have been during the first part of 1939] I had met Mornard [Ramon Mercader was using the name Jacques Mornard at the time he met Frida] the one who killed him [Trotsky] and he went around insinuating to me that I should take him to Trotsky's house. 'Not me, because I am quarreling with Old Trotsky', I told him. 'I only ask you, please to find me a house near there' 'Well look for it yourself I am too sick to look for houses for anyone and I cannot give you lodging in my house, nor can I introduce you to Trotsky. I never will introduce him to you'. But his girlfriend [Sylvia Ageloff] came and introduced him.

Mercader and Ageloff became lovers in Paris but it is not so well known that Mercader actively pursued Frida at about the same time — that is before the assassination. One of Ageloff's closest friends in Paris was Maria Craipeau who wrote an article in which she relates the story of the relationship among Frida, Ageloff and Mercader (Mornard). Craipeau wrote that Mornard (Mercader) told her about his meeting in Paris with

Frida and gave an embarrassed little laugh every time he told the story. He said the whole incident with Frida was silly.

Mornard (Mercader) claimed that he had learned from the newspapers about Frida's arrival in Paris. He knew she was the wife of Diego Rivera in whose house Trotsky was hiding in Mexico City. He claimed he bought a large bouquet of flowers and tried to give them to Frida at the opening of her art exhibit at the Paris gallery. (Frida's exhibit ran from the first to the fifteenth of March 1939.)

When Frida rejected his proffered flowers, he went back into the street and gave the flowers to the first woman he encountered. The woman was so startled by this mad gesture that she fled screaming and the flowers ended up in the gutter. Maria Craipeau asked him why he had been so intent on meeting Frida. He replied that it would have amused him to have met her.[474]

In September of 1939 the Riveras entered the final stage of a divorce. This was much against Frida's wishes but she went along with Rivera's desire to be rid of her.[475] If she had wanted to, Frida could have contested the divorce with a countersuit on the grounds of Rivera's adulterous conduct with other women. During this time Rivera was painting a portrait of Paulette Goddard, a motion picture actress from Hollywood who had gained notoriety as Charlie Chaplin's mistress. It was the gossip of the art world in Mexico City and Hollywood that Rivera and Goddard were having an affair. Strangely Goddard and Frida became good friends.

Despite their divorce the Riveras continued to see each other, take their meals together and frequently spend the night together. Rivera continued with his many mistresses and casual trysts. Frida was devil-may-care, exotic, vibrant and took her own lovers as she found them. One of the Spanish communist refugees from the Spanish Civil War was Ricardo Arias Vinas. Frida had met Arias when she was in Paris and they continued their affair when he came to Mexico.

Aftermath

When Frida learned the details of the Siqueiros raid on Trotsky's home of 24 May, she said of the attackers,

> They acted like a bunch of firecracker makers. They killed a gringo called Sheldon Harte and buried him in Desierto de los Leones and then fled,. Naturally the police caught them. They put Siqueiros in jail but Cardenas [President of Mexico] was his cuate. [Cuate means a very close friend, almost a brother but not of the same blood. Frida was implying that President Cardenas would eventually absolve Siqueiros of his responsibility for the raid.]

Regardless of Frida's perceptive comments (for that is exactly what happened and Cardenas did find a way to allow Siqueros to escape from any serious consequences for leading the 24 May debacle) the Riveras came under suspicion by the Mexican police who were investigating the raid. This suspicion was partially based on the well known fact that the Riveras and the Trotskys had quarreled.

Almost immediately after the raid on Trotsky's home, the Mexican police cordoned off Rivera's San Angel studio. Paulette Goddard, Rivera's mistress of the moment, happened to see the police start their approach to Rivera's studio from her hotel room across the street. She telephoned a warning to Rivera.

Another one of Rivera's women, Irene Bohus, who was with Rivera in his studio at the time stuffed him onto the back floor of her car and covered him with used painting canvases. Bohus then drove through the police cordon of Colonel de las Rosa and his thirty-odd men who had surrounded Rivera's studios.[476] Rivera managed to sneak out of Mexico and went to live with Bohus, and occasionally with Goddard, in San Francisco.

> Rivera escaped the attempt on his life by a crowd of Stalinists and Nazi agitators and stayed in the home of an American friend. Rivera said that it was claimed by Mexican officials that Sheldon Harte was kidnapped in Rivera's car and put the blame for the attempt on Trotsky's life on him. According to Rivera, Vicente Lombardo Toledano is the head of the Fifth Column, anti-American, pro-Stalin and pro-Nazi activities. It was Lombardo Toledano and his group who were responsible for the attempt on Trotsky's life.[477]

Soon after the first attempt on Trotsky's life in the 24 May raid, but completely unrelated to the raid, Frida became very ill. Rivera's departure to the States and his dalliance there with Bohus and Goddard exacerbated her poor health. Although the relationship between Ramon Mercader and Frida was somewhat murky to the world, Frida recognized Mercader as a charlatan and shallow but amusing conniver.

However, Frida did continue to allow him to visit her during her illness. These visits were rare and Frida only allowed him inside her home to help her escape the boredom of her loneliness. She said that Mercader told such obvious lies and fanciful stories about himself that they amused her and helped her bear her illness.

Three months after the attack by Siqueros and his gang of incompetents, Mercader succeeded in murdering Trotsky. The news that her friend, Mercader, had murdered her former lover, threw Frida into another intense illness and prolonged period of depression.

Frida called Rivera in San Francisco and told him, "Listen, they have killed old Trotsky this morning, stupid! It is your fault they killed him. Why did you bring him here?"

Rivera was working on the Treasure Island mural in San Francisco in August 1940 when Frida told him of Trotsky's assassination. He was alarmed at the political ramifications and the possible consequences to his personal safety from Trotsky's followers who may think he had something to do with the murder. In order to be on the safe side, Rivera hired a full detail of bodyguards to sit beneath his scaffolding and protect him while he worked.

It was well known to the Mexican police that Frida had met Mercader in Paris and had invited him to dine with her on occasions at her home in Coyoacan. As a result, Frida was under suspicion of complicity in the assassination. The Mexican police, true to their reputation for ineptness and venality, ransacked Rivera's house, looting as they went. They stole Rivera's clock, some of his paintings and everything else they could carry.

Frida said that, after the sacking of her home by the Mexican police, that she knew the police would search Rivera's house but she fooled them. She had gathered all of Rivera's political papers and thrown them into the cellar under the house where the police would not think of looking.

Cristina and Frida were held in jail for two days of constant interrogation. They were released when the police were unable to produce any evidence that they had anything to do with either the 24 May raid or the 20th of August murder.

In February of 1941 Ramon Mercader's interrogation was completed. He had not named Rivera as an accomplice. Further, he had specifically denied that the Riveras had anything to do with the Trotsky assassination. Only then did Rivera pack his bags and feel safe to return to Mexico City.

When Rivera was still in San Francisco he sent for Frida because he thought that Dr. Leo Eloesser could properly treat her continuing ailments. On his fifty-fourth birthday, Rivera remarried Frida on 8 December 1940 in San Francisco.[478]

Rivera's repeated efforts to be readmitted to the PCM were unsuccessful until 1954. Frida, on the other hand, because she had never publicly embraced Trotskyism was welcomed back into the PCM in 1948. During the mid-forties Frida's health deteriorated. Partly to ease the pain in her deformed leg and the repeated operations on it, Frida started to use drugs.[479]

There is no viable evidence to support the historical rumor that the Riveras had anything to do with planning or carrying out Trotsky's assassination. In later years Rivera bragged that he had invited Trotsky to come to Mexico so that the GPU would have a better chance to assassinate him. This represented Rivera's attempt to ingratiate himself with the PCM in order to gain readmission. There is absolutely no basis for Riveras bragging!

In the latter part of 1952 Rivera became infatuated with the Mexican motion picture actress, Maria Felix. Rivera's wanton public conduct with this actress was reported almost daily in the lurid Mexico City newspapers. Frida could not help but know of Rivera's blatant rejection of their marriage and specifically of her. All of this increased her physical decline.[480]

While she was ill, Frida was visited by Dolores del Rio, the most beautiful woman in all of Mexico, and her husband, the charro singer, Jorge Negrete. Maria Felix, Rivera's mistress who had become a personal friend of Frida's was a constant visitor. Teresa Proenza, the secretary to the Mexico President and Maria Elena Vazquez Gomez, a highly placed functionary in the Ministry of Foreign Affairs, all came to visit her.[481]

During her last illness Frida's allegiance to the PCM never faltered but became only more intense.

Safely ensconced in the welcoming arms of the PCM Frida lashed out at Trotskyism. Frida accused her former lover, Trotsky, of all the sins of omission and commission the PCM could dredge up for Frida to spew out for the edification of the press. In an interview published in one of Mexico City's influential newspaper. Frida explained the Trotsky asylum in Mexico this way,

> Diego said to me, 'I am going to send for Trotsky' and I said to him 'Look Diego, you are going to make a tremendous political error.' He gave me his reasons and I accepted them., My house had just been fixed up. 'El Viejo' Trotsky and 'La Vieja' Trotsky arrived with four gringos. They put adobe bricks in all the doors and windows. He went out very little because he was a coward. He irritated me from the time that he arrived with his pretentiousness and his pedantry because he thought he was a big deal.
>
> When I was in Paris the crazy Trotsky once wrote to me and he said to me, 'Diego is a very undisciplined individual who does not like to work for peace only for war. Be good enough to convince him to return to his party.' I told him that I can't influence Diego at all because Diego is separate from me. He does what he wants to do and so do I. What's more you [Trotsky] robbed me, broke into my house and you stole fourteen beds, fourteen machine guns and fourteen of everything. He [Trotsky] only left me his pen. He even stole the lamp. He stole everything.

It was around this time that Rivera confided to Frida that he had cancer of the penis. He told her he was terribly worried about losing this vital appendage. He was relieved when the cancer responded to radiation treatments and his penis did not have to be cut off, or so he feared the cure would have to be.[482] Frida still worried about him, his immaturity and what foolish things he might do next.

Frida was invited to Russia and desperately wanted to go. She knew she was close to death and since she was now a rabid communist she wanted to see the promised land of communism. She did not want to go without Rivera but he was not invited because he was not as member of the PCM.

Frida was keenly anticipating the celebration of her and Rivera's silver wedding anniversary on 21 August 1954 when they would have been married twenty-five years. Frida agreed to participate in a protest against the ouster of the pro-communist President of Guatemala, Jacobo Arbenz. Allegedly Arbenz had been deposed by the American Central Intelligence Agency's puppet, Castillo Armas.

The 2nd of July was a dreary, rainy day, cold and sooty, as only a dank day at that altitude can be. Frida got out of her sick bed, and with Rivera pushing her wheel chair, took part in her last public communist demonstration. Many thousands of her admirers rushed into the street to touch her, give her a hug, stroke her hair and in many other ways pay her homage.

On Tuesday, 13 July 1954, Frida Kahlo died of a pulmonary embolism brought about by her exposure to the wind and cold in the anti-CIA demonstration. Frida's last words in her diary probably best summed up her bitter philosophy of life and her disgust with the relationship with Rivera when she wrote, "I hope the exit is joyful ... and I hope I never come back. Frida."

Many people came to pay their respects to Frida's body, passing in mourning by her death bed. One friend noted that Frida's body was still warm. As the friend bent over the body to kiss it, goose pimples sprang up on Frida's still warm skin. The woman became hysterical and screamed that Frida was still alive.

Frida's body was placed in an open coffin in the rotunda of the Palace of Fine Arts in the center of Mexico City. All her jewelry had been removed and her body dressed in the simple Tehuana Indian costume she habitually wore. When Rivera saw the hair on her arms stand up he was convinced she was still alive and was not satisfied until the doctors opened Frida's veins and showed him that there was no blood in her system.

When Rivera had received permission from Dr. Andres Idurate, the Director of the Palace of Fine Arts, to have Frida's body laid out in the rotunda, he had promised that he would not use the occasion for a political demonstration. When the first honor guard was stationed, Rivera reneged on his promise and draped a Soviet flag of the hammer and sickle over Frida's coffin. There was subsequent hell to pay but no one removed the flag while the coffin remained in the rotunda. It is said that the Soviets were so pleased by Rivera's political ploy that they ordered the PCM to readmit Rivera into the party.[483]

Frida had always expressed a desire to be cremated. Her body was taken to the Pantheon Civil de Dolores in Mexico City. Shortly after noon on 14 July 1954 Frida's body was carried into the ante-room of the crematorium. John Pablo Saenz, a member of the Central Committee of the PCM eulogized Frida. He took the opportunity to outline for the avidly interested mourners, the communist party's views on the world situation from a Soviet point of view. At 1:15 p.m. Rivera and others lifted Frida's now stiffening body out of her coffin and placed it on the cart that would roll the body into the flames of the crematorium oven.

The assembled crowd, tightly packed into the little crematorium room, sang the "Internationale." At 1:50 p.m. the doors to the flaming, hot oven opened. Darts of flame leapt out of the open doors and the crowd instinctively shrank back. The cart bearing Frida's body was slowly wheeled toward the gaping doors that were filled with flames.

The crowded on-lookers threw themselves on Frida's body, tearing at her clothes, ribbons and all other adornments for a souvenir. They grabbed at everything on the body, anything at all as a memento of their fondly remembered, beloved Frida. There was much pushing and shoving toward the body on the cart. Some mourners fell on the body when they were pushed from behind.

As Frida's body slid closer to the open doors of the oven, the flames seemed to reach out toward her body. Suddenly, her stiffening muscles contracted from the heat and Frida sat bolt upright on the cart. At that instant the flames touched Frida's hair, igniting it and wreathing her head in a bright, blazing halo. It was all too sudden, completely unexpected and extremely frightening. The packed-in mourners, screamed, panicked and stampeded, wildly stumbling over each other in their rush to escape. The uncontrollable, screaming horde crashed through the outer doors of the crematorium, almost ripping them from their hinges, streaming into the open, shouting that Frida was alive!.[484]

Rivera and Siqueiros remained in the little crematorium room apparently immune to the panic around them and unperturbed by the pandemonium. Siqueiros later remarked that as he looked at Frida's peaceful face, surrounded by the brilliant flaming hair, that she seemed to be smiling back at the world she was glad she was leaving. Rivera died less than three years later on 24 November 1957.

SYLVIA AGELOFF

...it poisons the air we breath....

Julian Gorkin (Former Comintern official referring to the Soviet secret intelligence service the GPU.)

Long before the final phase of putting the plan into action there were the primary and secondary phases. The primary phase was to define the parameters of the problems, and consider the various options of action necessary to resolve them. The secondary phase was to flesh-out and put assets onto the tentative plan to determine if they all fit and could be efficiently coordinated. This would include field testing certain aspects of the plan and moving personnel and supplies into the staging area ready for the final phase of implementation. The readying of personnel in the secondary phase would include training such as that experienced in the Spanish Civil War.

The GPU recognized that, crucial to every one of the plans, intimate access to Trotsky to complete the assassination was essential. It could be by a commando-type raid, planting a bomb near enough to accomplish his death or putting an agent into a position where the agent could imbed an Alpenstock in his head. This meant finding someone with current access to Trotsky who could accomplish, preferably unwittingly, the introduction of the murderer technician into Trotsky's entourage.

Louis Budenz was a native of Indiana and a recent convert to the CPUSA. He had been recruited into the ranks of the communist conspiracy from his former position as the secretary of the Conference for Progressive Labor Action (CPLA). He was chosen by the GPU, together with others, as one of the spotters of likely access agents who might have an entre' to Trotsky.

In December of 1936, which was early in his career with the CPUSA, Louis Francis Budenz, was the labor editor of *The Daily Worker*, the newspaper of the CPUSA. He was called to the ninth floor of the party headquarters in New York City. Jack Stachel, a powerful member of the Political Bureau of the CPUSA and a Soviet espionage agent, had telephoned down to Budenz's office on the eight floor and told him to report upstairs for a meeting.[485]

Budenz met with Stachel and Jacob Golos, who was the chairman of the Control Commission of the CPUSA and a principal agent for the GPU. Golos worked under cover as an official of the World Tourist, Inc., a Soviet cover company. There was another man at this meeting whom Budez knew only a Michaels. He assumed Michaels was a Soviet intelligence officer since the others were deferential and obsequious to him.

Stachel told Budenz that he must go with Golos to meet a very important friend from abroad. Budenz was reluctant to leave his work as he was preparing an important editorial for *The Daily Worker* against the CIO (Congress of Industrial Workers, a powerful American labor union). In addition Budenz had a speaking engagement for that evening. Stachel impatiently deprecated Budenz's objections and told him that "nothing could be more important than this assignment." He told Budenz they could get a substitute for his speaking engagement and another writer would finish his editorial lambasting the CIO.

Budenz went with Golos to a Topps restaurant on East Fourteenth street that faced on Union Square not far from the offices of *The Daily Worker*. In a poorly lighted back corner booth, Golos introduced Budenz to a man called Richards. Budenz was immediately aware that Richards was a fictitious name. The man's heavy Russian accent emphasized the fact he was a foreigner unlikely to have an Anglo name such as Richards.[486]

Richards ordered Budenz to gather information about the Trotskyites, supposedly in order to negate their plot to kill Stalin and to invade the Soviet Union. This was during the great purge trials in Moscow. Budenz naively believed the contrived evidence from the trials indicating that Stalin and the Soviet Union were truthfully in grave jeopardy from the Trotskyites. He agreed to help, believing that whatever he could do would further the aims he strove for in the international propagation of communism.

From then on Budenz met Richards in various Childs' restaurants around New York on an average of three times a week; mostly in a Child's restaurant at 86th and Broadway. Budenz, through his many contacts in Trotskyite groups, easily obtained rosters of their membership. He also collected information on the "dangerous left," as Richards referred to the socialists who were followers of Norman Thomas, the persistent candidate for president of the United States on the Socialist Party ticket.

One time when Richards was angry over some trivial incident (Richards frequently threw childish temper tantrums according to Budenz) he threatened that he would throw Stachel out of his position as head of the CPUSA Political Bureau and put Budenz in his place. Richards did not think that Stachel was doing all he could to penetrate the Trotskyites and Socialists.

Not wanting to incur Stachel's wrath, Budenz rejected Richards offer of the promotion. He later learned that both Earl Browder, titular leader of the CPUSA and a GPU agent, and Stachel himself, knew of Richards's offer to Budenz. They called Budenz to a special meeting and Budenz surmised they were curious whether he would accept Richards's offer. The two connivers were overly affable and asked Budnez if there was anything they could do to help him in his work for Richards. Budenz realized that Richards, as a Soviet intelligence officer, was all powerful over all the Americans who were the supposed leaders of the CPUSA. In essence, he realized that Richards could order them to do anything. More importantly, as the leaders of the CPUSA knew, they had to obey Richards's orders.

Budenz continued to meet with Richards according to the pre-arranged plan that was established at each previous meeting. The place for the meeting was pre-set but the time could always be changed by a telephone call using an open code that a stated time automatically meant two hours earlier or later as agreed upon. Richards would call Budnez at his office at *The Daily Worker*, using a fictitious name and make certain Budenz could change his schedule in order to meet with him at the stipulated time.

In the spring of 1937 Richards introduced Budenz to another Soviet intelligence officer whose name was Michaels. (This was not the same Michaels as previously introduced to Budenz by Stachel and Golos.) Both Richards and Michaels impressed upon Budenz that they, all three of them including Budenz, were engaged in a desperate effort to stop the Trotskyites from succeeding in their alleged diabolical plotting against Stalin and the Soviet Union. Budenz believed them and intensified his efforts. He collected all the information he could about the activities of the

Trotskyites, the movements of their clandestine couriers, and the secret relations between the Fourth International with other left wing socialists.

One of Budenz's best reporting sources inside the Trotskyite factions was Bill Reich. In reality, and originally unknown to his Trotskyite companions, Reich was an ardent Stalinist. Reich subsequently revolted against his double agent role because he could not stomach his own betrayal of his Trotskyite friends. He therefore openly announced his Stalinist loyalties in order to discourage his Trotskyite friends from trusting him with their secrets. His mea culpa may have done wonders for his soul but the Trotskyites did not take him seriously. Most of the gullible Trotskyites continued to confide confidential information to him about the activities of the Fourth International which, reluctantly but dutifully, he continued to pass along to Budenz.

In mid-1937 Richards, and the new Michaels, introduced Budenz to yet another Soviet intelligence officer, who was clearly their superior, whose name was Roberts. The Soviet toadies deferred to Roberts on every decision without discussion and took orders without question from him.

Roberts impressed Budenz as being intelligent and almost paternal in his manner. He wanted to meet the loyal Stalinists who were penetrating the Trotskyite organizations on Budenz's instructions. He also wanted to meet any potential defecting Trotskyites who might be turned around into good defectors-in-place before they openly broke with the Trotskyite group and put themselves beyond the possibility of remaining as ostensibly loyal Trotskyites. His intention was for them to progress up the leadership ladder, so that one day they might be leading the Fourth International while actually working for him and the GPU.

Budenz later identified the GPU senior agent he knew as Roberts, as Dr. Gregory Rabinowitz, who was working under cover in the United States as the chief of the Russian Red Cross. Rabinowitz was a physician and an accomplished surgeon as well as a skillful intelligence professional.

Rabinowitz ordered Budenz to introduce him to Sylvia Franklin. She was also known as Sylvia Caldwell and her maiden name had been Sylvia Kallin.[487] Budenz had been sent to Chicago by Rabinowitz who believed it would be easier for Budenz to recruit young communists in Chicago who were not yet known to the Trotskyite organization in New York. The ultimate intention was to penetrate the New York headquarters of the Fourth International which was, in essence, also the headquarters of the worldwide Trotskyite organization.

Rabinowitz had told Budenz to contact Jack Kling, the head of the Young Communist League, as soon as he got to Chicago. Budenz asked Kling to find someone in Chicago who could move to New York and infiltrate the Trotskyite international offices. Kling introduced Budenz to Sylvia Franklin (using the alias of Helen) who was a social worker for the City of Chicago. Her husband, Irving Franklin, had fought in Spain with the Abraham Lincoln Battalion in a secret capacity for the GPU. After that he had been sent to Canada as an espionage agent to help the Soviet intelligence service there.

Rabinowitz came to Chicago to join Budenz and meet Sylvia Franklin. Rabinowitz stayed at the Stevens Hotel (now refurbished as the Conrad Hilton Hotel). It was at the Stevens hotel, when Rabinowitz signed the register, that Budenz saw that his real name was Dr. Gregory Rabinowitz. (Budenz was later able to verify this name with Elizabeth Bentley, the FBI undercover agent in the Soviet intelligence network, who certified the correctness of the identification.)

Sylvia Franklin suggested that a secure place to have dinner with Rabinowitz and Budenz would be the Drake Hotel, since none of her social worker clients would probably be there. The Drake Hotel was, and is, one of Chicago's better hotels.[488]

Rabinowitz arranged for Sylvia Franklin to go to New York. He obtained cover employment for her there with a woman doctor who was also a GPU agent in Rabinowitz's network. Rabinowitz also arranged for Franklin's husband to go to New York and rent a separate apartment in the Bronx so that Franklin could visit him from time to time for connubial purposes.

The reason that the married Franklins had to live apart in New York was because Rabinowitz knew that the Trotskyites preferred that only single women work at their international headquarters. This was a good, but very untypical, example of the thoroughness of the planning by the GPU planners. Sylvia Franklin was provided by Rabinowitz with expense money to rent and furnish an apartment of her own in Manhattan.

Sylvia Franklin went to New York and immediately volunteered to do secretarial work, and every other menial chore they had in the headquarters of the Fourth International, cheerfully and without pay. The Trotskyites were delighted to get such a good worker at no cost to them. Franklin worked hard and gradually made herself indispensable to James Cannon, the executive director of the Fourth International.[489] She became

Cannon's confidential secretary with no security investigation or background check of her past activities in Chicago.

At that time Cannon had an office with two clerks in the front room of his apartment at 126 West 11th Street, on the seventh floor. Reba Hansen, Joe Hansen's wife, took Sylvia Franklin's place as Cannon's secretary when Franklin left in 1947.[490]

Rabinowitz later confided to Budenz that there was not a thing that went on either in the Trotskyite movement in the United States or in Trotsky's own home in Mexico City, that Franklin did not immediately relate to the GPU. Franklin borrowed copies of Trotsky's mail and other important documents for the GPU to copy.

There is no doubt of Franklin's complicity. The GPU agent named Lucy Booker, later identified Sylvia Franklin as the person who brought documents from Cannon's officer to her apartment to be photographed and the copies delivered to Jack Soble, later convicted as a GPU agent.[491]

The Trotskyite headquarters in New York (the same headquarters as for the Fourth International) was also penetrated through the unwitting services of a well-known anti-communist newspaperman named Leigh White. Before he left New York for Chicago, Budenz, under the guise of being a fellow newspaperman, simply asked White whether Trotsky's mail from Mexico concerning his book about Stalin was sent to his publisher in the New York area. White, who was working for Trotsky's publisher at the time, blithely told Budenz that all of Trotsky's mail from Mexico was sent to Sara Wolf in New Jersey. Wolf had been Trotsky's secretary and was now the housewife of a professional man in New Jersey.

Trotsky's publisher, Maxim Lieber, was the founder, together with Whittaker Chambers and John Sherman, of the American Feature Writers' Syndicate, which was a branch of the GRU, the Soviet military espionage service. Lieber was a Soviet agent. On orders from his superiors in Soviet military intelligence he maintained his relationship as Trotsky's literary agent.[492]

Budenz passed this information to Rabinowitz who worked some GPU magic on the way the mail was delivered (probably by bribing the postman). The net result was that the GPU was able to intercept, read and copy all of Trotsky's mail before it was delivered to Sara Wolf and long before Trotsky's publishers saw it.

The GPU also had a technical penetration, or telephone tap, on the home telephone of James Cannon, the head of the Fourth International and the leader of the Socialist Workers' Party.[493] Rabinowitz had recruited Floyd Cleveland Miller, a radio soap opera writer, to monitor the telephone tap. Another GPU agent named Joseph Katz, known to Miller as Joe Gordon, had installed the tap on Cannon's home telephone. Katz ran the wire from the tap in the junction box on the telephone pole into Miller's apartment.

This tap lasted about one year during which time all of Cannon's telephone conversations were transcribed. Miller subsequently joined the Socialist Workers' Party. He was rapidly promoted to military affairs writer for the SWP's magazine, *Fourth International*, and wrote about military matters under the name of Michael Cort.

Budenz introduced Rabinowitz to a New York woman named Ruby Weil. Budenz had known Weil before when they both worked for the Conference for Progressive Labor Action (CPLA). Budenz had been the national secretary of the CPLA before he joined the CPUSA and Weil had been a member of the CPLA. He knew that Ruby Weil had secretly joined the CPUSA. Instead of being assigned to an overt cell of the CPUSA, Weil was sequestered to a secret school for infiltration training operated by Bernard Schuster.

Bernard Schuster was a notorious underground communist agent.[494] He was in charge of communist infiltration of the National Guard of New York and New England. His mission was to prepare the communist cadre within these American military organizations to be a Soviet fifth column to fight for the Soviets in the event there was a war with the United States. This far sighted plan by the Soviet Union was largely ineffective since it never attracted sufficient personnel from among the National Guard to make it a viable possibility.

Schuster trained Weil in infiltration techniques. Ruby Weil had been chosen because she was a close friend of Hilda Ageloff, the sister of Ruth Ageloff, whom Budenz believed had worked as Trotsky's secretary during the Dewey Commission. This was in error, since Ruth had never been a secretary to Trotsky. She had been in Mexico at the time of the Dewey Commission hearings and when they needed someone to type and file she had helped out.

This was as close to being Trotsky's secretary as Ruth Ageloff ever got. It was through this error of GPU operational intelligence that the GPU thought they could get close to Hilda Ageloff because Weil already knew Hilda.[495] Then, the plan called for Weil through Hilda to get close to Ruth whom they believed had been Trotsky's secretary.[496] All three Ageloff sisters, Hilda, Ruth and Sylvia, were members of the Trotskyite Socialist Workers' Party. Their importance to the Trotsky movement appears to have been overestimated by the GPU.[497]

Trotsky had considerable affection for Ruth Ageloff. She had not lived in the Blue House while working for the Dewey Commission but for several weeks had visited almost daily to share the life and work there. Trotsky had a very favorable impression of Ruth and any sister of hers would be well received by him.[498]

Rabinowitz discovered that Sylvia Ageloff, a Brooklyn social worker was a sister of Ruth and Hilda Ageloff.[499] He also learned that Sylvia usually took her vacations to coincide with needed secret courier work for the Fourth International.[500] Budenz introduced Rabinowitz to Weil as John Rich.

Rabinowitz gave Weil a large sum of money through Budenz so she could dress up for her new assignment of getting acquainted with the Ageloff sisters. He also gave Weil money to pay her arrears of telephone service so that Rabinowitz could call her to arrange meeting times and places. Rabinowitz told Weil that her new assignment would help stop the Trotskyite plans to murder Stalin.

A rumor reached Rabinowitz that Sylvia Ageloff might be planning a vacation in Europe. He instructed Weil to convince Sylvia Ageloff that Weil should go to Paris with her. The Trotskyite International Congress, the Fourth International, was to be held in Paris. Rabinowitz had received information that Sylvia Ageloff might attend this conference.[501] Weil succeeded in convincing Sylvia that Weil should accompany her to Europe. Weil's ostensible reason for wanting to go to Europe was to visit her sister, Corine, in England.

Weil balked at one point during the preparation for her journey with Sylvia and Rabinowitz believed that the operation might be finished. Budenz, by then, had moved to Chicago to be the editor of the *Midwest Daily Record*, a communist newspaper directed toward midwest readers. Rabinowitz ordered Budenz back to New York to try to convince Weil she must continue with the operation.

Budenz talked to Weil in the Grand Central Terminal in New York between trains. He had only from 10:30 to 11:30 p.m. to convince Weil of the righteousness of her going to Europe with Sylvia. After Budenz picked her up at Peoples' Press where she worked they returned to the station. Budenz was able to persuade Weil to continue with the plans to accompany Sylvia Ageloff to Paris. Budenz was then able to catch the 11:30 p.m. train back to Chicago.

This was poor planning on the part of the GPU. They should have had another case officer already introduced to Weil who would have established the rapport with her so that he could talk to her like a father. This is a good example of the necessity of prior planning of agent handling. There should not have been any reason that Budenz, once eased out of the case officer/agent relationship, should have to be brought back to talk with Weil.

Ruby Weil's reluctance to continue the cultivation of the Ageloff sisters resulted from the fact that some of her friends, who had known her as a overt member of the CPUSA, had seen her with the Ageloff sisters. These loyal comrades had reported Weil to the party's disciplinary committee for consorting with non-party people which was tantamount to being disloyal to the CPUSA. If she was dedicated communist, she obviously would not be associating with Trotskyites. This was a crime that called for severe censure and probable expulsion from the CPUSA.

Weil did not want the charge of disloyal behavior on her party record. Budenz had foreseen this possibility when he was in charge of the operation. He had told the section organizer of Weil's cell that she was to be engaged in secret party work and therefore exempt from routine discipline. This word had apparently not filtered down to the lower ranks where it would have protected Weil from any charges of disloyalty.

Rabinowitz sent Weil to see a member of the underground CPUSA by the name of Gertrude Allison who lived in Greenwich Village.[502] The purpose of the visit was for Weil to know, and be able to recognize Allison when she met her again in Paris. Weil was told that when she and Sylvia arrived in Paris, Allison would introduce her to a male agent whose task it would be to charm Sylvia.

Gertrude Allison worked at a Marxist bookstore in Greenwich Village named the Jimmy Higgins Bookshop. Allison was married to Eduardo Machado, one of the founders, together with his brother Gustavo Machado, of the Communist Party of Venezuela. In the 1920s Eduardo Machado had been very active in the Anti-Imperialist League in New York City which was a GPU front organization. Eduardo had twice been deported from the United States for subversive activity.[503]

Unable to return to the United States he and Gertrude had gone to the Soviet Union in 1932. The couple stayed in Russia for several years during which time Gertrude made several secret courier runs back to the United States for the GPU. Later Gertrude lived, off and on, in Paris in 1938 and 1939.[504]

In the days before she sailed for Europe with Sylvia, Gertrude Allison gave Weil detailed instructions on how to prepare Sylvia for the meeting with the man in Paris. Weil was to harp on Sylvia so that she would be anticipating an adventurous, exciting and romantic sojourn in Paris. Allison then proceeded Weil and Sylvia across the Atlantic and was waiting for them in Paris.

A few weeks before they were to leave for Europe, Weil had told Sylvia that her sister, Corinne, who lived in England, had sent the money for her passage. Since she was free and temporarily out of work, she would therefore be able to go to Europe on the ship with Sylvia as planned. Sylvia, as she later testified before Congress, was not overjoyed at the prospect of having Weil as a traveling companion but reluctantly acquiesced. They left on the S.S. Normandie on 15 June 1938.

When they arrived at Southampton, England, in June of 1938, Corinne met the ship. Weil and Corinne went up to London by car but Sylvia preferred to take the boat train. Sylvia had told Weil what hotel she would be staying at. In a few days Weil and Corinne contacted Sylvia at her hotel. They told Sylvia that she was very lucky since they had both decided to travel with her to Paris. Again, Sylvia was not too happy with this arrangement but saw no harm in it, anticipating that she could leave Weil and Corinne to their own plans in Paris.[505]

When they arrived in Paris, they all stayed at the Hotel St. Germain d'Pres. Weil, full of surprises, then told Sylvia she knew a comrade in Paris named Gertrude who was supposed to know all about how two young New York girls could have a good time in Paris. Sylvia later testified before Congress,

> This Gertrude knew somebody in Paris that she had been friendly with who was a young student who used to visit Gertrude at her apartment in Paris. She was going to get in touch with him and have him come over. So she called him and he came around to the hotel.

This charming gigolo, Jaime Ramon Mercader del Rio Hernandez, would eventually be Trotsky's assassin. At the time he was introduced to Sylvia he was using the name of Jacques Mornard.[506]

Mornard/Mercader began immediately to romance Sylvia.[507] He launched an intensive campaign to sweep her off her feet. She succumbed to this elaborate GPU plan and they became bedded lovers.[508]

Jacques Mornard was an ideal escort for a lonesome American girl in Paris. He was tolerably handsome, with GPU financing he was impeccably tailored.[509] He also had money to take Sylvia to the best restaurants, plays and could buy her expensive gifts. He studiously avoided discussing politics with an indifference that aroused no suspicions. Sylvia, ordinarily a thinking, politically-oriented person, was too blinded by adoration of her suave lover to notice his camouflage of political naivete.[510]

Mornard/Mercader affected the style and trappings of the son and heir of a wealthy aristocratic Belgian family. His nonchalance gave the impression of a man-about-town with time on his hands and no reason to worry about anything except where his next divertissement was coming from. He also had an apparent passion, ostensibly at least, to fulfill Sylvia's every dream of what a prince charming should be.

There was one thing about Mornard's/Mercader's life style that did bother Sylvia a little! The cover that the GPU had contrived for Mornard, obviously without adequately briefing him on how to live it, or having taken steps to properly backstop his cover, made Sylvia vaguely suspicious. Mornard's cover, in addition to being a man about town, was supposed to be that he was studying journalism at the Sorbonne University. He was also supposed to be writing free lance articles about sports in his spare time to augment the bountiful allowance provided to him by his wealthy parents.[511]

What made Sylvia suspicious was that while he was supposed to be writing about sports, he never took her to any sporting events. Normally he should be covering athletic events and interviewing athletes if he was to report on them in his free lance articles. However, he never went to any sporting events alone. He was always with Sylvia, wining, dining and courting her from early morning to the next early morning and sometime spending the night with her. They never went to any cafe or other place where the sporting crowd congregated or where sporting events were discussed or bet upon. Mornard never spent any time writing, much less doing any research for his articles on sports.[512]

The preparations for the Fourth International meeting in Paris were proceeding apace. The principal organizer for this meeting was Rudolph Klement. A Jewish emigre from Germany, Klement had been Trotsky's secretary at Barbizon in France and now held the position of organizational secretary for the forthcoming meeting.

In the middle of these important preparations to which he was devotedly dedicated, Klement mysteriously vanished from his home in Paris. The police speculated that he left his lodging on his own, or was kidnapped, just before breakfast since this meal was prepared on the table but untouched.

About two weeks later, Trotsky in Mexico received a letter ostensibly written by Klement from Paris but unaccountably postmarked from New York, The letter was signed with an old pseudonym of Klement's that he had long since discarded in correspondence with Trotsky. If Trotsky needed any evidence to condemn the letter as a forgery, and a very poor one at that, the use of the out-dated pseudonym convinced him.

This alleged letter of Klement's denounced Trotsky's alleged cooperation with Hitler and the Gestapo and vehemently proclaimed his break with Trotsky. The letter contained so many mistakes of fact that the real Klement would have known to be false. Trotsky concluded the letter had either been written by an imposter or by Klement under duress and torture. Copies of the letter had been sent to several Trotskyite politicians in Paris. Since the letter was thus receiving wide circulation Trotsky had to take a public stand on it and denounce it.

Trotsky therefore proclaimed loudly and stridently, "Let Klement, if he is still alive, come forward and state before the judiciary, the police or any impartial commission everything he knows. One can foretell that the GPU will in no case let him out of their hands."

A few days later, on 16 July 1938, Klement's body was found, maimed and mutilated, with the head removed from the torso with surgical precision. The body was found in Paris washed ashore on the banks of the Seine river. The arms and legs had been meticulously severed from the trunk by someone with considerable knowledge of anatomy and surgery.

The attitude of the writer of the alleged Klement letter was that of a disillusioned follower, breaking with Trotsky over principles. This was the same attitude adopted by the writer of the letter some two years later that was found on Trotsky's assassin immediately after his arrest.

Sylvia Ageloff was not an official representative to the Fourth International Conference that Klement had been organizing. She did attend several sessions as an interpreter. Sylvia spoke flawless Russian, French, Spanish and, of course, her native American English.

Jacques Mornard usually affected a nonchalance bordering on boredom concerning things political but begged Sylvia to take him with her to the opening meetings of the delegates to the Fourth International.[513] At these preliminary social functions Mornard met many of the more important delegates such as James Cannon, the staunchly loyal Trotskyite who was the American leader of the Fourth International and the Trotskyite Socialist Workers' Party.

Ironically Sedov assigned Mark Zborowski, the GPU/NKVD penetration of his entire operation, to escort the delegates to the meeting of the Fourth International that was supposed to be kept secret from the NKVD. Zborowski later admitted that he had kept the NKVD fully informed on all delegates and deliberations of this "secret" meeting.[514]

Mornard/Mercader also met Marguerite and Alfred Rosmer, two of Trotsky's closest personal friends and both prominent literary personages in France. Mornard would again meet the Rosmers in Mexico when they were houseguests at Trotsky's home in Coyoacan but then he would be using the name of Frank Jacson.

There have been accusations that the Rosmers, Alfred in particular, should have been suspicious of Jacques Mornard, especially when he met him the second time, when they were so intimately together on long automobile trips. Trotskyites claim that Alfred should have been suspicious of Mornard's claim of Belgium origin. Mornard did not speak French with a native fluency that someone raised in Belgium would ordinarily have.

Moreover, the Rosmers accepted Mornard, at least into the fringes of the highly confidential conference being held in their home on the outskirts of Paris, without any question or attempt to investigate his bona fides. Their personal acceptance of Mornard in Paris, and later as Jacson in Mexico, may have been the thing that partially allayed any suspicions that Trotsky might have had about Mornard/Mercader's bona fides. The

GPU would have hoped for this gradual acceptance of Mornard/ Mercader by Trotsky's friends on the fringes of his intimate circle. Thus, Mornard would later be accepted into the center of the inner circle of Trotsky's friends which would put him well within striking distance for the lethal blow.[515]

Shortly after Sylvia and Mornard became lovers[516], Mornard had to leave Sylvia, this would be in late July of 1938. Mornard claimed that he must return to Belgium because his mother had been in a serious accident.

On 26 July 1938, Mornard wrote a letter, supposedly from Belgium, to Sylvia explaining everything,

> Sylvia darling, As I told you, my brother had given me a very bad version of my mother's accident. The thing happened like this: my father and mother were coming from Ostend to spend a day in Brussels. They were in a car driven by a chauffeur and just getting to Brussels, my father had the car stopped for a necessary reason, and it was then that the car was hit by a ten-ton truck, killing the chauffeur and very seriously injuring my mother, My father, not having been in the car, was unhurt. My mother was operated on again yesterday (the second operation in three days) and the doctors say the greatest danger is past but they cannot be sure for a few days.

While Mornard was supposedly visiting his injured mother in Brussels, Sylvia took a sightseeing excursion to Prague. She intentionally returned to Paris through Brussels where she had made arrangements by letter with Mornard to meet him there. They had planned to travel back to Paris from Brussels together.

In place of a love struck Mornard welcoming Sylvia with outstretched arms and a bouquet of flowers, a very cold business-like woman met Sylvia in Brussels at the time and place her lover had specified in his letter. The woman told Sylvia that Mornard had been suddenly called to England. Disappointed and bewildered Sylvia returned to Paris alone.

She received several letters from Mornard after she returned to Paris. She was about to give up the whole Mornard episode as a bad learning experience and return to America. Suddenly, in early September,

Mornard reappeared. He explained to Sylvia that he had actually been in Brussels at the time of her visit but had been under military arrest for refusing to serve in the Belgium army.

Sylvia accepted Mornard back into her life. The couple continued to live together in Paris for some time until Sylvia's work-ethic conscience broke the idyllic reverie. Sylvia announced to Mornard that she must return to New York to look for work. She reminded him that she had resigned her job as a social worker for the city of New York when she left on her European vacation.

The GPU arranged for Mornard to tell Sylvia that she need not return to New York to get a job. (The GPU wanted to keep Sylvia in Paris in order to give the romance more time to become settled so that Mornard would be accepted without question as Sylvia's common-law husband and thus more acceptable to her Trotskyite friends.) Mornard proudly told her that he had obtained work for her right there in Paris with his publishers, the Argus Publishing Company. Therefore, he reasoned with her, they could continue to live together in Paris. Mornard did receive correspondence from the Argus Agency who published a magazine, *La Auto*.

The Argus enterprise was also the company that was allegedly buying the nebulous articles on sports that Mornard was supposedly writing. The arrangement for Sylvia was that the Argus editors would pay her 3000 French francs a month for each story of undetermined length each week as long as it was about some aspect of psychology. Mornard said he had suggested the subject because he knew that psychology had been her major at Columbia University under the famous Dr. Hook.

Sylvia could choose the topic of her weekly article and, within each week's loose deadline could write at her leisure. Mornard added that there were some conditions. First. the Argus editors did not want any direct contact with Sylvia and secondly, she would never know what publications were to print her articles.

Strange as it may seem, blinded by her desire to remain with her lover in Paris, Sylvia never questioned these preposterous conditions. Had she applied a modicum of objective reasoning to these proposals she would have realized that she had no marketable reputation as a professional writer. The Argus editors had never read anything she had written. Further, why were these hard-nosed business editors willing to pay her, an unranked, unheard of, amateur psychologist 3000 francs a month? This was more than any editor would pay the leading French psychologist for an authoritative article on a subject which would command genuine reader interest.

Sylvia did not want to be sensible. She was in love and wanted only to believe those things that would make that love endure and be reciprocated. She therefore believed that a commercial publishing house would not want to meet their writer and would not want to offer their writer guidance and constructive criticisms. Students of this episode have asked why Sylvia did not ask a librarian or the Economic Attache in the American Embassy for information on the Argus Publishing Company. The answer is, that subconsciously she probably did suspect the proffer was bogus but she did not want to know the truth.

After an idyllic period of living together in Paris, Mornard surprised Sylvia with the announcement that he had obtained a job as an American correspondent for a Belgian newspaper. He wanted her to proceed him to New York. He promised that he would meet her there as soon as he completed the details about the job. Mornard was vague as always, with no specific newspaper mentioned, but once again Sylvia accepted his story at face value.

Accordingly, in February of 1939 Sylvia left Paris and sailed back to New York expecting Mornard to follow her in a few days. Instead she received a cable from him telling her his departure from Europe had been delayed. His follow-up letter explained that he was unable to get a United States visa. Sylvia commiserated by letter with her lover over this unexpected obstacle to their happiness together and once again she believed his story.

Of course Mornard's story was all a lie! If Sylvia had been at all perceptive she would have realized there was absolutely no valid reason why the United States would refuse to grant a visa to the son of a Belgian diplomatic family, unless he was a criminal or a communist. Mornard had told Sylvia on many occasions about his noble lineage. He claimed that his father had been the Belgian Ambassador to Teheran during World War One.

The reason that Mornard tarried in France instead of immediately following Sylvia to the United States was because his real mother, Caridad Mercader, had arrived in France from Spain. In 1939 the Spanish Civil War had come to a bitter end for the Republicans. With their defeat there had been a general exodus from Spain to France of thousands of communists who had lost their identities and all their worldly possessions along with losing the war. Some were trying to reach Moscow. Others, some of whom were part of the GPU assassination team, were mingling with the legitimate refugees on their way to resettlement in Mexico via Paris.

When the Ile de France sailed into New York harbor on 9 September 1939, a few days after the outbreak of World War Two, Ramon Mercader, the former Jacques Mornard was aboard with yet another identity as Frank Jacson.[517]

As soon as Jacson/Mercader cleared through the United States Customs he presented himself at Sylvia's home at 50 Livingstone avenue in Brooklyn. He explained his new identity by telling Sylvia that he had to flee Europe to avoid being conscripted into the Belgium army. He claimed the only way he could evade the emigration and military authorities in Belgium was to use a false identity.

The two lovers resumed their common law marriage and moved into an apartment in Greenwich Village in New York that was loaned to them by a Trotskyite friend of Sylvia's.[518] There they enjoyed connubial bliss while Jacson/Mercader searched for a job.[519]

The passport Jacson/ Mercader had under the name of Frank Jacson, was a genuine but altered Canadian passport, No. 31377. It had been issued originally by the Canadian government in Ottawa on 22 March 1937 to one Tony Babich, a Yugoslavian-born naturalized Canadian citizen. Babich had been born on 13 June 1905 at Lovinak, Serbia and naturalized as a Canadian citizen in 1929. As far as is known, Babich never had any connection with the GPU while he was alive. He had made the mistake of going to Spain to fight with the communist Republicans in the International Brigade.

Babich had left his Canadian home on the pretext of going to visit his family in Yugoslavia but instead went to Spain. At that time the Canadian government prohibited its citizens from traveling to Spain so Babich had to lie in order to leave Canada legally for Spain.

In Spain Babich had dutifully surrendered his passport to the GPU at their insistence as they made all members of the International Brigade surrender their passports. According to the official reports of the Republican government, Babich had been killed in action at the front in the early part of 1937, this meant he had been killed almost the day he arrived in Spain.

Unfortunately Ramon Mercader's Canadian passport as Frank Jacson had been badly botched in the alterations. An inept, typically unworldly, GPU automaton technician who knew nothing of the English language, but only the cyrillic alphabet, made a crucial phonetic error. The passport should have been altered to Jac*k*son with the K but this did not seem right to the technician who was grounded in the cyrillic and he decided to omit the K.

The GPU, if it realized the mistake, had to proceed with the passport altered with the name Jacson. They knew they had to use either a United States or Canadian passport in order to avoid having to apply for a United States visa where there might be other more compromising questions.

Having arrived in the United States undetected, Jacson had completed the first stage of his journey. He was now close enough to his Mexican designation to be able to move quickly when his orders came from the GPU. He told Sylvia that he had purchased the Frank Jacson passport on the black market in Paris for $3500. He said his mother, who was equally fearful of his being called into the Belgian army, had given him a total of $10,000 to buy the passport and pay the expenses of his trip to the United States.

The original GPU plan had been for Jacson/ Mercader to stay in the United States, specifically in the New York area, for several months before going to Mexico. He was to re-establish himself with Sylvia and, in general, pick up area knowledge and build genuine cover that would be backstopped.

The GPU believed they had a communist contact in the personnel office of the New York World's Fair who could get Jacson/Mercader a job there as a guide or driver. However, something had gone wrong with this plan, so the GPU decided to speed up their plan and move Jacson/, Mercader to Mexico ahead of the scheduled time.

On GPU instructions Jacson/Mercader told Sylvia he had been offered a good job in Mexico. The alleged job with the Belgium newspaper that had prompted the whole idea of the move to the United States was now forgotten by Sylvia. The job in Mexico was allegedly purchasing critically needed war materials for shipment to Britain and France to aid them in the war effort against Nazi Germany. His employer in Mexico, he told Sylvia, was a Peter Lubeck who had agreed to pay him

$50 per week. Lubeck, according to what Jacson/Mercader told Sylvia, was the manager of a large British import firm.

Jacson/Mercader spent about a month with Sylvia in their Greenwich Village apartment. He gave her $3000 to keep for him of the money he had supposedly received from his mother. He told her to take the money to him when she came to visit him in Mexico.

Again we are confronted with Sylvia's incredible gullibility. If Jacson/Mercader's passport story was not sufficiently unbelievable, read on about some of Sylvia's doubts. Sylvia testified before a Congressional Committee on this aspect of her relationship with Jacson/Mercader,

> When he told me that on the boat he was telling everybody what was Lower Manhattan, and so forth, I said, 'How could you tell them, you didn't know?' He said he had been so interested he had studied pictures and everything. When I thought it over, I figured her must have been in New York before, although he said he had not.

On 6 October 1939 the Mexican Consulate in New York issued Jacson/Mercader a tourist card. He entered Mexico at Nuevo Laredo, across the Rio Bravo from Laredo, Texas on 12 October. He signed his name, as he had been told to by the GPU who did not want to give the border crossing authorities any more than the barest information, merely as "F. Jacson." He gave his home address as 1260 St. Davis street in Montreal, Canada and listed his occupation as a mechanical engineer.

Sylvia became tired of waiting for Jacson/Mercader to get settled in Mexico and send for her as he had promised. She did not want to spend any of the $3000 he had entrusted to her for safe keeping but she still needed living expenses. She took a job with the New York Home Relief Bureau, as a social worker.

While Jacson/Mercader was biding his time in Mexico waiting for the GPU to decide on his business cover and where his office should be located, he took up his favorite sport of mountain climbing. He had been told by the GPU to acquire local area knowledge while waiting for their decision. He toured Mexico City and the surrounding area like a typical tourist.

During these rounds he became friendly with a Sr. Patino of the Pan American Airways office in Mexico City. Patino was an accomplished mountain climber with considerable expertise. He listened to Jacson brag of how he had climbed and conquered Mont Blanc in Europe and other impressive heights.

In November Patino organized a climb of the local mountain climbing club to go up Popocatepetl. This was one of the beautiful snow covered peaks, together with Iztacihuatl that could, in those days before the smog, be clearly seen from Mexico City. Jacson appeared for the climb in full climbing regalia. According to Patino he was overloaded with useless equipment as an amateur would be who did not know much about mountain climbing. With this load Jacson tired long before the group reached the summit. He had to be left at a rest station midway up the slope in a completely exhausted condition.

In December, Leonid Eitingon, the GPU case officer told Jacson/Mercader that he could purchase an automobile since the group might need one for later in the Trotsky operation. On 7 December 1939 he purchased a 1937 four door beige colored Buick sedan from the Hernandez Agency at the corner of Morelos and Iturbide streets.

Not being able to wait any longer, Sylvia arrived in Mexico City in January of 1940 eager to resume their relationship. Jacson feigned delight in having her with him again and they took up residence in an apartment at Hamburgo 159 in Colonia Juarez in the center of Mexico City. Eitingon had instructed Jacson/Mercader to resume the common law marriage. Sylvia's presence in Mexico represented one of the few remaining entres to Trotsky's inner circle that the GPU could depend upon, if the planned 24 May raid by Siqueiros failed.[520]

There was a slight change in Jacson/Mercader's attitude regarding politics in general after Sylvia's arrival in Mexico. He grudgingly showed a beginning awareness of some things political for her sake.

As part of his new political indoctrination Jacson/Mercader affected a liberalist's inquisitiveness for all different political views. He still strongly espoused, for Sylvia's edification, a loyalty to Trotsky's basic communist philosophy. To demonstrate the strength of his resolve to Sylvia and his invulnerability to opposition views, he insisted that Sylvia accompany him to an anti-Trotsky public meeting of the Mexican Communist Party.

James Ford, who was the vice presidential candidate on the CPUSA ticket in the United States, and the most prominent token black in the party, was to speak at this PCM rally. Ford's speech was vituperatively anti-Trotsky. Sylvia sat through the meeting with clenched teeth, hating it more each minute and feeling ashamed that she was listening to such filth about her idol Trotsky.

Trotsky had prophesied that for Stalin to order his murder, he must have the complete subservience of the local communist party. The Mexican Communist Party was headed by Hernan Laborde, an anarchist and poet of some small repute. Laborde was generally regarded as a gentlemen, a man of his word and a humanitarian. Laborde enjoyed affecting the demeanor of the popular image of the Bolshevik revolutionary. He appeared at party meetings armed with a brace of pistols, in a long black cape and looked very much like the arch-typical conspiratorial Bolshevik. As a crusty individualist he resented Stalin's interference with the PCM through the Comintern representatives assigned as watchdogs.

Laborde had guided the PCM into accepting the popular front philosophy, originally decreed from Moscow. Therefore the PCM had enthusiastically given their support to the leftist regime of Mexican President Lazaro Cardenas. A concomitant of accepting Cardenas's ideology was to avoid at all costs a break with his government. This included giving continuous lip-service to Cardenas's principle of political asylum in Mexico for all those who sought it, and that included Trotsky.

The GPU knew that in order to have a cooperative communist party supporting Trotsky's liquidation instead of his political sanctuary, it was necessary that Laborde and his like-thinking cohorts be purged from party leadership.

James Ford, the black leader of the CPUSA was a member of the purge commission sent by Stalin to oust Laborde's group from the PCM leadership. Another member of the purge commission was Vittorio Codovilla, the Comintern's principal trouble shooter in South America. The presiding judge of the purge commission was Vittorio Vidali, who was then operating under the alias of Carlos Contreras, previously noted as the GPU's arch assassin.

The PCM held a secret meeting in March of 1940 to consider the issue of "The Struggle Against Trotskyism and Other Enemies of the People." It was a one-sided debate since the purge commission had rigged the decision. Laborde and his followers were expelled from the PCM. A new leadership was announced by the purgers that had no scruples against the extermination of Trotsky, regardless of the Mexican president's assurances to him of asylum and sanctuary.

Sylvia had introduced Jacson/Mercader to the Rosmers at their house in Paris at the time of the meeting of the Fourth International.[521] The Rosmers were in Mexico visiting the Trotskys on a purely social basis as old friends of the family. The Rosmers had escorted Trotsky's grandson, Seva Volkov, to Mexico from Paris the previous summer. The two couples, the Rosmers and Sylvia and Jacson/Mercader, renewed their acquaintanceship and dined together on several occasions. They also took long drives in the country in Jacson/ Mercader's newly acquired Buick automobile.

As soon as she arrived in Mexico City, Sylvia confided to the Rosmers that the man they had known in Paris as Jacques Mornard was in Mexico using a false Canadian passport under the name of Frank Jacson. She told them the whole story of why he had to travel under a false passport because he was a fugitive from the military conscription in Belgium.

Sylvia asked Marguerite Rosmer to check on room 620 in the Ermita building to see if that was really Jacson/Mercader's office. Sylvia's sister had gone to room 820 which Jacson had originally told Sylvia was his office and learned that no room 820 existed in the Ermita building. Jacson had temporized that his office was not in room 820 but in 620.

Marguerite Rosmer went to room 620 and was greeted by an office boy who told her that the room was in fact Jacson/Mercader's office. The boy neglected to tell Marguerite that room 620 was also the office of David Alfaro Siqueiros.[522]

This cross-contamination of cover offices was a severe breach of operational security by the GPU. It was a prime example of typically insecure GPU planning. Regardless of whether or not Eitingon, or anyone else had or had not known about the cross-use of the office. The agents should have routinely been instructed that under no circumstances should they undertake such a commitment without prior permission. (Eitingon should have been more on top of things like this!)

Having told the Rosmers that her "husband" was in Mexico on a false Canadian passport, Sylvia was concerned that they would tell Trotsky. She decided that before Trotsky heard about it from the Rosmers that she had better tell him the entire story.

The next time she was in Coyoacan, she told Trotsky about Jacson/Mercader's trouble with the Belgium military authorities and his consequent need to use the false passport. She promised Trotsky that she would bring her husband to the house to see him. Sylvia was concerned that if Jacson/Mercader became known as being on friendly terms with Trotsky, and he was subsequently discovered to be illegally in Mexico, that guilt by association could cause Trotsky serious consequences.[523]

Sylvia had to return to her job as a Brooklyn social worker and this meant leaving Jacson/ Mercader alone in Mexico. Therefore, to preserve Trotsky's innocence from being contaminated by Jacson/Mercader, Sylvia made him promise that while she was away, he would not visit the Trotsky house. Jacson/Mercader promised Sylvia that, while she was in the United States, he would not go near the Trotsky house.

DIGRESSION TO JACK SOBLE AND FRIENDS

> Intelligence is like kippers. Individually each tiny bone amounts to little but all together are of immense importance to the kipper.... nothing we find and put together is valueless. All should be carefully documented and studied."
>
> Anthony Glees, *Secrets of the Service*, p.303

This is a digression — a detour for a moment from the main story of Trotsky but still very intimately related to the topic of Trotsky's life. The tentacles of the theme, the newly turned up tangential avenues that must be investigated for completeness, all bear an importance to the body of the story about Trotsky.

What is of major importance to the counterintelligence aspects of the Trotsky case is the later use of GPU assets, i.e., were not there prohibitive security ramifications against using a GPU agent again, who had been blown in the Trotsky operation?

In order to illustrate these considerations, the life and metamorphosis of Abraham Sobolevicius to Jack Soble, and the people he touched, presents an instructive chronicle. Although this chapter is labelled "Digression" it continues to add substance to the Trotsky story and should not be skipped.[524]

In 1927, Abraham Sobolevicius became a GPU agent. In 1931, under the name of Abraham Senin, he penetrated the inner circles of Trotsky's home on Prinkipo Island. He came at Trotsky's invitation because he had inveigled his way into the German Trotskyite organization and Trotsky thought him to be a loyal follower. As a GPU agent living by Trotsky's side he reported to his superiors in the GPU the most intimate details of Trotsky's life, his motivations and his plans for future operations inside the Soviet Union.

Almost a quarter of a century later this same man, now known as Jack Soble, was arrested on 25 January 1951 at his apartment in

Manhattan. At 6:30 a.m. on that date seven FBI agents arrested Soble, supposedly a respectable businessman selling pig bristles, and his wife, Myra, for espionage against the United States. The Sobles were each held in lieu of $100,000 bail.

The two brothers, Abraham and Ruvin Sobolevicius were Lithuanian Jews who, using the names Abraham Senin and Roman Well, first appeared as members of a German Trotskyite group in Leipzig in 1929. It was not known until later that both men had been recruited by the GPU in 1927 and specifically trained to infiltrate Trotskyite groups.[525]

On 26 April 1930, Abraham Senin wrote to Trotsky that he was taking a vacation with his wife away from his work with the Trotskyite group in Berlin. With admirable candor, Senin told Trotsky that his wife, who was a Soviet citizen and a member of the Communist Party was employed by the Soviet trade commission in Berlin. (The GPU felt this forthright admission of Senin's wife's party affiliation would appeal to Trotsky's sense of honesty.)

Senin added, in order to clinch Trotsky's attention with his feigned honesty, that his wife was not known to be the wife of a Trotskyite. She would no longer be able to keep her job with the Soviet trade commission if the truth about her husband's political affiliations were known.

The two brothers had risen rapidly in the German Trotskyite organization and they helped Trotsky's son, Lyova, in Germany with the dissemination of the *Bulletin of the Opposition*. In August of 1931 they both went to Prinkipo Island where Trotsky had his exile headquarters to visit with him. The Sobolevicius brothers (the future Jack Soble and Dr. Robert Soblin) had access to Trotsky's codes, secret inks and cover addresses used to correspond with Trotsky's supporters — all betrayed to the GPU together with the lists of Trotsky's supporters inside the Soviet Union.[526]

After they left Prinkipo the German Trotskyite movement, under the brothers' tutelage, became disorganized and ineffectual. In January of 1933 they published a false issue of the German Trotskyite newspaper, asking that the organization return to Stalinism.[527] Just before the Moscow purge trials in 1936 the brothers disappeared from active GPU operations in the west. About that time, late 1935, Senin was in Moscow helping the GPU in their reprisals against the relatives of Trotskyites still living in Russia.

Subsequently, in 1940 Abraham Sobolevicius (Senin) was sent to the United States by the NKVD as an espionage agent under the name of Jack Soble. He came to the United States on the eve of Pearl Harbor by way of previous service for the GPU in Japan. He arrived in America only a few weeks ahead of his old collaborator Mark Zborowski.[528] Soble held the rank of colonel in the Soviet hierarchy and had served with the GPU on the Republican side in the Spanish Civil War. His wife Myra was his co-worker in his espionage activities for the Soviets. Their son, Larry, had been named in honor of Lavrenti Beria.

Lavrenti Beria, the chief of the NKVD, and the number two man in the Soviet government, sent Soble to the United States to infiltrate the Trotskyite, and Jewish and Zionist organizations in the New York City area. Beria had told Soble, "Comrade Stalin remembers your name and the service you performed regarding that dastardly enemy of the state, Trotsky."[529] Soble was able to save the members of his huge family and bring them out of Russia and get most of them to the United States. Robert Soblen, his brother who was also a GPU agent, came to the United States about this same time.

In the United States Soble worked directly for Vassili Zubilin, ostensibly the Third Secretary of the Soviet Embassy in Washington but, in reality, he was the chief of NKVD operations for the entire United States.[530] He told Soble that there were three Trotskyite groups in the New York area and that the NKVD already had an agent as a leader of each group.

Soble said he met the three Trotsky agents, each at different restaurants, to get their reports on the names, addresses of members, their activities and trends of the Trotskyite movements.[531] Soble not only knew the Trotskyite sympathizers in the United States but he also learned all of Trotsky's followers in Russia.[532]

Floyd Cleveland Miller was an admitted Soviet espionage agent who turned informant for the FBI. Miller identified Lucy Booker as his cut out to Jack Soble, identified by Miller as a Soviet principal agent. Elizabeth Bentley was told by Jacob Golos that Lucy Booker had done work for the Russian espionage system in Germany and the United States. Soble told Miller that Booker was thoroughly reliable for confidential work.[533]

In order to appreciate the extent of Zubilin's activities, the story must jump ahead to 29 March 1943 (well after the Trotsky assassination) when a scientist working at the top secret Manhattan project at the radiation laboratory at Berkeley, California visited the home of Steve Nelson.[534] An honor graduate of the infamous Lenin Institute in Moscow, Steve Nelson was a member of the National Committee of the CPUSA and a veteran of the Lincoln Battalion of the International Brigade of the Republican side of the Spanish Civil War.[535]

The scientist told Nelson that the envelope he was giving him contained a complicated formula that revealed the top secret military use of atomic energy. Nelson was next observed giving an envelope, believed to contain the top secret formula, to one of his regular contacts, Peter Ivanbov, the vice consul of the Soviet Consulate in San Francisco.

About ten days later, on 10 April 1943, Nelson received another visitor who gave him $50,000 and was heard to say to Nelson the payment was for "an assignment well done." The FBI surveilled Nelson's visitor on the train across the United States from California to New York where the visitor entered the Soviet Consulate General in New York City. The visitor was later identified as Vassili Zubilin.

In 1944 Zubilin directed Jack Soble to take over the operational direction of the Boris Morros Music Company of Hollywood and New York. This company, although overtly legitimate, was a cover company for Soviet espionage. The Boris Morros Music Company had an extensive clientele of prominent recording artists who were unaware they were being represented by a Soviet commercial cover company.[536]

Morris was introduced into Soviet espionage directly as a result of Leon Trotsky. In 1933 Morris was a booking agent for show business acts when a promoter offered to get him Leon Trotsky for a specialty act at the New York Paramount Theater. Morris had this offer under consideration when Gregory Melamed, a representative of Amtorg Trading Company, a Soviet commercial cover company for espionage, contacted him. Melamed first asked Morris if his relatives, who were still in Russia, could use any help. The quid pro quo for Melamed's help would be to stop any chances of Trotsky gaining any stage in America from which he could speak his opposition to Stalin.[537]

Boris Morros, the owner of the music company (on paper at least) that bore his name, briefed Soble about the music business. He stressed how it would have to be operated in order to leave enough time for the agents to handle their espionage assignments and still appear to work at the legitimate business of the company. Morros told Soble that Alfred

Kaufman Stern, a loyal communist, had put up $130,000 of his own money to finance the company.

Stern, Morros said, wanted more than token espionage work for his money and had been given a title of vice president and put in charge of the New York office. Morros, who deprecated Stern's intelligence, hastened to assure Soble that Stern was merely a front.

According to Soble, differences were constantly arising between Morros and Stern.[538] Soble was spending an inordinate amount of time mediating these petty squabbles. Soble visited the Sterns in what he later described as "their palatial apartment" on Central Park West on one of his many peace making attempts between the two.

Soble knew that Stern was wealthy in his own right, having inherited money from his family.[539] In addition, he received a considerable sum in the divorce settlement from his first wife, Marion Rosenwald, the daughter of the former president of Sears Roebuck. Stern was known as a "patsy" by the CPUSA because he was always good for a healthy financial contribution for any communist cause. He served on the Central Committee of the New York chapter of the CPUSA during the 1940s.

Stern had re-married a woman whom Soble described as "the vivacious, flirtatious, hard-drinking Martha Dodd Stern."[540]Stern constantly pestered Soble for more espionage work but Soble refused to commit himself. Soble was convinced that Stern, although a loyal communist who would do anything he was told, was so dominated by his kinky wife that no one could predict what she might do.[541]

The Boris Morros company was eventually dissolved because Morris and Stern could not work out their business differences. Morros remained in Soble's espionage network as a trusted agent being given more and more responsibility. This was despite the repeated dire warnings from both Sterns that Morros could not be trusted. Soble ignored the Sterns' warnings attributing them to petty spite and jealousy.

Unfortunately for Soble, this time the Sterns were correct. Morros had been an FBI informant for years and continued as a FBI penetration of the Soviet espionage network for the next ten years. Ironically, when the Soviet headquarters became suspicious of Soble they requested his return to Moscow through Morros. Soble refused to return to the Soviet Union and was arrested by the FBI before the GPU could arrange his kidnapping. (Soble was convinced after Mark Zborowski's arrest that it was only a matter of time before the FBI arrested him.)

Morros was able to report that Martha Dodd Stern had recruited several agents, among them one Jane Foster Zlatovski (code name SLANG) into the Soviet espionage network.[542] Jane had been born to wealthy parents who lived in the exclusive Nob Hill area of San Francisco. When Soble first met Jane she had just completed a three year tour with the Office of Strategic Services (OSS) the wartime espionage arm of the United States. Jane was preparing to transfer to Vienna, Austria where she would be in charge of a classified American radio broadcasting project where she would have access to sensitive material.

According to Soble, Jane was a valuable source of intelligence for the Soviets. Jane was married to a hard-drinking lout and "ne'er-do-well" named George Zlatovski who was a lieutenant in United States Army intelligence service. George had been born to lower middle class parents in Duluth, Minnesota but had worked himself up through military service to where he was an "officer and gentleman" by virtue of an act of Congress that granted him a commission in the army.

Martha Dodd Stern told Soble that Jane Zlatovski badly wanted to do more important espionage work for the Soviets. Jane herself told Soble that she and her husband were prepared to "go to any lengths" to be useful to communism and the Soviet Union. To show the extent of her cooperation with Soviet espionage, at the first meeting she had with Soble, she briefed him on her classified work with OSS in Indochina.

Emphasizing her dedication to the Soviet cause of espionage against the Western world, Jane gave Soble a six page secret report on Indonesia.[543] She told him that the Soviets would find the report very useful. The Soviets at that time were negotiating with the Indonesian government on a very important matter and Jane's report spelled out for them exactly what they needed to know in order to conclude the bargaining to their advantage.

Later, when they were stationed in Austria and France, Jane and George Zlatovski, continued to provide the Soviets with reports on classified American intelligence and diplomatic matters.

Although Jane and George were each aware of what the other was doing for the communist cause and Soviet intelligence, they quarreled incessantly over each other's extra marital sexual encounters with many other people. They fought constantly, Jane scratching George's face and tearing off his clothes. George blackened Jane's eyes and threw her out of their apartment without clothes. She went to Soble in this battered condition to complain about George's treatment. Both Jane and George drank to excess, regularly finishing a bottle of liquor at bedtime and taking a tumbler full of liquor upon awakening to get going in the morning.[544]

More damaging in the long run, were the reports the Zlatovskis provided with the details of the sex, drinking, gambling and other personal quirky habits of the American diplomatic, military and intelligence personnel stationed there. This information could later be used by the NKVD as blackmail against these people many years later after they had been promoted to sensitive positions.

George Zlatovski, through his normal work with military intelligence, was privy to the resettlement plans for refugees and defectors from the Soviet Union. These refugees were crowding into the displaced persons' processing centers operated by the American government in Europe after World War Two. Zlatovski reported to the Soviets what he learned from the debriefing dossiers, including a refugee's place of eventual resettlement.

In this manner many Russian refugees were located by the NKVD after they had settled into their new homes in the United States, Canada or Western Europe. The NKVD would force the refugees to re-defect back to the Soviet Union thereby achieving a propaganda coup to embarrass the west and discourage future defectors. Alternatively, the NKVD would force the refugees to accept espionage assignments against their new homelands. Neither of these demands were subtle but were accompanied by direct threats against the refugees' relatives still living in the Soviet Union.

Jane and George Zlatovski were indicted 8 July 1957 for conspiring with the Soviet Union against the United States. They were living in Paris at the time and refused to return to the United States. French law treats espionage as a political offense and refused to extradite them.[545]

Martha Dodd Stern, eleven years younger than Alfred Kaufman Stern, was according to Soble, always seeking new adventures. She enjoyed bragging to Soble, over very dry martinis, about the thrills of her sexual escapades and the allegedly perilous intelligence assignments she had performed for the Soviets. She was born in 1908 and educated at some of the country's best schools by her father, William E. Dodd, noted educator and diplomat. Martha attended finishing school in Washington, D.C., went to the University of Chicago for three years and did her final year at the Sorbonne in Paris.

One time Martha bragged to Soble about how she had spied on her own father to gather information for the NKVD. When her father had been the United States Ambassador to Germany from 1933 to 1937, Martha had a passionate love affair with an official of the Soviet Embassy in Berlin. (There is speculation that Martha's NKVD lover in Berlin was Vassili Zubilin.) The Soviet was an official of the NKVD and at his behest Martha stole classified files from her father's office safe in the American Embassy.[546]

Martha's husband, Alfred Stern, continually complained to Soble that he wanted to do more for the Soviets. In an attempt to keep him content, Soble introduced him to Anatole Gromov, the first secretary of the Soviet Embassy in Washington but also the senior NKVD officer in America. Gromov was to find some espionage work for Soble to do to keep him happy.[547]

Gromov's code name was Al. He was the same NKVD officer to whom Elizabeth Bentley turned over the secret material she collected from the Soviet spy ring in the American government.[548]

The Bentley network was closely connected to the Trotsky case. Jacob Golos, Bentley's Soviet case officer, was one of the principal American case officers for the Trotsky assassination. Bentley routinely picked up mail received by Rose Arenal, a Brooklyn school teacher and the wife of Luis Arenal. This mail cutout operation was set up by Maxim Lieber, a GPU agent handled by Jacob Golos. Lieber was a literary agent who handled Trotsky's writings. This professional link with Trotsky enabled Lieber to maintain an insight into Trotsky's activities which Lieber passed on to his GPU superiors. Later on, inexplicably, on orders from his superiors, Lieber dropped Trotsky as a literary client.[549]

The Arenal brothers, Luis and Leopoldo, had been the leaders, together with Siqueiros, of the 24 May raid on Trotsky's house in Coyoacan. Further Angelica Arenal, their sister, was married to Siqueiros, the leader of the 24 May raid.[550]

The Arenal brothers are believed to be the murderers of Robert Sheldon Harte, Trotsky's guard who was kidnapped and murdered by the 24 May raiders. Harte's body was found, a month after the raid, buried beneath the dirt floor of the kitchen of a shack in the Desierto de los Leones, a distant suburb of Mexico City.

Bentley would pick up the letters that had been sent to Rose Arenal by her husband and take them to Golos for delivery to Gromov. Before Trotsky's assassination the letters contained requests from the Arenal brothers and Siqueiros for instructions for the raid and pleas for more money. After Trotsky's assassination, with Mercader, the assassin, in jail, the letters dealt almost solely with plans for his escape.

After Soble's arrest and conviction for espionage, and after he had become a witness for the government against his former subordinates, the Sterns decided to leave the United States. They managed to flee to Mexico one step ahead of the FBI. With the Sterns in Mexico, the United States Department of Justice had to file a formal extradition request with the Mexican government.

It was on 17 December 1953 that the Sterns, together with their adopted son, Robert, who was then seven years of age, decided to move permanently to Mexico City. They did not have the proper documentation to enter Mexico when they arrived at the border. They conned and bribed their way past the legal technicalities with the help of a good "coyote" (expediter). They showed their birth certificates and marriage license and with a hefty "mordida" (mordida translates as bite but means bribe) in American money, they passed into Mexico. (Later, when the American government was attempting to extradite the Sterns, the Mexican government claimed they could not serve the subpoenas on them because they were in Mexico "illegally" and there was no record of their ever arriving in Mexico.)

Alfred and Martha Stern established a financial holding company called Mercantial Latino Americano, S.A. and another front for their business activities called Negociadores, S.A. Both of these fronts were located at the same address in Mexico City at Reforma 104 on the fifth floor. The Sterns acquired a lot of money making loans (to other members of the American Communist Group in Mexico-ACGM) at high interest rates.[551]

There was in Mexico at the time the Sterns arrived, a semi-formal group of communist expatriates loosely referred to by the FBI as the American Communist Group in Mexico (ACGM).[552] The group included a very few former NKVD agents, some wanna-be NKVD agents and some hard line former members of the CPUSA. Mostly, however, the people that grouped together for comfort in a strange land, were merely former liberals who had been frightened by the McCarthy era witch hunt.[553] Regardless of their stripe, the ACGM welcomed the Sterns who were delighted to be so warmly received.

The Soviet Embassy in Mexico City encouraged the propaganda value of the ACGM and made use of some of the members for actual espionage assignments. The Soviets gave covert support to the aspirations of the Movimento Pro-Independencia de Puerto Rico (MPIPR) represented in Mexico by the Dona Laura Meneses de Albizu Campos, wife of the founder of the MPIPR, and the Secretary of State of the MPIPR, Juan Juarbe y Juarbe.

Pedro Albizu Campos, the leader of the MPIPR was languishing in an American jail being bombarded, he claimed, by death rays emanating from the White House. Albizu Campos had been convicted of masterminding the abortive attempt to assassinate President Truman and of actually firing sub-machine guns into the House of Representatives in an attempted revolt for Puerto Rican independence.

The day of reckoning dawned for the Sterns on 12 February 1957. This was the day when agents of the Mexican Gobernacion, the Ministry of Justice, and officers of the American Embassy, served subpoenas on them at their luxury apartment in Colonia Polanco.[554]

The former American Ambassador to Mexico, William "Big Bill" O'Dwyer, had remained in Mexico when his stint as ambassador was over and had opened a law office. He was quickly retained by the Sterns to fight the extradition. Within minutes of the subpoenas being served, the Runyunesqe hustler from New York, telephoned the American Embassy complaining that the subpoenas were improperly drafted and therefore deficient for service on the Sterns.[555]

The Political Section of the American Embassy worked many long hours and redrafted the subpoenas that were validly served on the Sterns on 27 February. Both Martha and Alfred were each given, along with the subpoenas, $468 American dollars for expenses to cover the trip to New York to appear before the Federal grand jury.

The Sterns, as every thinking person could have predicted, failed to appear in New York and subsequently refused to acknowledge any further communications from the American Embassy or the Mexican Gobernacion. Diligent officers from the American Embassy, accompanied by Mexican officials from the Direcion Federal de Seguridad (DFS — the

Mexican equivalent of the American FBI) staked out the Sterns's Polanco apartment, Alfred's office on the Paseo de la Reforma and their new house in Cuernavaca in order to attempt to serve them again with subpoenas but to no avail.

The Mexican government offices involved promised that they were doing everything possible to arrest the Sterns and honor the American extradition request. The Mexican government promised they would throw them over the center line of the bridge across the Rio Bravo (Rio Grande) between Nuevo Laredo, Mexico and Laredo, Texas into the waiting arms of American justice, just as soon as they could be located

This was a time-honored custom of cooperation between the two countries. The Mexican police would "deport" the undesirable "gringo" who was wanted by the FBI or some other law enforcement organization. Deportation in this sense amounted to the gringo being literally pushed across the center line of the international bridge which meant he was pushed out of Mexican territory onto American territory. He was therefor automatically eligible for a legal arrest on the American side by American authorities who had been alerted to the deportation by the Mexican police.[556]

The Sterns were each fined $25,000 for refusing to appear before the New York Grand Jury. As an additional slap on the wrist they were adjudged to be in contempt of court. They remained in Mexico, keeping a very low profile. The Sterns were able to continue to dodge the process servers. It was alleged that it was costing the Sterns upwards of $15,000 per month in the form of "mordidas" (bribes) to Mexican officials to permit them to continue to "elude" the process servers.

On 7 June 1957 American Ambassador White delivered personally to Mexican President Ruiz Cortines a verbal message from President Eisenhower, requesting cooperation in a matter of importance to the security of the United States which was to assist in the extradition of the Sterns to the United States. On 24 June Ambassador White spoke again to President Cortines about the matter. On 26 June Ambassador White spoke with Mexican Minister of Gobernacion Angel Carvajal (equivalent to the American Attorney General) and Finance Minister Carrillo Flores, both of whom promised to look into the matter, but nothing further was heard from them or anyone in the Mexican government; such was the power of the Sterns' money.

A change of the Mexican administration and a little more pressure from a new American Ambassador eventually resulted in an improved system of serving subpoenas and processing the extradition.[557] The Sterns, recognizing the proverbial handwriting on the wall decided to leave Mexico. (If everything had been done properly, once served the Sterns would have been taken into custody and flown to New York within an hour after their arrest.)

The Sterns chose to escape to the glorious peoples' Republic of Czechoslovakia. They had Boris Kolomyakov, the Second Secretary of the Soviet Embassy and ranking NKVD officer in Mexico, help them sneak away. Kolomyakov realized the new president of Mexico meant business and the Sterns had better leave Mexico immediately. They were smuggled aboard a KLM plane departing Mexico at 1:00 p.m. on 12 July 1957 headed for Amsterdam via Montreal and then on to Switzerland.[558]

Just before their flight to Czechoslovakia Alfred sold shares of stock in American companies for cash that were valued at $15,000,000. Also, in preparation for their departure both Sterns renounced their United States citizenship and bought (in the truest sense of the word) Paraguayan citizenship. They purchased Paraguayan passports Nos. 30 and 31.[559]

The Sterns settled in Czechoslovakia and Martha wrote to her friends in the ACGM about life in the glorious peoples' republic. She realized that her letters were censored and paid obligatory obeisance to them with flowery phrases extolling the benefits of life in the glorious peoples' republic and how bountiful life was behind the so-called Iron Curtain. Martha stressed what a cornucopia flowed from the state controlled economy and that they had more than enough of everything they needed.

Incongruous as it may seem, Martha closed every letter with an eloquent plea for some of the basic necessities of life. She begged her friends to send her soap, any and all types of canned food, toilet paper, toothpaste, needles, thread and practically any item the average American or Mexican could buy in bulk but were unavailable in the glorious peoples' republic.

The Sterns continued to plead their innocence. Congressman Don Edwards, (Democrat — California, 10th Congressional District) was prompted to take the Sterns's case by Sylvia Crane. For many years Crane had been a member of the National Committee Against Repressive Legislation which was cited as a extreme leftist group. Crane claimed close and long time personal friendship with the Sterns. She traveled to Prague on several occasions to coordinate their campaign to get the espionage indictments against them quashed.[560]

Crane convinced Edwards to send a letter to Attorney General Griffen Bell, on 30 June 1977, stating that the indictment against the Sterns had been based upon the testimony of Boris Morros who, Edwards claimed, "files indicate that Morros was a person of doubtful integrity." Actually the indictments against the Sterns were based on the sworn testimony of many witnesses including Soviet spy master, Jack Soble. Soble testified in detail to the Sterns's activities in espionage when they were working for him.

Edwards's letter continued, that a friend of the Sterns (possibly Sylvia Crane) had recently visited them in Czechoslovakia and found them both anxious to return to the United States. (Remember there was a shortage of basic necessities in the country they had chosen to live!) If there was a dry eye left in the Department of Justice after that plea, Edwards took care of it next. "It is my understanding," he wrote to the Attorney General, " that Mr. Stern believes he is close to the end of his life and would like to spend the remaining months or years back home."

Edwards letter was answered by Deputy Assistant Attorney General, Robert L. Keuch, who wrote that the espionage indictments remained outstanding against the Sterns. He reminded Edwards that the Sterns's lawyer, Victor Rabinowitz, had been in contact with the Department of Justice and "any discussion with persons other than the principals or their attorney would be inappropriate."

Not deterred by this polite and legal rebuff, two weeks later Edwards wrote to President Jimmy Carter. He addressed the letter to White House Counsel Robert S. Lipshutz, asking that the indictments against the Sterns be quashed. This request flies in the face of the evidence against the Sterns. There is sworn testimony that the role of the Sterns included the recruitment of new espionage agents, one of whom was employed by the Office of Strategic Services and by the Central Intelligence Agency. The Department of Justice had always maintained that the two fugitives from justice would have to cooperate with the FBI's counterintelligence service before there could be any consideration of leniency.

Edwards replied that "Justice [the Department of Justice] is 'going by the book' and that would include these old sick folks returning, being interviewed by the FBI, court proceedings, etc." Edwards continued, that this major spy ring "wasn't much to begin with more than 20 years ago," and concluding, "sometimes justice includes being kind and generous."

Benjamin Civiletti, Assistant Attorney General for the Criminal Division, making an obvious concession to Congressman Edwards, who incidentally was the chairman of the Congressional subcommittee that had jurisdiction over appropriations for the Justice Department, replied that he had asked for a reevaluation of the entire Stern matter.

Not yet content, Edwards applied pressure at another point. Addressing his letter to Presidential Assistant Anne Wexler. He addressed this letter to "Dear Anne" on 12 June 1978 and wrote that "to arrest and try these two old people for events which arose out of the McCarthy era would not serve the cause of justice in any sense. The White House could show great compassion by indicating to the Justice Department an interest in allowing the Sterns to return home unmolested."

Edwards persisted and in January of 1979 wrote to Civiletti's replacement as head of the Criminal Division, Phillip Heymann, that he "would be especially interested in learning of your projection of when the indictment [against the Sterns] might be actually dropped." Continuing, Edwards revealed his lack of knowledge about the case, "My own research into the case indicates that the evidence which led to the indictment originally was questionable and because the case is so old, there is no way it could be successfully prosecuted." Unfortunately Edwards was probably correct about his last statement. After more than twenty years in hiding and avoiding facing the indictment, it would have been impossible to resurrect the original witnesses, most of whom were dead.

Irwin B. Nathan, deputy to Phillip Heymann, on 20 March 1979, wrote to Edwards that the Assistant Attorney General had signed a motion to dismiss the indictments for espionage against the Sterns. Sylvia Crane, a member of the Advisory Board of *Counter-Spy*, a magazine which had taken a leading role against intelligence and counterintelligence programs of the FBI and CIA, told a *Washington Post* reporter that, "the persecution of the Sterns was part of the McCarthy dirty tricks campaign."

In summation there may be a bond of brotherhood of government officials of Mexico, Paraguay and the United States when it comes to doing business for profit with the Sterns. It should be noted that Congressman Don Edwards of California revealed (as he was required by statutory disclosure laws) that, on 24 October 1979, he received a sizeable contribution for campaign expenses from one Alfred K. Stern!

In the Mexican edition of a major American newspaper of 24 June 1986 the following article, reported from Prague, Czechoslovakia, appeared:

> Alfred Kaufman Stern, 88, former Illinois Housing administrator who was charged in 1957 with espionage for the Soviet Union died of cancer on Monday [23 June 1986] in Prague exile, his wife reported.
>
> Mr. Stern and his wife, the former Martha Dodd, left the United States in 1953 after being accused of subversive activities. They were indicted in absentia on espionage charges on 9 September 1957. The indictment charged them with conspiracy to act as Soviet agents, receiving American military, commercial and industrial information and transmitting it to the Soviets.
>
> Mr. Stern was born on 29 November 1897 in Fargo, North Dakota into a wealthy family. In 1938 he married Martha Dodd, daughter of the late William E. Dodd, United States Ambassador to Germany before World War Two.
>
> Mr. Stern served as chairman of the Illinois Housing Commission and later as vice president of the National Association of Housing Officials and Directors and as Chairman of the Citizens Housing and Planning Council in New York. He also founded the Institute for Psychoanalysis in Chicago.
>
> The Sterns went to Mexico and visited Prague in 1957. They stayed in the Soviet Union for about a year before settling in Prague in 1958.
>
> In 1963 the Sterns went to Cuba and lived there until 1970 before returning to Prague. In Prague Mr. Stern worked as a consultant to the Ministry of Construction using his expertise in housing and construction.
>
> Besides his wife, he is survived by a son, Robert.[561]

Another member of the Soble spy network was Esther Rand who was an unindicted co-conspirator with Soble. She was targeted against

Zionism. Zubilin told Soble that the Soviets regarded Judaism as an enemy of communism and that he had a penetration in the offices of the United Jewish Appeal he was turning over to Soble by the name of Esther Rand. She furnished Soble with copies of all correspondence between the United Jewish Appeal and the State Department regarding efforts to rescue the Jews from Nazi Germany. She also furnished information about other Zionist groups. Esther Rand was born in Russia of Jewish parents in 1907 and came to the United States as a child.

With the indictment against her quashed, Martha could have returned to the States. She claimed she was afraid to, citing her fear that she was being tricked and would be prosecuted if she did return to U.S. jurisdation. She died in Prague, a lonely old woman, in the 1980s.

Robert, the Sterns adopted son, was reported in the 1990s as living in Cuernavaca, Mexico.

THE 24 MAY RAID

...if the facts were not so strange and exciting in themselves....

Colonel Leandro Sanchez Salazar, Chief, Mexican Secret Service, referring to the lack of planning and inefficient execution of the Siqueiros raid on Trotsky's compound.

A symphony orchestra might begin with a pianissimo, a barely perceptible rumble on the kettle drums. Then gradually build to a fortissimo roll, taken up and amplified by the large bass drums with the entire orchestra crashing at full volume for the finale. Thus did the rumblings of the anti-Trotsky GPU propaganda crusade begin. The original Soviet plan was to incite mass world support for the expulsion of Trotsky from Mexico. Unfortunately for them, this GPU blitz did not sway world opinion except to increase the fervor of the already committed hard core Communist Party members.

When the oil companies (American and other foreign held oil interests) were expropriated by Mexican President, Lazaro Cardenas, the Stalinist press charged that Trotsky was their "representative." Before the Stalin-Hitler Pact, the Stalinist press charged Trotsky of being an agent of Nazi Germany. After the Stalin-Hitler Pact, they accused him of being an agent of American and English oil interests.[562]

In essence, the anti-Trotsky propaganda campaign failed miserably. One of its principal purposes was to convince President Cardenas of Mexico to accede to what the GPU hoped he would interpret as overwhelming public outrage against Trotsky being allowed to settle in Mexico. The communist press assumed more strident tones, accompanied with the implication of more drastic reprisals to come, if the Mexican government did not do something to banish Trotsky from his sanctuary in Coyoacan.

The Trotskyite Party of Mexico had regrouped and changed the name from the LCI to the Partido Obrero Internacional (POI). They published a newspaper *Lucha Obrera* (Worker's Fight) which did all it could to counter the Soviet propaganda campaign to defend Trotsky but its circulation was minimal.[563]

Trotsky wrote of this noticeable change in the tone of the communist press and commented that it reminded him of the "way people write who are preparing to change the pen for the machine gun." Trotsky could envision no other purpose for the GPU propaganda barrage against him except, "they plan to commit a terrorist act but it is indispensable to prepare public opinion [first] for the deed."

> I know that Stalin often admitted that my deportation abroad was his great error. Only a terrorist act could correct that mistake. [Leon Trotsky, Coyoacan, Mexico, circa January 1938.][564]

At the end of April 1940 Trotsky had written a message to the Soviet workers, peasants and soldiers entitled, "You Are Being Deceived." The message was dynamite which further incensed Stalin against Trotsky. Stalin perceived that all his fears about Trotsky's formidable power were justified and he could not permit Trotsky's power to continue calling for insurrection against his regime.[565]

Thus, on May Day 1940, the PCM staged a giant parade through the center of Mexico City. They bussed in from the boondocks,and scoured the city barrios, to marshal some 20,000 semi-uniformed, faithful men and women to march and shout anti-Trotsky slogans. Some carried banners that proclaimed, among other more virulent pieces of hatred, "Throw Out the Most Ominous and Dangerous Traitor Trotsky."

Trotsky occasionally reminded himself of how time has a habit of repeating itself. The acute sense of deja vu reminded him of the methods police agents in the Soviet Union used to entrap unwary innocents. Trotsky vividly recalled when he was about sixteen years-of-age and living at the home of an old apple grower named Shvigovsky in the town of Nikolayev.

Trotsky had just joined his college classmates in a revolutionary organization called the Narodniks. This small group of well-intentioned, idealistic revolutionaries were enraptured by their newly discovered panacea for all the ills of the world. They were determined to convert a poor, down trodden laborer who worked for Shvigovsky to throw off his shackles and be free by revolting against his employer. Trotsky earnestly tried to convince this uneducated worker of the benefits of the Narodnik revolutionary ideals only to discover he was a police agent who jailed Trotsky.

The communist newspapers, *La Voz de Mexico*, *El Popular* and *Futuro* were controlled by Vicente Lombardo Toledano, the perennial Mexican socialist political candidate and obsequious GPU toady. They accused Trotsky of reneging on his promise to President Cardenas that he would not become involved in Mexican internal politics. Trotsky countered by describing Lombardo in a derogatory manner:

> Lombardo's job is to function in the trade unions as a mask for the GPU activity and as an exponent of Stalinist policy without holding a membership card in the PCM.[566]

The communist press accused Trotsky of establishing a center for espionage in Mexico; of being an advance agent for the oil interests of the capitalists countries of the United States and Great Britain; of actively cooperating with the "fascist" Dies Committee of the United States Congress for "thirty pieces of silver" and of attempting to subvert the Mexican form of government by an alliance with fascist General Franco of Spain. They also accused Trotsky of actively supporting thc anti-Cardenas Mexican opposition party of General Almazan and, for good measure, of intervening in Latin American affairs as an agent of the "fascist, capitalist imperialist powers."

This libelous vitriol, repeated daily in the communist press, was occasionally picked up and replayed in the non-communist newspapers of the capital. Such gems appeared as follows: "Trotsky, the old traitor, demonstrates to us that the older he grows the more cur-like he becomes" and "What a slippery fish is the little old traitor." This was in reference to his being called to testify before the Dies Committee of the United States Congress. "The dirty Judas," was in reference to the thirty pieces of silver previously noted which referred to the fact that Trotsky would be paid expenses to testify before the Dies Committee.

Charles Curtis, an American Trotskyite in Mexico, had attempted to reorganize a dysfunctional Trotskyite party. The new group published a pro-Trotskyite magazine *Clave* (*Key*) which did all it could to defend Trotsky and counter the broadside PCM propaganda blast against him.[567]

Trotsky was being made ready for the kill from all sides. Unbeknown to Trotsky, in addition to the propaganda barrage against him, Jacson/Mercader was casing his home in preparation for the Siqueiros raid. There was yet another part of the preparation taking place, literally, almost within sight of his front door.

The GPU felt that some diversionary tactic should be arranged to reduce the effectiveness of the Mexican soldiers on duty guarding the front of Trotsky's home. The alertness of the guards had to be diverted to allow the Siqueiros raiders an unimpeded chance at entering the front gate.

Siqueiros ordered Antonio Pujol, an avante guard Mexican painter, to establish an apartment close to Trotksy's home. Pujol was to bait the apartment with sexy, provocative women who could entice the guards away from their posts with a promise of something better to do than guarding an old Russian exile.

Siqueiros and Pujol found two willing candidates who admirably filled all requirements of being sexy and pretty and loyal communists. They were Julia Barrados de Serrano and Anita Lopez Chavez. Pujol set them up in a yellow painted, two story apartment building close to Trotsky's house at 85 Abasolo street in apartments 11 and 13.

Julia Barrados de Serrano was the wife of David Serrano Andonequi who lived at 85 Violeta street in Mexico City. David Serrano fought in the Spanish Civil War. He was a major with Siqueiros in the Republican army. While in Spain he had bigamously married a Spanish communist woman without informing Julia that their marriage was ended. Serrano and his Spanish bride had a daughter whom they named Sovietina, in honor of the glorious Soviet Union.

Serrano was a member of the Politburo of the PCM. He had been trained in Moscow with the GPU and Leonid Eitingon in Spain for the preceding six years.

Siqueiros wanted to make certain that things were going according to the plan with the girls in their newly installed "casa chicas," so he visited them frequently. He was driven there in his American La Salle sedan that had Texas license plates which caused a commotion every time it arrived in the neighborhood. Was this a breakdown in the GPU's security for the operation?

The girls' assignment was to seduce the guards posted in front of Trotsky's home. They were to elicit information about events inside the compound and the nature of the construction of defensive fortifications around the perimeter of Trotsky's house. (Most of the heavy construction took place after the Siqueiros raid but there was always some building going on inside the compound.)

The girls had sexual intercourse with the guards. Julia, the more dominant partner, had sexual relations with most of the guards. One of them, Rudolfo Fregoso, bragged about the number of guards who had sex with Julia.

Anita Lopez had a boyfriend named Mariano Herrera Vasquez, a 26 year old Mexican electrician who had been a member of the PCM since 1934. On 5 May Anita introduced Herrera to Pujol, who told him to go each morning at ten o'clock to the corner of Tampico and Chapultepec streets and wait for a contact.

On 10 May Siqueiros picked up Herrera at the contact point and told him he would be paid ten pesos a day if he wanted work. Siqueiros drove him to London street in Coyoacan where another of the Siqueiros conspirators, Luis Mateo Martinez, was staying. Herrera stayed with Martinez until 13 May when another Siqueiros cohort, Juan Zuniga Camacho, came and paid him the wages due thus far. Herrera was delighted to be paid for doing nothing. Especially since during this time he was able to visit his girlfriend, Anita, in her new apartment on Abasolo street.

On the 17th of May, Herrera was told to meet Siqueiros at the Tampico/Chapultepec meeting site. Promptly at the appointed time of five o'clock Siqueiros, Siqueiros's wife, Angelica Arenal de Siqueiros, and Antonio Pujol, picked up Herrera for a shopping spree. They drove to the center of Mexico City where they purchased a folding cot, food and some artists supplies such as an easel, brushes and tubes of oil paints.[568]

The quartet of shoppers drove to the village of Santa Rosa that was situated between the villa of former Mexican President Alvaro Obregon and the Desert of the Lions. There Angelica and Herrera took the purchases to a isolated shack, a short walk from the dirt road where they had parked the car. Angelica told Herrera to stay in the house with her brother, Luis Arenal, and the other two men already there; one of the men was called Narciso and the other was called the Invalid.

On the 22nd of May Siqueiros came to the shack in the Desert of the Lions to check on the men and pay their wages. Narciso and the Invalid requested permission to leave for the afternoon to buy some shoes. Siqueiros granted permission but warned them to return in time for night watch duty. He exhorted them to keep a careful eye on the shack and the few things he had put in it.

Narciso and the Invalid returned to the shack on the evening of the 22nd of May as they had promised. The next morning, the 23rd of May, Herrera was restless and resented the privilege to leave that had been given to the two other caretakers but not to him. Herrera was miffed and took off.

He went into Mexico City and got drunk on pulque — an alcoholic beverage made from the maguey cactus. He arrived back at the shack the following morning and found the Invalid there alone. The Invalid told Herrera that on the previous night, Antonio Pujol had been very angry that Herrera was absent but took Luis Arenal and Narciso with him. Pujol left the Invalid as the only guard of the shack and supplies inside.

On the 25th of May, the day after the raid on Trotsky's home, Luis Arenal returned to the shack to pick up the Invalid. He told Herrera to clear out and go someplace where he could not be found. Arenal gave him fifty pesos of hush money. Later, in his testimony, Herrera said he did not know what he was supposed to remain hushed about for the fifty pesos!

Herrera was in a quandary and did not know what to do. Since he had been hired originally through his girl friend, Anita Lopez, he hurried to Abasolo street. He wanted to ask Anita or her girlfriend, Julia Barrados, if they knew what he was supposed to keep hushed about.

At 85 Abasolo Herrera found the doors locked and the apartments apparently abandoned. Herrera then hurried to the house of Anita's parents at 134 Liberty street. There he learned that the Trotsky home had been raided the night before. Anita's parents told him that Juan Zuniga Camacho had told the girls the night before to abandon their apartments as soon as they heard shooting near to Trotsky's home.

The previous night, that is the night of the raid, the girls had had an exceptionally wild party. Using the pretext that they were going to leave their apartments in Coyoacan to live in the center of Mexico City, they suggested to Trotsky's Mexican guards that they all have a wild "despedida" (farewll party) at their apartments. They promised the guards there would be lots of pulque, mescal and tequila, good music from the radio and the implied opportunity of a romp in the hay with one, or maybe both, of the hostesses. This was the kind of an offer very few of the poorly paid Mexican guards could refuse.

Other Preparations

About four o'clock in the afternoon of 17 May 1940, Luis Mateo Martinez, made a strange request to the magistrate of the 11th Police Delegation in Mexico City. He told the magistrate that he wanted to borrow three complete police uniforms for some friends. The friends, Martinez explained, wanted to disguise themselves as policemen in order to raid the headquarters of a henchman of presidential candidate General Juan Andrew Almazan.[569]

Almazan was the reactionary candidate and not favored by the entrenched bureaucracy which was clearly the party of choice of the magistrate as part of that entrenched bureaucracy. The convoluted story that Martinez used was obviously designed to appeal to the survivor instincts of the politically attuned magistrate.

Martinez told the magistrate that his friends had learned that a large cache of arms and ammunition had been hidden at the home of this Almazanista political leader. After his friends grabbed the arms as evidence they planned to denounce Alamzan to the Public Prosecutor as a radical traitor to the peaceful electoral process of Mexico. Either because of these political blandishments, or because Martinez had given the magistrate the traditional mordida (bribe), the magistrate agreed to give Martinez the three police uniforms.

The magistrate took Martinez on a tour of the 11th Delegation to locate the storeroom where the uniforms were kept. Unfortunately, the uniforms could not be found that day. The magistrate asked Martinez to return in a few days when he would have the uniforms ready. Martinez went back on two different occasions but the uniforms could not be found. Finally, the magistrate asked Martinez to come to his home at ten o'clock the following morning and he promised to have the uniforms there for him.

During the night the magistrate apparently had second thoughts and realized for the first time that what Martinez was planning was illegal. (At least that is what he claimed to the Secret Service officer who investigated the circumstances leading up to the raid when it was over.) Whatever the magistrate's motives were, he stayed away from his home the next morning and Martinez came and went without the uniforms. There was some subsequent benefit to the Mexican authorities from this contact by Martinez. It provided the thread of what would later be the unraveling of the crime of the raid on Trotsky's home.

The break in the investigation started with a tip about Martinez's attempts to get the uniforms from the 11th Delegation. Fortunately, in the ensuing investigation, the magistrate knew Martinez's true identity and exactly where he lived. The Mexican Secret Service was able to arrest him almost immediately after the investigation of the raid began.

Luis Mateo Martinez was a twenty-six year old Mexican national and a primary school teacher. He was a member of long standing in the PCM. He readily admitted to the Mexican Secret Service that he had been trying to procure the police uniforms for David Serrano Andonequi. Serrano was an overt member of the PCM and, as noted, had been a major in the Republican army in the Spanish Civil War.

Serrano lived at No. 85 Violeta street in Mexico City which was a boarding house for Spanish Civil War refugees. His attractive wife, Julia Barrados, younger looking than her thirty years, with an aggressive Latin personality, also lived part-time with him there; that is, when she was not servicing the guards from Trotsky's home in her apartment on Abasolo street.[570]

Concurrently preparing for the same raid was Nestor Sanchez Hernandez who lived at Calle Guatemala No. 54 in one of the older sections of Mexico City. Sanchez Hernandez and Felix Guerrero Mejia were contributors to Lombardo's magazine *Futuro*. Sanchez Hernandez had written an article specially attacking Trotsky.[571]

Sanchez had been a captain in the International Brigade in the Spanish Civil War. The Mexican Secret Service later ascertained that, at this time (before the raid), Sanchez was corresponding with someone living at the Majestic Hotel in Mexico City.[572] Sanchez obtained the police uniforms from Antonio Pujol that Martinez had failed to get from the magistrate at the 11th Delegation.

Pujol obtained the uniforms from an unknown source and gave them to Sanchez who had the cloth patch of the 7th Police Delegation sewn on them. Sanchez was a young Mexican man of 23 years of age, dark complexion, with an open, peasant face, a broad forehead and an almond shaped pair of black eyes. He was a law student at the time of the raid.[573]

Sanchez was a close friend of Siqueiros through their service together in the Spanish Civil War. Siqueiros had asked Sanchez in April if he would like to participate in an affair of exceptional importance. Siqueiros had hedged and blustered but had not told Sanchez exactly what the important affair was to be. Sanchez soon deduced that the event Siqueiros was alluding to had to do with some, as yet unspecified, method of eliminating Trotsky.

As Sanchez testified later, " I had always been told by Siqueiros that Trotsky was a dangerous counter-revolutionary, an enemy of the Russian revolution and of our own Mexican revolution." Siqueros had assured Sanchez that he had worked out a plan to eliminate Trotsky. He bragged to Sanchez that when the moment came he would have all of the resources and money he needed to successfully carry out his plan. Siqueiros finally revealed his plan, Sanchez demurred on several points. He had almost decided not to join Siqueiros because he thought the plan was too vast, vague and cumbersome to be carried out effectively by a small group of untrained, unrehearsed mercenary buccaneers.

Sanchez was under no illusions concerning who was giving Siqueiros orders. He later claimed that he was not fully cut-in on all the details of who was sponsoring this great event but he knew it was not Siqueiros. He noticed that on every important issue in the preparations for the raid, Siqueiros would have to leave the safehouse to attend a secret meeting. Sanchez knew that in those secret meetings Siqueiros was getting his detailed instructions on what he should do next.

Sanchez later said that, "They [the foreigners directing the raid] made use of Siqueiros because he was Mexican, to recruit the necessary people and for the material preparations for the attack."

On 23 May Siqueiros met with Sanchez at the corner of Chile and Cuba streets. Siqueiros led him to a house nearby on Cuba street. There they waited until close to midnight when they were joined by several men bearing large suitcases. The newcomers unpacked the suitcases as if for inspection by Siqueiros. The suitcases contained police uniforms, several Thompson submachine guns, four revolvers, two Thermos bottles made into bombs, rubber gloves and a large quantity of rope in case they had to scale any walls.[574]

At Siqueiros's order the men tried on the police uniforms. The uniform Sanchez received fit him well. Pujol put on a Mexican army lieutenant's uniform while the others donned the police uniforms. They laughed at themselves and at each other as if they were going to a masquerade costume party.

During his questioning by the Mexican Secret Service after the raid, Sanchez maintained that, at this point in the preparations, he did not know where they were going to raid. Several times he had decided that they were really going to raid the arms cache of an Almazanista political leader.

Siqueiros left the house on Cuba street for a little while and returned about two o'clock wearing the uniform of a Mexican army major. Siqueiros did not tell the others where he had obtained the major's uniform and the men did not press him for an answer. In addition Siqueiros wore a rain coat, eye glasses and a false mustache. In point of fact, the men thought he looked ridiculous.

Sanchez carried a .45 caliber Star pistol with two additional clips. Pujol carried a Thompson submachine gun, and the others in this Mexican group had two revolvers among them. Noting their disappointment, Siqueiros told them that the other units that would be joining them at the target would supply all of the fire power they would need so as to be in complete command of the battle situation. The Mexican men were somewhat mollified by Siqueiros's assurance of the ultimate superiority of the combined fire power. Some still grumbled that they wanted to have a personal weapon for self defense during the fire fight.

A car arrived at the entrance to the house on Cuba street. It was a large, dark grey, four-door sedan of American manufacture. The men silently loaded into the car. They were careful to conceal their weapons even though there was no one at that time of early morning to see them in the dark, poorly lighted street. They started off for their destination (still not knowing it would be Coyoacan) in high spirits but tense and subdued.

During the drive, Siqueiros, from the front seat, gave them the details of the plan for the first time. There would be other groups joining them at the target area he told them. The groups were to be kept separate until they were brought together going through the gate at the target. His group must overcome the police guards in front of Trotsky's house but without firing a shot. Siqueiros assured the men in his car that overwhelming the guards at the gate would be easy.

Siqueiros implied that he had already made arrangements that would overcome the guards at the main gate without resistance. Under no circumstances, he assured them, would an alarm be raised before they got inside the compound. After they were inside the compound all hell would break loose so it would not make any difference if an alarm was raised then. Siqueiros gave each man an envelope with 250 pesos as their pay for the night's work.

The target for the raid was Trotsky's home at the corner of Viena and Morelos streets in the Mexico City suburb of Coyoacan. In May of 1940 neither street was paved nor asphalted. These dirt streets were a muddy quagmire during the rainy season with a choking fine dry dust during the summer.[575]

Trotsky's new home was of simple brick and stone construction with little ornamentation. The house was basically a T shaped building surrounded by a wall of twenty feet on one side, the house wall on a second side, the guards' rooms were a wall on a third side and common high wall with the adjoining house on the fourth side. There was a massive wooden gate of double doors reinforced with iron sheeting that led from the street into the garden. The dank ivy climbing in spots up and over the walls gave the house an air of a prison rather than the comfortable dwelling that it actually was.

The garden inside the compound was completely surrounded by the house and high walls. The house itself was airy, spacious and bright; altogether it was a pleasant sanctuary in a fortress like compound. Large tress, many of them decades old, with luxuriant foliage, shaded parts of the garden, making for some quiet alcoves.

The lawn was well cared for and plants and flowers brightened the sides of the brick and tiled walkways and floral beds. Cacti intermingled with the primitive wild plants, and vines that Trotsky so loved to collect and nurture abounded in attractively wild but controlled profusion. This lovely oasis made some visitors reflect that it was a little corner of paradise surrounded by the sinister nether regions of the resident exile's personal hell.

The flower scented night air was clear on Viena street in the hills above the valley of Mexico City in the early morning hours of 24 May 1940. No haze obscured the brilliant moonlight. The stars appeared to be close at hand and seemed to add light to the serenity of the dusty street, that was quiet in the night's slumber.

The Mexican police, who were back from Gloria's and Julia's were bedded down on their straw mats with serapes wrapped around their bodies for warmth. Two of their comrades, barely awake, were sitting slouched against the trees across the street from the large double doored gate at the entrance to the house. The muffled cough of the Trotsky guard on the roof by his rifle could not be heard in the street below. The distant sounds form the city in the valley made up a barely perceptible background hum but did not intrude on the pleasant tranquility that enveloped the sleeping Trotsky household.

At the top of Viveros de Coyoacan, a park a few blocks from Trotsky's home, there were three men waiting for Siqueiros's car. One of these men was Juan Zuniga Camacho but Sanchez did not know the other two. One might have been Eitingon (who was later to be known to the raiders as Felipe, French Jew) and the other man was probably George Mink or Carlos Contreras. They had gone ahead with Zuniga to make sure that everything was in readiness for the other groups of Spaniards who were scheduled to arrive a short time after Siqueiros's car.

Siqueiros spoke briefly with Zuniga and the other two men and returned to his car with an exhilarated attitude of expectancy. He drove slowly to the street parallel to Viena street. Silently the car eased to the side of the road and he motioned for the men to get out. Siqueiros exhorted the men in a hoarse whisper to move quickly and not to talk. The men said later that Siqueiros was the one who was doing all the talking. He signalled the men to assemble in a group behind some trees in a small depression behind the guards' shack in front of Trotsky's home in an attempt at concealment. Siqueiros checked his watch to make certain they were coordinated with the other groups. He than gave the order to silently move toward the shack where the guards were sleeping.

At the beginning of Trotsky's exile in Mexico the guard complement of the Mexican police and soldiers furnished on orders of President Cardenas, was thirty-three men. This number provoked protests from the PCM. Especially vociferous jibes came from Vicente Lombardo Toledano, Mexico's preeminent socialist, communist dupe and GPU toady. Lombardo petulantly claimed that the guard contingent for the "Russian refugee," as he referred to Trotsky, was costing Mexico too much money.

President Cardenas had no fear from Lombardo since he had been responsible for Lombardo's rise to power. In 1936 President Cardenas had backed a giant national labor confederation with Stalinists, Trotskyites and Marxists on the first executive committee. The new labor group took the name of Confederation de Trabajadores de Mexico (CTM). It was with Cardenas's specific approval that Lombardo became the Secretary General and virtual dictator of this new labor organization. Therefore, Lombardo owed Cardenas a favor.[576]

After all, Lombardo contended, what could possibly happen to Trotsky in peaceful, well controlled and socialistically advanced Mexico? Lombardo contrasted the benign political climate of Mexico where Trotsky should need no guards with the capitalist, violent and dangerous environment of the United States where Trotsky would certainly need many guards.

As a result of the hostile PCM demands, the Mexican government had reduced the number of guards in front of Trotsky's home to eight men. Eventually, after equally vehement demands for the assurances of Trotsky's safety came from thousands of his supporters from all over the world, the guard contingent was settled at ten. The guards were split into two sections, each of four or five going on duty for twenty-four hours shifts. In addition, to their side arms, the guards were given five guns with five bandoliers of ammunition with a reserve of 250 rounds. Jesus Rodriguez Casas, an amiable Mexican who became a close personal friend of Trotksy's, was the chief of the guards. At the time of the raid he had occupied the post as chief for the entire time Trotsky had been in Mexico, i.e., about three years and four months.

Augmenting the Mexican guards outside, were Trotsky's personal guards inside the compound. They were Charles Cornell of Fresno, California; Harold Robins from Detroit, Michigan; Jack Cooper of Minneapolis, Minnesota; Walter Ketley of New York City; Robert Shelton Harte also from New York City and Otto Schuessler from Germany. Schuessler's wife was living with him inside the compound.

Siqueiros checked his watch again and peered searchingly into the dark for the other groups., He noted several individuals from the other groups approaching also dressed as policemen. Sanchez's group, that was led by Siqueiros, was in front of the other groups and was the first to reach the police shack.

Siqueiros's men, shouting "Viva Almanzan!" pounced on the policeman in the shack and quickly subdued them. The captured police were trussed up before they could offer any resistance. Even those policemen who were not napping had been lulled into false security when they first became aware that the men approaching their post wore police uniforms. In order to divert suspicion away from the real sponsors of the raid, someone shouted, "Long live Almazan!" The men said it was difficult to determine in the dark and confusion but it appeared that it was Siqueiros who gave the shout.

Two other captured policemen were brought into the shack by the other raiders. All policemen were disarmed and securely bound with the ropes Siqueiros had brought. Sanchez and another raider, El Provencial, were told to stay in the shack to guard the trussed-up policemen.

The raiders from other groups quickly assembled by the police shack to join with Siqueiros's group. Most of the newcomers were dressed in dark civilian clothing that bore a superficial resemblance to police uniforms in the dark.

The accents of the new arrivals made the Mexicans immediately aware that they were foreigners. These newcomers probably included Codovilla and the rest of the Spanish communists from the International Brigade.[577]

The combined raiding party, now commanded by Siqueiros, crossed quickly to the large fortified gate to Trotsky's compound. They paused for a moment to exchange a few words with the guard on duty inside, probably Robert Sheldon Harte. The gate was opened and the raiders passed quickly inside.[578]

Sanchez claimed he saw, and later deposed in his statement to the Secret Service, that it was Robert Sheldon Harte, who opened the gate.[579] He recalled that during the ride from Cuba street, Siqueiros had repeatedly assured the men that nothing could go wrong with the plans for the raid. He strongly implied that one of Trotsky's guards had been bought off and would cooperate with the raiders by opening the gate.

During Jacson/Mercader's earlier visits to the Trotsky compound he had become friendly with the new guard, Robert Sheldon Harte.[580] This friendship appeared to spring full blown almost as soon as Harte arrived from New York. Jacson/Mercader let Harte use his car, of which he was inordinately possessive, for personal outings. It was probably poor security for these two GPU agents to become so friendly, especially when Jacson/Mercader took Harte whoring in Mexico City.

The combined raiding groups went quickly and noiselessly through the open gate and swiftly spread out inside the compound. They opened fire at random, spraying the inside of the compound indiscriminately with withering sustained bursts from submachine guns and pistols. The initial prolonged bursts lasted only a few minutes but seemed much longer to those inside the compound. Sporadic firing continued as real and imaginary targets of opportunity appeared in the shadows of the garden shrubbery or in the elusive play of moonlight along the side of the house.[581]

Inside the house when the raid began were Trotsky's three servants: Carmen Palma, the cook, Helen Estrada, the general housework maid and Melquiades Benitez, the gardener. The guards on duty were: Harold Robins, the shift chief, Otto Schuessler, Walter Ketley, Charles Cornell, Jack Cooper and, the latest arrival from the United States, Robert Sheldon Harte.

Also in the house were Alfred and Marguerite Rosmer, the French Trotskyites who were visiting the house as guests of the Trotskys. The Rosmers were old friends of Trotsky whom he had known in Paris during the First World War before he had been expelled from France.

Alfred Rosmer was a liberal leftist revolutionary, confirmed Trotskyite and journalist of international renown. Rosmer and his wife had arrived from Paris only a few months before the raid bringing Trotsky's grandson, Esteban Volkov, affectionately called Seva by his grandparents. The Rosmers' visit with the Trotskys was almost finished and they intended soon to return to France. Also in the house at the time of the raid were Natalia and Leon Trotsky and their grandson, Seva Volkov.

When the raiders struck, the guard, Otto Schuessler, became instantly wide awake, grabbed for his machine gun but the breech was blocked and he could not return the raiders' fire. The overall precautions that Trotsky had adopted for the security of the house, plus the police on patrol outside the compound, made the house an almost impregnable fortress. It was generally conceded by armament experts that the system of defense was good, that is, unless someone opened the gate and let in the enemy!

The basic wooden gate, had two panels on either side, with the total width more than adequate to allow a car to pass through. These two halves had been reinforced with metal plating on the outside. This reinforced gate had a locking mechanism on the inside augmented by two large wooden bars that were kept across the center of both halves. By lifting one of the wooden bars the gate could be opened a few inches. The guard on duty inside could then look out to see the person waiting outside. Outside, directly above the middle of the gate entrance, was a strong electric flood light which illuminated the space in front of the gate. The whole area in front was bathed in light to prevent anyone from approaching too close to the gate unnoticed.

Charles Cornell was asleep in the guards' dormitory when the raiders struck. He became quickly wide awake when he heard the sounds of machine gun fire. Initially he thought the heavy firing was from his fellow guards repelling some attackers. Cornell's recollection after he woke was a voice shouting from the garden area in English, "Don't move and nothing will happen to you!"

There was no question in the minds of Trotsky's American guards that the voice that shouted this instruction to stay down and they would not be hurt was definitely an American voice and was not one of them.[582]

Cornell later deposed that he scrambled frantically in the dark trying to find his pistol. He knew the voice that had just shouted in English did not belong to anyone living in the compound. Only then did he remember that he did not have his pistol. It was a common practice of the guards to lend their weapons to one another. The night before the raid Cornell had loaned his pistol to Harold Robins.

Cornell stumbled toward the door of his room. As soon as Robins, from his own vantage point saw him in the dim light, he yelled, softly but with suppressed urgency, "Put your head down, Charles! Don't let them see you!"

Cornell peeked cautiously out of his partly opened door in the direction of the guard post at the gate. He saw three shadowy figures; two men dressed in dark suits and the third man dressed in a suit of lighter color. Cornell quickly turned back into his room to get a rifle. At the same time, amid all the confusion, Cornell could only think of arming himself and getting out of his room into the courtyard below to repulse the invaders.

He remembered that he was wearing light colored pajamas, in contrast to the dark garb of the invaders that offered protection in the darkness. He put on a dark top coat over his pajamas. "I might get shot," he thought, "but at least I won't give them an easy target!" Shrugging his arms into the coat he turned to leave the room when Robins called again, "Don't show yourself, Charles. Keep you head down!"

Robins, at the door of his room, could see the man firing a submachine gun.[583] He stayed alert and ready with his rifle at the crack of his door. Robins deduced that the confusion among the invaders kept them from mounting any concerted attack. He doubted if they were yet in full control of the entire compound. He saw a furtive figure run across the courtyard and threw a quick shot at him but missed. He was certain it was one of the invaders because he knew intuitively the physical conformation and running gait of everyone in the house.

About 4:00 a.m. on 24 May 1940, a staccati noise like the rattling of rocks in a tin can broke the stillness inside Trotsky's home. Trotsky, tired and drowsy, barely awake, thought he was hearing the sounds of a typical Mexican celebration with fire crackers. He smelled gunpowder and became immediately wide awake. He realized in a flash that "they" had come to kill him!

Natalia, a little quicker and more alert than Trotsky, jumped out of bed, pushed Trotsky under the sturdy wooden bed frame and shielded him with her body. She contorted their positions until she had Trotsky wedged between the bottom of the heavy headboard and the bedroom wall. The firing stopped and still they lay, tightly huddled and motionless not knowing what might come next and fearing the worst.

The firing resumed, criss-crossing through the windows and doors, almost two hundred shots in all; rapidly, randomly fired into the bedroom. About one hundred shots were fired into the walls close to their hiding place and into the bed where they had just been sleeping.[584] Natalia raised herself to peer through a slit in the bed frame but Trotsky dragged her back down.

Trotsky's own description of the 24 May raid bears relating in full.

> Natalia and I were sleeping in our bedroom, which is between the study and our grandson's room, when we were suddenly awakened by a brisk firing. My first impression was that it was the noise of firecrackers or fireworks, such as are used here [in Mexico] for religious feasts. But I soon realized that they were reports made by guns. We quickly threw ourselves under our bed. Natalia pushed me towards

the southeast corner of the room and forced me to lie flat on the floor., She remained standing for a few moments, as though to protect me by her body until I made her lie down beside me. I am certain that is what saved us both. We were able to see that they were shooting towards the interior of the room from the door of Esteban's [Seva's] room, the door of my study, from the bathroom and from the bedroom windows. All at the same time. A real cross-fire. We remained stretched out on the floor until it had ceased, and then saw they had set an incendiary bomb at the door which communicated with our grandson's room, We thought for a moment that the little boy had been killed.

The assailants numbered about twenty and were armed with machine guns. Doubtless they thought they had killed us. When the firing ceased, my wife and I hurried to extinguish the fire which was already burning the door and floor boards where they had placed the thermos bomb containing incendiary matter. From my grandson's room they had also thrown another, but it fell outside the window which give onto the garden and burnt part of the lawn. There is not doubt the attackers intended not only to kill us but to set fire to the house.

With the object of getting rid of all traces and getting rid of my archives. They had already tried to do the same thing in Paris, by melting down a steel door with a blow pipe [torch] and destroying papers which weighed 150 pounds. This time they had even greater reason to destroy my records for everyone knows, and the GPU knows better then anyone, that I am writing a biography of Stalin. That is why they had brought several incendiary bombs with them.

Our first thought after the attack was for Esteban [the grandson] whom we feared must have been killed but he ran towards us almost as soon as the firing ceased. The child had been surprised like us, by the attack from the assailants but nothing had happened to him because he too had thrown himself on the floor and instinctively hid under the bed.[585]

Seva had in fact screamed to Natalia and Trotsky from his room next to theirs, "Grandma! Grandpa!." Natalia recalls that this scream chilled them to the bone.

Trotsky later wrote that his first thoughts were that Seva had been kidnapped. Before he could move out from under the bed, a man's silhouette appeared at the open doorway. The figure was backlighted from the flames in the courtyard and the flare from an incendiary bomb that was sputtering in Seva's room. The details of the front of the silhouette were sufficiently illuminated to reveal the curve of a military helmet, shining buttons, epaulets on the shoulders, a hawkish aquiline nose and an elongated face with prominent cheekbones. (This description by Trotsky is very close to that of David Alfaro Siqueiros, dressed in a garish army officers's uniform.)[586]

This man stopped at the doorway between Trotsky's and Seva's rooms. He was apparently looking for any signs of life. Although he could not discover any challenges to his deadly authority, he still fired a volley of shots at the beds in both rooms. He then turned and ran towards the courtyard.

The courtyard sounded alive with firing, curses, shouts and warnings from both sides to their comrades to take cover or run or shoot or hold their fire. Seva's room was on fire but he was not in his room. Natalia remembers it best,

> Then it was all quiet. Unbearable silence. Where could I hide him [Trotsky] for safety? I was losing my strength from tension and hopelessness. I thought that at any moment they would return to finish him. Where were all the members of the household, the Rosmers, the secretaries, the guards, the police? Were they all killed? We felt the stillness of the night, like the grave itself, like death itself. And suddenly there came again the voice of our grandson; but this time it was from the patio and sounded quite differently. Ringing out like a staccati passage of music, joyously, bravely, 'Alfred! Marguerite!' It returned us to the living. Seva had saved himself also by hiding under the bed. Even before the shooting had ceased and thinking that his grandparents were dead he went looking for the Rosmers even though he had

been slightly wounded by a bit of flying metal in his left big toe.

Seva remembers his grandfather (Trotsky) dashing out into the garden and taking pot shots with his pistol at a figure on the canal bank above the level of the wall at the back of the compound. This figure turned out to be an old woman. Trotsky had totally missed with his shots.[587]

After the raiding party had fled, taking Trotsky's two cars, the household members assembled on the patio. The Ford was found at the corner of San Pedro and Viena streets and the Dodge on Merida street in Colonia Roma. There was burglary paraphernalia, guns and other incriminating items found in both cars that had been abandoned by the raiders.[588]

Fortunately no one had been killed or seriously wounded. Trotsky himself was cut about the face by flying shards of glass.[589] This lack of serious casualties was to be the basis for the investigators' subsequent skepticism. The Mexican Secret Service suspected the attack was not bona fide but was really a hoax engineered by Trotsky on himself.

The raiders obviously intended to murder Trotsky and his family because they aimed only at their bedrooms. Never, during the entire few minutes of the raid, except for a few shots in what they perceived to be self defense against the guards, did the raiders fire at any targets other than the bedrooms.

Convinced that Natalia, nor Trotsky nor Seva could have survived the concentrated, withering fire, the raiders withdrew. They threw an incendiary bomb into the garden and house as they retreated. For good measure they left a powerful demolition bomb (which failed to explode) on the patio. They departed in Trotsky's two cars which were always parked in the garage inside the compound behind the main gate with the keys left in the ignition.

Robert Shelton Harte had disappeared with the raiders. The Mexican police outside the compound who had seen the raiders leave, maintained that Harte had left willingly with the raiders and was obviously not putting up any resistance to going with them. Two of the raiders had been guiding Harte by the arm but he definitely did not appear to be under any form of compulsion or restraint.[590]

Their miraculous escape was a source of great joy to Natalia and Trotsky. They were safe, Seva was safe, none of the guards nor their dear friends had been hurt. Most of all, their arch enemy, the villain Stalin, had once again been foiled. Trotsky gave every outward appearance of being amused that such a gigantic raid, obviously so painstakingly planned and so laboriously executed, could have failed so miserably.

On the other hand could the whole thing have been a gigantic charade? Could the raid and all it seemed to imply been nothing more than a warning? In the years to come that is exactly what Siqueiros tried very hard to make everyone believe.[591]

The simple tragedy of the raid from Stalin's point of view was that Siqueiros had badly botched the job; so badly in fact that the GPU wanted to divorce themselves from all responsibility for the raid. The Mexican authorities were keeping watch on a Soviet merchantman anchored in the port of Manzanillo. It was feared the ship might be used as a means of escape for Siqueiros, his wife, his two brothers-in-law and other people wanted by the police.[592]

The GPU calculated, after the debacle of the raid was sorted out, that the task of murdering Trotsky would now be infinitely more difficult. Trotsky had been alerted and would doubtless increase his defenses. Further, world opinion had been piqued and would more strenuously denounce the perpetrators when the killing was subsequently accomplished.

For a short time after the raid there was a contrary opinion. A few staunch comrades opined that the raid might have been an independent maverick action by Siqueiros and the PCM. They speculated, encouraged by the GPU to take the heat off them, that Siqueiros and the new hierarchy of the PCM, thought they would please Stalin by doing what they thought he wanted done.

> Twenty-one persons believed to have been involved in the attack have been arrested. There are four who so far have failed all attempts at capture. According to information in the hands of the police the attack was financed by the PCM. The organizers were Siqueiros, his brother, Alfredo, Antonio Pujol and Pedron Zuniga Camacho. They belong to the PCM and all fought in the Spanish Civil War. Most of the information came from Nestor Sanchez Hernandez, who was

> paid to take part in the attack. He said each assailant received 250 pesos. (The Mexican peso was quoted as being worth .20 cents US or about five pesos to the dollar.)[593]

The conduct of the raid itself raised many questions. Who besides Harte could have been a penetration inside Trotsky's compound? Who gave the raiders the detailed information about the interior of the compound they seemed to have? Someone inside must have been supplying information before Harte arrived. If so, who? Why had Harte left with the raiders? It would have been better GPU planning if Harte had remained behind to report on the aftermath of the raid. How had the raiders avoided all of the electric alarm wires and booby traps throughout the garden and compound? Had someone inside conveniently short-circuited or cut the wiring of the alarms? (Harte had asked Trotsky for the key to the alarm central the evening of the raid.)

Why had Harte been taken with the raiders when he would have been free of suspicion if he had remained? Trotsky's comment in defense of Harte is part of the answer. Trotsky often replied to those who suspected Harte might have been a GPU penetration, that such speculation was nonsense. Trotsky pointed out that if Harte had wanted to kill him he could have done so many times before the raid. He could have simply stabbed Trotsky in his sleep or while he had his back turned tending to his rabbits in the garden.

After the raid, except for the exuberance of being alive, the foreboding and deep doom that hung over the compound was pervasive and stifling. Another attack was expected imminently. Nerves were taut and frazzled. Trotsky could only explain that his own and Natalia's narrow escape was a freak of luck. Trotsky always gave Natalia's courage all the credit for saving his life by pushing him under the bed.

On succeeding days, when they arose from the night's sleep, Trotsky would play a little game with Natalia. (This little game was not enjoyed by Natalia as much as Trotsky seemed to enjoy the macabre humor of his charade.) Each morning when he got out of bed, Trotsky would stretch and say to Natalia, "You see, they did not kill us last night after all and you are still dissatisfied." On occasion he would add plaintively, as an endearment to her, "You see, Natasha, we have been given a reprieve!"

Trotsky refused to cringe in terror waiting for another attack or to work less than he had before. He remonstrated that there had been so much to do before the raid and now even more, in order to fight the injustice of the attack that had endangered his wife and precious grandson. He wrote replies to the never ending lies that were printed about him in the communist press.

Trotsky wrote for publication in the world press that those who were guilty of the political preparation for the attack were such organizations as "The Friends of the USSR," an organization affiliated with Vicente Lombardo Toledano's Confederation de Trabajadores Mexicano (CTM).[594] Also taking part in the propaganda vilification of Trotsky were the newspapers controlled by Lombardo such as, *El Popular*, *La Voz de Mexico* and *El Nacional*. Colonel Sanchez characterized Lombardo as "Moscow's principal instrument in Latin America."[595]

During the second week in June a group of Trotskyites from Minneapolis came to Coyoacan to discuss the Fourth International with Trotsky. They were James Cannon, Farrell Dobbs, Sam Gordon and Antoinette Koniakov. Jacson/ Mercader took them for a ride in his Buick sedan to Toluca. While at Trotsky's the four Minneapolis comrades discussed whether there could be a realistic revolution in the United States to seize power. Although this was ridiculous, Trotsky still maintained that the Fourth International would change the world. Trotsky wanted them to try a small test revolutionary uprising in Minneapolis.[596]

This group begged him to allow them to smuggle him into the United States. They assured him he could live there peacefully, well protected and unafraid. Trotsky refused, saying that he could not sulk all his life nor could he do his best work if he was living furtively.

Reluctantly he did yield to his American friends, and the many concerned Mexican authorities, who urged him to strengthen the defenses of the compound. Accordingly, new guard towers were erected at the corners of the walls, new reinforced concrete was used to patch-up crumbling masonry, armored doors and steel shutters were installed. Trotsky viewed these embellishments with distaste. He commented to one of his secretaries, Joseph Hansen, that all of the fortifications reminded him of the first jail he had been in, "The doors make the same sound," he said.

One day Hansen caught Trotsky looking at the new turrets with their gun slits, with what might have been an expression of incredulity. Trotsky turned to Hansen and commented bitterly, "Highly advanced civilization — that we must make such construction." Hansen added that Trotsky seemed like a man waiting in the condemned cell for the fatal day of his own execution.

Trotsky seemed determined to make the most judicious use of every hour left to him. Fortunately his keen mind, sense of irony in viewing the future realistically, and a bubbly good humor kept Trotsky going on an even keel.

He developed an unique philosophy because of his virtual imprisonment. He believed the world was relapsing from what might have been an age of progress and triumph for humanity, back into the savage cruelties of the barbaric ages. Even he, surrounded as he was, by guard towers, ramparts and seige walls, was somehow, he thought, involved in this general back sliding.

Despite his neo-fatalism, Trotsky refused to wear a bullet proof vest that his American friends had sent him.[597] He gave the vest to his guards to wear. He refused to permit visitors to be searched for weapons even though his guards had introduced this procedure into the compound's regimen. Trotsky also made his dislike obvious when any of his guards or secretaries sought to be present when he was talking to visitors.[598]

INVESTIGATION OF 24 MAY RAID

A little rebellion now and then is a good thing.

Thomas Jefferson

In subsequent evaluations by some historians, Trotsky's guards have been blamed for not being more forceful in repelling the raiders on that fateful 24th of May. In all fairness, the guards were surprised but did all they could in the face of the overwhelming firepower of the raiders.

Lt. Jesus Rodriguez Casas of the Mexican police guard in front of Trotsky's house and five of his men were detained for questioning. Rodriguez was off on the day of the raid but the police took him in anyway; they wanted to know why he had taken that particular day off.

Colonel Leandro Sanchez Salazar, a former Mexican army officer on detached service to the Mexican police as Chief of the Secret Service, assumed command of the investigation of the 24 May raid on Trotsky's home.[599]

The Chief of Police of the city of Mexico, General Manuel Nunez, was in charge of the investigation of the raid for the city but Colonel Sanchez was in charge of the Federal investigation. There was apparently no jurisdictional disputes during the investigation of this case.[600] However, Colonel Sanchez describes General Nunez as his superior.[601]

At the beginning of his investigation Colonel Sanchez counted seventy-three bullet holes in Trotsky's and Natalia's bedroom. This included two lethal holes directly through the bed exactly where Trotsky had been lying.[602]

On the banks of the Rio Churubusco, close to the intersection of Viena and San Pedro streets, near Trotsky's home, Colonel Sanchez found Trotsky's Ford motor car stuck in the mud of the river bank at the ford over the river. This was one of the get-away cars the raiders stole from Trotsky's garage. Near the car was a rope ladder with a grappling hook at one end, apparently abandoned by the raiders. Not far from the car the police found an electric saw, an iron jimmy-pry bar (picturesquely referred to in Mexican street slang as the "holy child"), two Mauser hand guns that had been taken from the police guards, some .45 caliber cartridges and a loading tool for Thompson sub-machine gun clips.

The second get-away car that had been stolen from Trotsky's garage was a Dodge. It too was found abandoned some distance from Trotsky's home on Merida street in Colonia Roma, almost to the center of Mexico City. Inside the Dodge were some civilian clothes of a dark blue material which could have easily passed, in the dim light outside the compound, as police uniforms. Also found were two black leather belts (the kind customarily worn by the police), some 35 mm cartridges and an axe.

Colonel Sanchez at first believed that Trotsky had engineered the raid against himself to arouse world sympathy for his exile predicament but he soon discarded this notion. Narciso Bassols was the former Mexican Ambassador to Moscow and France, serving as a prominent lawyer in Mexico City for communists and ultra-liberals. He did all he could to convince the police that Trotsky had engineered the raid on himself to garner sympathy. The PCM, not to be outdone, issued a statement that the raid had been organized on Trotsky's behalf by the Dies Committee of the United States House of Representatives.[603]

The good Colonel Sanchez continued to believe someone inside the house had cooperated with the raiders. He, of course, suspected that Harte was at least one of the raider's penetrations inside the compound but he could not garner enough immediate evidence to prove it. He reasoned that others, from the close-knit fraternal group that lived inside the compound, might be able to help identify other penetration agents.

Sanchez knew he must interrogate members of the household for leads and this was bound to arouse resentment in this mutually loyal group. Accordingly, he choose those least likely to withstand his interrogation despite how well they may have been coached by Trotsky, if in fact he had planned the raid as a diversion.

All of the insiders wished to preserve the family secrets from the prying eyes of the Mexican police. The insiders felt that anything the Mexican police learned about the raid and what went on inside the compound would subsequently be obtained by the GPU through their informants within the Mexican police. Colonel Sanchez also wanted to get information from insiders who were not part of the politically loyal group that demanded silence among Trotsky's group of foreigners. Colonel Sanchez therefore arrested the Mexican cook and Mexican parlor maid.[604]

He took them to a jail called "El Pocito." (This means The Little Well and is aptly named. Rumor has it that the name is derived from the fact that after an interrogation inside that jail, the prisoner becomes a veritable fountain of information.)

The cook told the Colonel that she had observed the guard, Otto Schuessler, go past her room in the servants' quarters below the guards' dormitory with a revolver in his hand immediately after the shooting started. Schuessler had taken a position by the kitchen door where he could observe the melee in the court but, according to the cook, he had not fired a shot. Both the cook and the parlor maid had observed Schuessler and Charles Cornell as being very nervous the evening before the raid. They both opined that these two guards had shown increased signs of nervousness ever since the raid. Colonel Sanchez promptly ordered the arrest of the two nervous guards and had them brought to El Pocito.

The two guards readily admitted their actions during the raid. As soon as the shooting stopped they had both run to the door leading into Trotsky's bedroom but it was locked. They had then run into the garden and discovered the two cars and Harte missing.

Trotsky complained to President Cardenas that his two loyal guards and servants were innocent of any complicity in the raid and had been unjustly arrested by Colonel Sanchez. The maligned chief of the Secret Service, who was merely trying to do his duty, was reprimanded by his superiors. President Cardenas forced him to apologize to Trotsky for arresting the servants and guards.

The point that Trotsky was making, and he stressed this with the Colonel, was that he was not too concerned about the arrest of his people. He could, however, tell from the arrests that the police, as directed by the Colonel, were on a fruitless quest and needlessly expending investigative energy. Trotsky persistently maintained that the real culprits like Siqueiros, and the other GPU hirelings, should be the ones the Colonel should concentrate his investigative assets upon. Trotksy insisted that when all the investigative leads were uncovered they would ultimately lead directly back to the arch villain, Joseph Stalin.

United States Foreign Service Officer, Robert G. McGregor, Jr., had a conversation with Trotsky regarding the raid:

I [McGregor] mentioned [to Trotsky] statements attributed to him which appeared in the [Mexican] press implicating Narcisco Bassols. [Bassols was a leading Mexican Stalinist who had been the Mexican Ambassador to France in the 1930s. He had also held several Mexican government posts. He had visited Moscow on many occasions and was rumored to be a willing tool of Stalin in the Western Hemisphere. He was leading the propaganda campaign in Mexico to force President Lazaro Cardenas to expel Trotsky from Mexico.] Trotsky said his statement had been misinterpreted and referred to his declaration immediately following the attempt a month ago, to the effect that David Alfaro Siqueiros and Narciso Bassols would be able to give much information concerning the attack. I told Trotsky that I heard sometime ago [that] on one of his trips Bassols, when he was Mexican Minister to France, had endeavored to persuade President Cardenas of the desirability of expelling Trotsky from Mexico. Trotsky confirmed this information, stating that Bassols had undoubtedly received orders from the GPU agents in Geneva, Switzerland.

In a strictly confidential and private manner Trotsky told me he suspected the order for the attempt on his life came through the Soviet Ambassador in Washington, Oumansky who, according to Trotsky, is a GPU agent. Trotsky stated that lacking a diplomatic legation in Mexico, the Soviet officials in Washington maintain an agent here, who is none other than American citizen Harry Black. [Black was well known as a Stalinist and the managing editor of Vicente Lombardo Toledano's magazine *Futuro*. Black was often used, in order to make him feel he was a part of the GPU team, as a secret courier between the senior GPU representatives in Mexico and Washington.][605] He [Trotsky] pointed to Black's article in *The Nation* early in June [1940] entitled, "The Phantom Conspiracy" which endeavored to explain away the attempt on Trotsky's life. Trotsky said that Black works closely with Vicente Lombardo Toledanop and the CTM [Confederacion Trabajodores de Mexico - Mexcian Confederation of Workers - a Mexican labor union

> dominated by Lombardo and the PCM.] and while he does not go so far as to say that Toledano [sic] is a GPU agent or affiliated with the Stalinist group he did say that Black was an extremely dangerous man.[606]

At this same time, during the investigation of the attempt on Trotsky's life, the intrepid macho, Diego Rivera, fled Mexico to the United States. He claimed the reason he left for safer climes was because he had arranged for Trotsky's asylum in Mexico. Therefore he reasoned, the same people who were after Trotsky would soon be after him. This was either pure fear on Rivera's part or a chance to get some publicity to sell his paintings.

When Colonel Sanchez apologized for the arrest of Trotsky's staff, he told Trotsky what the servants had told him about Schuessler and Cornell being nervous before and after the raid. Trotsky explained to the Colonel that everyone in the compound often displayed signs of nervousness and worked under great stress. For example, Trotsky told the Colonel that the night before the raid his typist, Madame Fanny Yanovich, who usually worked only to three or four hours a day, remained the entire day to help Trotsky finish some important work. Trotsky also remembered that at about five o'clock in the afternoon of the day or the raid, Harte had come into his study to ask Trotsky for the keys so that he could check the alarm system inside the compound.

The alarm system connected sensors that were placed at strategic points around the house and garden to a central panel. On the brick and masonry wall that surrounded the property there was a network of electric contacts, placed so that it would be impossible for anyone to scale the walls without touching one of them. The slightest pressure on these contacts would cause a bell to ring at the central. At the same time a small light bulb on the control panel would blink off and on in a position corresponding to the location of the intrusion on the wall. This control panel not only signalled the danger but also the exact spot on the wall that was being breached.

As was bound to eventually occur, during the post-24 May investigation, Colonel Sanchez began to make arrests of the real culprits involved. One day Colonel Sanchez happened to stop at a bar at the corner of Arcos de Belen and Luis Moya about one o'clock in the afternoon, He overheard five "tramway employees" discussing the raid.

One said the police magistrate at Tacubaya told him how he had supplied police uniforms that were used in the attack. From this lead the facts of the raid were consequently revealed.[607]

Sanchez learned that a member of the raiding party, one Nestor Sanchez Hernandez, had been a captain in the International Brigade during the Spanish Civil War. When he arrested Serrano Andonequi, another one of the raiders, the Colonel found an envelope in his possession on Majestic hotel stationery addressed to Nester Sanchez Hernandez at 54 Avenida Guatemala in Mexico City. By 17 June Colonel Sanchez had arrested some two dozen persons who were allegedly involved in the raid, including four chauffeurs who had driven the attacking force to Coyoacan.[608]

Colonel Sanchez was impressed by the similarity of the backgrounds of the raiders as he uncovered their identities. He commented that it was abundantly clear that many former communist Republican combatants of the Spanish Civil War had been used in the attempt on Trotsky's life. He then began to realize that behind the communist veterans was the power of the GPU. By this time, the good Colonel had discarded his original theory that Trotsky had orchestrated the raid himself.

The Colonel discovered that Nestor Sanchez Hernandez had disappeared from 54 Avenida Guatemala. Continuing his search for others he discovered, among the papers left behind at Serrano Andonequi's house, the lone address of Corregidora 101. The door keeper at this address was Nestor Sanchez Hernandez's uncle.

The uncle told the Colonel that his nephew, Nestor, had stored a suitcase with him. A few days after the raid, the uncle guessed it might have been the 28th or 29th of May, at about half-past nine o'clock in the morning, nephew Nestor had brought a large suitcase with a blue label and white crosses on it (for identification) for his uncle to store.

Later nephew Nestor returned to remove something from the suitcase but the uncle could not see what it was. On one other occasion there had been a letter waiting for nephew Nestor at the uncle's apartment that had been forwarded by Nestor's father. Nestor read the letter in the uncle's presence but the uncle did not know what it contained. When he left, Nestor warned his uncle not to tell anyone that he had ever visited the uncle's apartment.

The Colonel opened Nestor's suitcase and found a police uniform with the insignia of the 7th Company, two bandoliers of ammunition and a Star automatic. There was no doubt that this evidence implicated Nestor in the raid on Trotsky's compound — not only because of the police uniform but the Star automatic had been taken from one of the Mexican police officers on duty guarding Trotsky's gate. The Colonel soon arrested Nestor Sanchez Hernandez.

Nestor Sanchez Hernandez was a youth of 23 years of age and a law student. He was not a member of the PCM but readily admitted participating in the raid. He claimed he had done so only out of friendship with David Alfaro Siqueiros.

Sanchez told the Colonel that he was convinced the individual used by Siqueiros to open the gate had been Robert Sheldon Harte. He pointed out that the attack took place at exactly the day and time when Harte was on duty at the gate. It was beyond a doubt that Harte had been bribed by Felipe (the French Jew), according to Sanchez. Further, he told the Colonel, it was evident that they had known each other before the raid because of the friendly way they talked and acted in the car after the raid.[609]

Felipe and Harte were in a get-away car that had been stolen from Trotsky's garage and Sanchez had been riding with them. During the drive he heard their conversation in English but could not understand what they were saying. He did note that the conversation was friendly and animated, strongly suggesting that Felipe and Harte were considerably more than casual acquaintances. Sanchez was not able to add more about the identity of Felipe but did hear Harte refer to him in an intimate form of address as "Felipe."[610]

J. Edgar Hoover, the Director of the American Federal Bureau of Investigation, informed Colonel Sanchez that the instigator of the attack on Trotsky had been a man called George Mink. Hoover claimed that Mink had arrived in Mexico from Philadelphia a few days before the raid. George Mink, according to Hoover, was one of the principal GPU agents in the world. He had already successfully accomplished killings and other important missions for the GPU in Spain, Japan, the United States and other countries.

> The Mexican police began a search today for an American named George Mink, in connection with the attempted

> assassination of Leon Trotsky.... It was reported that Mink and Herbert Pissman, a German-born visitor to Mexico, were being sought for questioning.... Pissman was reported to have arrived in Mexico ten days ago.[611]

Being a thorough investigator, Colonel Sanchez tapped the telephone lines of the principal suspects. He began this technical operation by tapping the telephone lines at Trotsky's compound.

In Robert Shelton Harte's room at the Trotsky compound the Colonel found a key to room 37 of the Hotel Europa in Mexico City. He learned that on the evening of the 21st of May, Harte had spent the night there with a prostitute. The prostitute told the Colonel that Harte had been very drunk that night and had a large sum of money with him. There was independent evidence from some of Trotsky's other guards that Harte had a large number of American Express Travelers' checks in his possession just before the raid.

Jesse Samual Harte, Sheldon's father, arrived in Mexico to help in the investigation of his son's disappearance.[612] Jesse Harte was the president of International Factors Corporation and apparently a politically influential person in the United States. He carried letters of introduction to Mexican officials and to the United States Ambassador from several prominent United States Senators and from J. Edgar Hoover — all of which meant absolutely nothing to Colonel Sanchez who was already doing his utmost to investigate a very complicated case.

Jesse Harte had some interesting tid-bits of information for the Colonel. He said his son, Sheldon, had never told his family he was working for Trotsky or that he had anything to do with the Trotskyite Fourth International. He had told his family that he was going to Mexico to work for a Mr. Williams but that he had had a falling out with him.

He claimed in his letters sent home that he was maintaining himself by giving English lessons to the daughter of the owner of the boarding house where he was staying. Obviously this was a trumped-up story by the GPU as a cover for Harte since he was really living in the Trotsky compound.

Harte had asked his parents to send letters to him care of the Wells Fargo office in Mexico City. This made it more difficult for anyone from home who inquired about Harte to learn he was working for Trotsky. He could have had his letter sent to him in care of Trotsky's address on Viena street in Coyoacan if he had not been trying to deceive his family.

After Harte's disappearance, Harold Robins, one of his fellow guards, was cleaning out Harte's room in the compound. Robins found a Spanish-English dictionary with the signature of David Alfaro Siqueiros inside the front cover. Harte's trunk in his room at Trotsky's house had a sticker from a Moscow hotel on it.[613]

Still more strange, for the die-hards who believed Harte's fundamental loyalty to the Trotsky cause, was his father's comment when he found out that Sheldon was working for Trotsky. The father reported that his son had no sympathy for Trotsky, or Trotsky's cause since he was a staunch Stalinist.[614] In fact, Jesse Harte explained, his son so revered Stalin that he had a portrait of Stalin hanging in a position of honor above his bed in his room in New York.[615]

Trotsky himself vehemently denied any possibility that Harte had, even remotely, been inclined to cooperate with the raiders.[616] Trotsky claimed, self-importantly, that he was a very good judge of character. Historians cite the Stalin picture above Hart's bed as evidence of Harte's true orientation, justifying his cooperation with the raiders. Others are equally convinced that Harte was fundamentally loyal to Trotsky and cite the fact that Harte could have murdered Trotsky many times while they were together in the compound before the raid.[617]

Trotsky may have had inner thoughts about Harte's basic loyalty. Shortly after Harte's arrival, during some heavy construction work that was being done on the walls and turrets of the compound, Harte was on gate duty. He had to open and close the gate every fifteen to twenty minutes to allow the workers with their wheelbarrows to go into the street to dump their loads and then back into the compound. At the same time, Harte was so engrossed in building a bird cage for his pets that he did not want to be continually interrupted. He handed the gate key to a workman. Trotsky saw this and reprimanded him, "Bob, you might be the first victim of your own carelessness!" This was about a month or so before Harte's disappearance.[618]

Soon after his arrest, Nestor Sanchez called the Colonel to his cell in the El Pocito jail to tell him that about three weeks before the raid he had been with Juan Zuniga Camacho, known as Pedro N. Mr. Zuiniga was a miner whom Siqueiros had brought to Mexico City from the town of Hostotipaquilla in the state of Jalisco, to take part in the raid.

Nestor Sanchez had taken Zuniga to a house where Spanish refugees and Mexican recruits were waiting for the raid to begin. At this house he met Mariano Vasquez Mateo. One of Vasquez's sisters was married to an Italian ex-combatant of the International Brigade. This married couple, Sanchez told the Colonel, lived at 38 Republica de Chile street in Mexico City. This address was also an assembly area for Mexicans and Spanish Republicans preparing for the raid, much like Serranos house at 85 Violeta street.

Nestor Sanchez knew that Vasquez was working for Siqueiros and living at another of Siqueiros's safehouses at 108 Londres street. Sanchez had accompanied Vasquez to the house on Londres about three times and on one of these visits he had met Felipe. Felipe gave Vasquez a telephone number to use in case of emergencies.

The Colonel quickly traced the telephone number to a house in Acacias. The house had recently been occupied by a Cuban woman and foreigner who spoke Spanish with a French accent. The Cuban woman was no doubt Caridad Mercader and the Spanish speaker with the French accent, who used the name Felipe, was no doubt Leonid Eitingon. The neighbors at Acacias told the Colonel that the couple had hastily departed the preceding day.

Inside the house on Acacias the Colonel found ladies underwear that had been bought in the Boulevard Saint Michel in Paris. The Colonel deduced that the couple had come to Mexico from Paris. This itinerary was compatible with Caridad's trip to Paris from Spain to see her son before he left for New York.

The Colonel circularized all Mexican police organizations, all military outposts and border crossing areas in the Republic. He admonished them to be on the alert for three men: David Alfaro Siqueiros, Antonio Pujol and Juan Zuniga Camacho. He added a fourth man later, Siqueiros's brother, Jesus Siqueiros who disappeared at the same time as his brother. There was no evidence to connect Jesus with the raid but his disappearance was enough to raise the Colonel's suspicions.

The Colonel raided an apartment on Calle Chile in Mexico City suspected of being the home of Juan Pujol. In the apartment which was occupied by Pujol's mother and brothers, the Colonel found a suitcase containing stylish women's clothes from America. Pujol's mother explained that the suitcase and the clothing belonged to an American woman who was a friend of her son.

As Colonel Sanchez was leaving Pujol's apartment, Mariano Herrera Vasquez walked into the hallway. Nestor Sanchez had identified Herrera as one of the miners Siqueiros had brought to Mexico City from Jalisco. Herrera was 26 years of age, an electrician and a member of the PCM since 1934. He admitted that he had been living with a woman named Anita Lopez, whom the Colonel recognized as one of the women from Abasolo street by Trotsky's house.

In the rooms formerly occupied by the girls on Abasalo, Colonel Sanchez found rough sketches of Trotsky's house. They were apparently work sheets which had been used in constructing the operational plan for the raid on Trotsky's house.[619]

Many years later, in 1972 before he died, Siqueiros gave his account of the 24 May 1940 raid:

> On 24 May 1940 we said that it was time to break from our inertia. I got hold of an army major's uniform and disguised myself as an officer. Twenty of my companions disguised themselves as soldiers. We took the police guarding the Coyoacan fortress by surprise and immobilized them. We captured the American, Sheldon Hartc, who was one of Trotsky's personal guards, and broke into the patio of the house. I confess that at that moment I was paralysed by the emotion. I had taken part in various clandestine operations and was used to danger. I had participated in political struggles in Central and South American counties. But despite this, I had never found myself faced with the necessity to kill anyone in cold blood.[620]

The PCM and Stalinist press of Vicente Lombardo Toledano stressed, at the time immediately after the 24 May raid, that Siqueiros was "mad" and "undisciplined." They categorically stated that when he was leading the raid on Trotsky's house, Siqueiros was strictly off on a frolic of his own without any connection to, or support from, the PCM.[621]

PREPARATION FOR MURDER

Truth may never die but it leads a tortured life.

Wall plaque in the office of a Chicago/Cook County prosecutor.

There was a universal image of Trotsky presented to the world by the non-communist press after the 24 May raid. He was portrayed as an impoverished, puny, old man facing tremendous odds — a victim of oppression, living virtually friendless and alone in far away Mexico City.

The image sometimes continued, revealing that he was ineffectively watched over by a bunch of inexperienced American college boys who had allowed him to be attacked and almost killed by a gang of professional Mexican thugs. Stalin and the GPU were suspected of complicity by a few more politically sophisticated journalists but, immediately after the raid, there was no general condemnation of the Soviet hand behind the gang of criminals.

The communist press and the GPU controlled media outlets, on the other hand, vociferously branded the raid as a self-inflicted ploy by Trotsky in an attempt to garner world sympathy.

The writings which poured from Trotsky's pen in rebuttal continued to reach out lucidly to those who would listen. Unfortunately they captivated fewer and fewer adherents. The diminishing ranks of Trotsky's disciples were listened to by fewer and less important people. The conflagration in Europe was soon to engulf the globe and make nonsense out of many of Trotsky's predictions. Trotsky was virtually alone and defenseless. Yet the most powerful man in one of the most powerful countries on earth would still go to enormous lengths to kill this man. The reason was, that even though Trotsky did not possess any coercive power he was still the most powerful force in the world as an anti-Stalin symbol and, as such, was by Stalin's definition, a profound enemy that must be silenced.

One hitherto unknown result of the 24 May raid may have been Trotsky's tentative peace offering to Stalin. This was a personal letter from Trotsky to Stalin written from Mexico two months before he died; this would be approximately mid-June of 1940. In the letter Trotsky rebuked Stalin for subverting the ideals of the world revolution. If Stalin would correct the situation, Trotsky wrote, he would be willing to cooperate again with Stalin without any stipulations. Allegedly, Stalin was furious and asked, "How long must we tolerate this scum?" Stalin thought Trotsky's attitude totally intolerable and within two months Trotsky was assassinated.[622]

Trotsky had always been Stalin's scapegoat for all of the disasters that occurred inside that Soviet Union, and after the Siqueiros raid the demonic symbolism became even stronger. According to Stalin there should not have been any calamities or disasters in the Soviet Union since his communist theory was scientific and therefore infallible. The failure of the wheat harvest, for example, could never be explained simply by inclement weather. The low industrial production must never be attributed to the ineptitude of Soviet workers, as that would be a betrayal of Stalin's communist ideology.

The culprits responsible for these disasters had to be found therefore outside of Stalin's communist system. Saboteurs from outside the Soviet Union were easily created, unable to defend themselves and made very believable to the people inside the Soviet Union. Trotsky became Stalin's arch saboteur, to be blamed for everything that went wrong inside the Soviet Union.

One of the tragedies of Stalin's campaign of vilification against Trotksy was the corruption of a wholly new generation of Soviet citizens, GPU officers and Soviet bureaucrats. There should be sorrow in the fact that some of the young GPU officers who were involved in the plan to murder Trotsky were too young to know that Trotsky had been Lenin's right hand man and at the center of Soviet power.

They did not know that but for a quirk of fate, Trotksy would probably have been the leader of all Russia instead of Stalin. It was ironic that the mighty had fallen from such heights. Trotsky had become the scapegoat for a Georgian peasant's failure to efficiently rule the country.

Stalin's propaganda also blamed Trotsky for the failures of the communist system outside the Soviet Union. According to Stalin, Trotsky was responsible for the defeat of the Republican army in Spain.

To the very end though, Trotsky's writing independently proclaimed him to be the consummate politician. Most historians agree, even Trotsky himself, that he was not a prophet. Many of his prognostications based upon dialectic materialism had not come to pass. He had, even up to the end, preached an irony of history that,

> ... a socialist program based on the internal contradictions of a capitalist society has petered out as a Utopia.

Trotsky cautioned in this regard, ever the humanist,

> ... however in this even [a classless society] a minimum program [socialist] would still be necessary to defend the interests of the slaves against the totalitarian bureaucratic system.[623]

The consensus among scholars who understand Trotsky is that he was the prototype for all fighters who show their faith in the basic goodness of the world by proclaiming that faith. Trotsky was a fighter for a true workers' state which, even though Utopian, had to remain his goal for mankind. He put his views forth in his writing for the world to read and ponder their own fate with humanism and rationality.

At the time of the raid Trotsky was sixty years of age. He had a broad chest, thin spindly legs and was of medium height. He had thick grey hair, a high forehead and dark blue eyes. His little goatee was probably his most recognizable facial feature together with his steel rimmed spectacles. He ordinarily parted his hair on the left side of his head. When he wanted to go incognito he would shave off his goatee and part his hair down the middle.

Trotsky usually worked twelve or more hours a day. After the raid he continued his work habits but did not leave the compound for his walks in the country. The only exercise he got was the relaxation of tending his rabbits and fowls. He spent a lot of time working out feeding formulas for his pets. He always asserted that their feed had to be the best and scientifically prepared.

After the raid an event occurred that did not break the routine, yet it was one of the most important events in the entire scenario. On 28 May, four days after the Siqueiros raid, the protagonists of this historical drama did meet face to face for the first time. Destiny was rapidly approaching a show down with fate!

Jacson/Mercader had promised Sylvia that he would not visit Trotsky's home while she was in New York. Despite this promise he had been worming his way into the compound by way of his friendship with Alfred and Marguerite Rosmer.

The occasion for this auspicious meeting between the great historical personage and the soon to be legendary arch-villain, was the scheduled departure of the Rosmers to Vera Cruz where they were to take a ocean liner to the United States. Fortunately the Rosmers had survived without injury from the Siqueiros raid and were eager to return to their home near Paris.

After he learned the Rosmers planned to sail from Vera Cruz on the first leg of their journey, Jacson/Mercader volunteered to drive them there. The prospect of the grueling day-long journey from Coyoacan to Vera Cruz over primitive, winding, mountain roads was gladly undertaken by Jacson/ Mercader. He thought chauffeuring the Rosmers would cement his relationship with them and, through them, he might gain a better entre' to Trotsky.

Jacson/Mercader explained his willingness to undertake such an arduous journey by saying he had business in Vera Cruz. He explained to the Rosmers, that he had to make the trip at least twice every month and he would combine a business trip with taking them to their ship in the Vera Cruz harbor.

Natalia had never seen Vera Cruz and decided she too would enjoy the outing with the Rosmers since this nice young man had so kindly offered to be their chauffeur. Trotsky was eager for Natalia to take this pleasant break from the gloomy routine in the compound, as she had been under considerable strain since the raid. Right after the raid, for the next four or five days, everyone in the compound had been under incessant questioning by the police and the reporters who had flocked to the house from all over the world.

Jacson/Mercader was overjoyed upon learning that Natalia was going to join them on the Vera Cruz trip. He assumed that once he had Natalia as a friend, the next step of getting in to see Trotsky whenever he wanted to, would be easy.

At precisely 7:58 a.m. on 28 May 1940, Jacson/Mercader was admitted to the Trotsky compound for the first time according to the log maintained by the guards. In addition, to Natalia and the Rosmers, Riva Hansen, the wife of Trotsky's new chief secretary and guard, Joseph

Hansen, also had accepted Jacson/Mercader's invitation. Joseph Hansen later reported this momentous occasion:

> Trotsky was in the patio and met Jacson for the first time. They shook hands. Trotsky continued with his chores about the chicken yard. Jacson retired and began speaking with Seva, Trotsky's grandson, to whom he gave a toy glider. Both Natalia and Trotsky noticed him in Seva's room and asked Seva what it meant. Jacson then explained the workings of the glider to them.
>
> Trotsky with his customary thoughtfulness for others asked Natalia if Jacson should not be invited in. Natalia responded that he must have already had his breakfast. At the table, however, as a matter of courtesy, he was invited to come in and have a seat. He took a cup of coffee. This was the first time Jacson sat down at a table with Trotsky.

The drive to Vera Cruz gave Jacson/Mercader an opportunity to play the congenial host. He exerted all his boyish charms on Natalia. However, Natalia developed some doubts about Jacson/Mercader's veracity when they reached the outskirts of Vera Cruz. He appeared to be lost. This was contrary to what he had bragged about on the way down from Mexico City, that he made the trip to Vera Cruz often and knew the way.

When the threesome of Natalia, Riva Hansen and Jacson/Mercader arrived back in Coyoacan, at 3:42 p.m. on 30 May, Jacson stayed in the compound talking with Natalia for thirty minutes. These thirty minutes enabled Jacson/Mercader to increase his knowledge of the interior of the compound, the way the guards handled visitors and their movements patrolling the interior.

On 17 June the Mexican Secret Service, working under the leadership of Colonel Sanchez, cracked the 24 May raid case wide open. They arrested some twenty-five of the persons who had been involved.

The PCM was glad and eager to escape responsibility for the 24 May fiasco. They therefore made a big propaganda show out of disowning Siqueiros, Contreras and all those known Party members who had been arrested or named as wanted by Colonel Sanchez.

David Serrano, the husband of Julia Barrados, who also represented the GPU on the Political Bureau of the PCM, was one of those arrested. Serrano swore that neither he nor Siqueiros had anything to do with the raid on Trotsky's home. Serrano tried to place all the blame on his former comrades who had been ousted from the PCM, Laborde and Campo. Regardless of these diversionary tactics and accusations it was firmly believed by Colonel Sanchez that Siquiros, the hard core Spanish Republican refugees and the communist trade union miners from Siqueiros's home state of Jalisco were the main participants in the raid.

In the global arena of public opinion there was no doubt that Siqueiros and the Spanish Republican mercenaries were the principal perpetrators. A predictable back-lash of the raid were ripples on the public conscience that condemned as a class all leftist artists because of the involvement of Siqueiros, Pujol. All those so-called free-thinkers (meaning those who espoused any kind of radical socialism) and, all those characterized as liberal leftist who had voiced any kind of support for this gangster operation against Trotsky, were also looked upon askance.

A small but discernible change came from the PCM. In an effort to bolster its flagging public image and striving for non-complicity in the raid, it wanted to put as much ideological distance between itself and Siqueiros as possible. The PCM openly denounced Siqueiros as an "uncontrolled element who is considered half mad."

Outside of Mexico the turmoil of world conflict continued to simmer threatening to erupt into a boil, taking the front page of newspapers away from the news of the raid. In May 1940 the Nazi blitzkrieg was trying to over-run Europe. They were spreading a form of socialism they called Nazism while their temporary allies, including the Soviet Union, were spreading their own brand of socialism that they called communism. The individual supporters of each form of socialism claimed theirs was different. Cynics and impartial observers saw them as identical and were therefore not surprised at the 1939 Nazi-Soviet Pact of Non-Aggression. Consequently, in his defensive planning Trotsky had to consider the possibility that the Nazi Gestapo might now be working with the GPU toward his elimination.

Trotsky's will to live was indomitable but the years of Stalin's relentless persecution had taken their toll on his reserve of energy and courage. He was more fearful of death after the 24 May raid than he had been before. Those closest to him observed small signs that he might be slightly more convinced of his own immortality.

Trotsky was convinced that Stalin and the GPU would try again to kill him. The only questions were when and how? He anticipated that the next attempt would probably coincide with some crucial event in Europe of predominant world interest. He prophesied the event would probably have to do with the war or some outrageous act of Hitler. Therefore it would render relatively insignificant the news that a lone Russian emigre, living in isolation in far away Mexico, had been murdered.

Trotsky was not a quitter. He may have resigned himself to the expectation of another attack but he by no means acknowledged that it would succeed. To help him in preparing his defenses, contributions from followers and humanitarians had poured into Coyoacan, especially from the Trotskyite party faithful in the United States. A specialist in military construction was retained to advise on how the already formidable compound fortress could be further strengthened.

A redoubt with bomb resistant ceiling and floor was considered; twenty foot high walls were erected at what were considered vulnerable points around the perimeter; reinforced steel doors with electric controls replaced the former wooden entrance gate; two towers were reinforced with fields of fire directed inwards toward the patio, the house itself and the area in front of the gate; barbed wire was strung along the top of the walls, over the gate and between the towers. Nets were strung over the courtyard and interior of the compound to prevent bombs or mortar shells from being lobbed over the walls.

In addition to strengthening existing security, Hansen and the other guards were concerned about more sophisticated attacks. They speculated that Siqueiros's miners might try to tunnel under the compound walls from a nearby location and dynamite the house; a small plane could dive over the residential area of Coyoacan unobstructed by tall buildings and bomb the compound; Trotsky could be ambushed on the highway (they tried to obviate this possibility by using conveys, decoy cars and split second timing); a time bomb could be delivered to the house in one of the many packages Trotsky received every day from all over the world; poison could be introduced through food brought into the kitchen for Trotsky's meals (the cook was alerted to this possibility and remained vigilant). Demonstrations, with overwhelming numbers of people, could be organized by the PCM which could then force their way into the compound and seize Trotsky for a secret trial. All of these were ever present considerations and ominous portents.

The major problem in defending the house was its fixed position. There was no way that this condition could be changed. If Trotsky's defenders were guerrilla fighters who could constantly change their base of operations they could disrupt the plans of the attackers. There were some hideouts at homes and ranches of friendly Mexicans where Trotsky could be taken in case of emergencies. Conversely, if the GPU discovered the hideouts, taking Trotsky there could be putting him into a more deadly trap which was not as well defended as his Coyoacan compound.

President Cardenas, outraged at Siqueiros's action in belying Mexico's traditional policy of sanctuary for political refugees, ordered the guard force around Trotsky's compound tripled.

Trotsky took an active part in all of the construction work, offering sensible suggestions for improvements. Occasionally, albeit rarely, he even helped with the mixing of cement with sand. He was not so tractable however with the suggestions for changing the security controls inside the compound. He adamantly opposed, in fact vetoed, a suggestion that everyone who entered the compound would have to searched for weapons.

He also swept aside as impossible the idea that he should never talk alone with anyone in his study. Although he would not live long enough to regret these words, wherever he is now he must regret them. He said, "Either we trust the people and admit them without a search or we do not admit them at all." This is reminiscent of that tired old joke about "... except for that one little mistake...."

Another bit of irony is that Trotsky said of the spartan and confining life he had to live in the isolation of his Coyoacan fortress that it reminded him of Kherson, the first prison he was in. He claimed that the doors made the same sound as the gate, a hollow thump, as they closed. He usually ended this reminiscence by gazing wistfully about the interior of his fortress with its high looming walls and saying, "This is not a home, it is a medieval prison."[624]

In a conversation one day with Jacson/Mercader, Trotsky asked him what he thought of the fortress under construction. Jacson/ Mercader glanced over the work being done on the walls and towers and commented with what history must concede is a supreme bit of irony. He told Trotsky, "In the next attack, the GPU will use other methods!." Joe Hansen and Charles Cornell were close by when this remark was made.

Hansen asked Jacson/ Mercader what other methods would the GPU would use to attack the next time. Jacson/Mercader, who already knew what other method would be used by the GPU, merely shrugged his shoulders and did not reply. Trotsky, not to be turned aside in his thinking about a subject that greatly interested him, prophetically supplied the answer as he replied to Hansen's question, "A single agent of the GPU who could pass himself off as my friend could assassinate me in my own home."

On Thursday, 8 August 1940, Jacson/Mercader and Sylvia Ageloff had a social tea with Natalia and Trotsky. They discussed, among other things, whether the defense of the Soviet Union in World War Two should be supported by members of Trotsky's Fourth International while Stalin was still in power? Trotsky supported this view. Sylvia was opposed to giving any help to the Soviet Union while Stalin was still in power. Natalia remained silent (she usually agreed with everything that Trotsky said so everyone knew her position) and Jacson/Mercader remained non-committal.

Jacson/Mercader again visited the compound on Saturday, 10 August 1940 and stayed for 37 minutes, leaving at 3:23 p.m.. This was a personal reconnoitering wherein Jacson/Mercader committed to memory every detail of the interior of the compound that might help him in completing the murder without interference and with his escaping afterwards.[625]

On that day he had carried all of the pieces of equipment he would use in the murder so that he would be thoroughly familiar with how to carry them and use them with natural ease. He had driven his Buick sedan to the door of the compound, parked it facing away from the wall toward the open street; this was the way he would park his car on the day of the murder in order to facilitate his escape. He then minutely examined every step he would take going into and coming out of the compound and house. He commented later that doing all this looking and making mental notes without being obvious to the guards was very difficult.

Jacson/Mercader had come to the compound to discuss with Trotksy an article he was preparing for publication. Jacson/ Mercader had asked Trotsky for this interview to critique his article. He explained that he really wanted to be more active in the Fourth International. To further this ambition he had written an article he hoped would be published. He realized, he said, that he was not a professional writer (apparently having forgotten that, when he met Sylvia, he had claimed he was a professional writer). He hoped however, that the Trotsky would read his article and be able to offer some suggestions that would make it more acceptable. Trotsky had agreed to see him only reluctantly but managed to put on a welcoming countenance. He genuinely liked Sylvia (Jacson/Mercader's common-law wife) and wanted to help her husband in anyway that he could.

This visit was really a dress rehearsal for Jacson/Mercader and lasted exactly eleven minutes. Trotsky was alone in his study writing when Jacson/Mercader entered. Natalia was not at home but out shopping. Jacson/Mercader carried his raincoat over his arm, concealing the pistol, knife and ice axe. He kept his hat on so his hands would be free to quickly draw his weapons. Jacson/Mercader had gone over the schematic of Trotsky's study, studying the plan of his attack countless times.[626]

He should not sit in the chair that Trotsky usually proffered to his guests. This chair was too far away from Trotsky who would be sitting at his desk. Instead Jacson/Mercader should sit on the edge of Trotsky's desk where he could easily reach Trotsky's head with the ice axe. Although this would be very disrespectful to Trotsky, and he might take umbrage, it had to be done to establish this as a precedent with Trotsky. During the next, and what would be the final visit, Jacson/Mercader would again have to assume this close position sitting on Trotsky's desk so he would be in a position to strike the fatal blow.

Trotsky told Natalia the day after Jacson/Mercader's dress rehearsal visit, in which he had affronted Trotsky's sense of propriety. "I don't like him. What sort of a fellow is he ? One should find out more about him." Trotsky continued talking to Natalia about Jacson/Mercader's visit, "Yesterday he did not resemble a Frenchman at all. Suddenly he sat down at my desk and kept his hat on all the while!" Natalia, genuinely concerned about Trotsky's agitation added, "Yes, it is strange! He [ordinarily] never wears a hat." Natalia subsequently stated that she felt that Trotsky noticed something else that was even more foreboding and sinister in Jacson's deportment that day. She also felt that Trotsky did not want to do anything that might offend Sylvia by snubbing her husband.

There has been speculation by some Trotsky historians as to why Jacson/Mercader did not take advantage of the dress rehearsal visit to carry out the murder of Trotsky at what appeared to be an opportune moment.

The answer is very simple, his escape mechanism was not in place that day. Eitingon and Caridad were busy with other matters. Therefore with absolutely no assurance that he would have any kind of a chance for success in making good his escape, Jacson/Mercader did not even consider trying to murder Trotsky during the rehearsal visit.

JACSON AND SYLVIA IN MEXICO CITY

Stalin and the assassin
In frank cooperation
Carried out their crime
With precision to its
Final destination.

Corrido sung in the of Mexico City after Trotsky's assassination.[7]

When Sylvia Ageloff left for the United States, Jacson/Mercader's role for the GPU was primarily to continue to gain access to Trotsky's compound solely to case it for the Siqueiros raid. At this juncture he was not the designated back-up killer.

Previously he had tried to obtain this casing information through adroit but circumlocutious debriefings of Sylvia, her sister Hilda and the Rosmers. He also tried to use dissimulating elicitation with the guards when he waited for Sylvia outside the compound.

The guard schedules were important in order to determine which ones were lazy or more alert and who was to be on duty the night of the raid. Detailed personality information about everyone in Trotsky's house, particularly the guards, was important to determine character flaws or personality weaknesses that might make them susceptible to blackmail.

In planning a semi-military operation such as the 24 May raid, all services and support systems in and out of the target are important — the placement of the guards, the route of incoming power lines, who delivers the bottled cooking gas, the telephone lines, whether the sewers are pump-powered or gravity flow to take the waste water from the house to a deeper main sewer, etc.

---

[7] A corrido is a popular ballad made up by the people spontaneously in bits and pieces in honor of a popular person, usually as a lament on the occasion of his death.

The GPU had arranged for a tap on Trotsky's telephone lines. Ovakimian had made arrangements to send an agent down from New York to monitor the tap.

The roster of the Mexican police outside the front gate of the compound was an important consideration in the comprehensive casing of the target. Most of the information about the police and their activities was being collected by another compartmented operation by the two women in a nearby apartment.

The activities of the servants were also a part of the Jacson/Mercader assignment, i.e., what time did they start their chores each morning, when did they go to bed each evening, when was their day off, where did they go, did they have a relative in the PCM who could be used to appeal to them, how best could the servants be neutralized during the attack? The GPU would also want to make certain that no Mexican servant was killed during the attack since the resulting adverse publicity would be disastrous.

Despite his promise to Sylvia not to go near Trotsky or the compound while she was in New York, fate intervened. In an unexpected manner Jacson/Mercader was given an excuse to renege on his promise. In all candor, he probably never had any intention of keeping his promise to Sylvia not to go near Trotsky; nor any other promise he ever made to her!

Fate's intervention came when Alfred Rosmer became ill. Jacson/Mercader immediately offered to help the Rosmers during Alfred's illness by running little errands for them and driving Alfred to and from the French Hospital in the Colonia de los Doctores on Nino Perdido street.

In order not to alienate Sylvia beyond the point of reconciliation should she learn he had broken his promise to her, Jacson/Mercader decided to take the offensive. Accordingly he told Sylvia when she returned from New York that, because of his past friendship with the Rosmers, they had imposed upon him to help during Alfred's illness. He had had no choice but to go to Trotsky's compound to help them or they would have thought he was rejecting their friendship.

There was no doubt that Jacson/Mercader was turning in casing reports to the GPU after each visit to the compound. In April of 1954, Vladimir Petrov and his wife Evdokia Petrov defected from the NKVD, the successor of the GPU. They had been assigned as NKVD officers at the Soviet Embassy in Australia. During his testimony before the Australian Royal Commission on Espionage, Vladimir Petrov testified that in 1948 he had perused the voluminous file on Trotsky in the NKVD headquarters in Moscow. He testified that it contained,

> Copies of instruction sent out from NKVD headquarters in Moscow to the NKVD residents in the Soviet embassies in all countries where Trotsky had lived at various times, including instructions to the NKVD resident in the Soviet Consulate in New York, who directed the assassination operation on the American continent. There was complete photographic documentation of Trotsky's life, from his first days in the Soviet Union, before his banishment in 1928, right up to his last day in Mexico. There were numerous photographs taken inside the fortified villa, perhaps by Mornard [Jacson/Mercader] himself showing guards, fences and courtyards, photographs of Trotsky with his wife, Trotsky sharing tea with friends.....[627]

On 11 April 1940 Jacson/Mercader moved into a tourist motel called Shirley's Courts. This tourist court was operated by an amiable, rangy Texan from the United States named Jim Shirley. The motel was located between Villalonquin and Sullivan streets just west of the intersection of Insurgents Norte and the Paseo de la Reforma. This is a major intersection in Mexico City.[628]

When Jacson/Mercader checked into the motel he told Jim Shirley that he was a road engineer working on the construction of the Chiapas highway. He said he had come to Mexico City because construction on the highway was bogged down due to the mud during the rainy season.

Shirley's Courts was composed of large comfortable tourist cabins with private bathrooms, surrounded by green lawns and flowering plants. It was considered a luxury motel that catered to the affluent North American tourists who drove their cars to Mexico City. It had ample parking space for cars next to each cabin and was secluded from the surrounding busy city by a high stone wall. It had a good restaurant where the food was prepared under rigid sanitary conditions to help protect the delicate stomachs of the North American tourists from catching the bugs of the dreaded "turista" disease, or better known as the "Aztec Two Step" and "Moctezuma's Revenge."

When he checked into the motel on 11 April 1940, Jacson/Mercader signed the register as Frank Jacson, a Canadian citizen and then left for five days. When he returned he was driving the Buick sedan with Mexican license plates he had purchased at the Hernandez Agency on 7 December 1939. He remained at Shirley's Courts until 12 June when he left for the United States. When he checked in, he had two large suitcases and a heavy trunk with him which he put into storage at the motel. This luggage could have contained some of the weaponry and equipment that Siqueiros used on the 24 May raid. On the night of 23 May, he took his Buick sedan out of the motel garage, checked out of the motel luggage storage room his two suitcases and the heavy trunk and loaded them into his Buick.

He had deposited $900 American dollars with Jim Shirley when he checked in. On the afternoon of 23 May he withdrew $100 and converted it to Mexican pesos.

Jim Shirley's son was an enthusiastic mountain climber. Several times during his stay at Shirley's Courts, Jacson/Mercader had discussed mountaineering with young Shirley. Four years before Jacson/Mercader came to the motel, young Shirley had purchased an ice axe (also referred to as an Alpenstock, piolet, pick-mattock) to use on his climbs. Jacson/Mercader stole this ice axe from Shirley's office where his son had left it after showing it to Jacson/Mercader. This was the ice axe subsequently used by Jacson/Mercader on 20 August to murder Trotsky. Instead of being a new, untraceable ice axe that could not be easily traced back through Shirley to him, this axe was eminently traceable.

The ice axe had inscribed on the side of the handle "Garantie Werkgen Fulpmes" which made it unique in Mexico and made subsequent identification easy. Security conscious GPU planning would have dictated the use of a clean, untraceable ice axe brought surreptitiously into Mexico from some place in American or Europe by a seaman courier or through the diplomatic pouch of a friendly nation.

If Jacson/ Mercader had escaped after murdering Trotsky, leaving the ice axe at the scene, it would have been only one piece of evidence against him. It would also have to be proved that he had been the only person who could have stolen the ice axe from young Shirley.

When Colonel Sanchez was investigating Trotsky's murder, Jim Shirley recalled that Jacson/Mercader had received several telephone calls from a woman. The switchboard operator was unable to give the Colonel any clues regarding the mysterious female caller. Jacson/Mercader always took his calls at the public telephone in the office of the motel and was careful that none of his conversation could be overheard by others in the office. Whenever he used the telephone, Jacson/Mercader placed his back to the wall, kept his voice low and his eyes on the entrance door.[629] One day two men came openly to visit Jacson/Mercader at Shirley's Courts. They talked together for about an hour and a half seated in the garden within view of the office where Jim Shirley could see them. They could not be further identified by Colonel Sanchez.

Jacson/Mercader appeared before the United States Consul in Mexico City on 12 June with his doctored Canadian passport as Frank Jacson. He requested a United States transit visa with his ultimate destination of Montreal, Canada.

On 12 June 1940 when he left Shirley's Courts, Jacson/Mercader, despite his concern about the FBI, gave Jim Shirley Sylvia Ageloff's address of 150 Livingstone in Brooklyn, New York as the place he would be staying.

Prior to his departure for the United States on 12 June, Jacson/Mercader came to see Trotsky. He told him that he was flying to New York to confer with the men who had hired him to work in Mexico. He complained that the business in Mexico had not turned out to be as profitable as they had originally anticipated. He told Trotsky he was leaving his Buick sedan for the guards to use. He bade a solicitous goodbye to Trotsky and drove to the Mexico City airport with Jake Cooper, one of Trotsky's guards. He left his car in Cooper's care.

Cooper waited with him at the airport until his plane left. During the wait of an hour or so, in the restaurant on the second floor overlooking the airport, the two had lunch. Cooper asked him how well he knew some of the leading members of the Fourth International. Cooper assumed he knew them well since he was always bragging about having met them on an intimate basis when he was in Paris with Sylvia.

Jacson/ Mercader rattled off some of the more important names of the Trotskyite leaders in the Fourth International from Europe and the United States. He claimed close personal friendships with most of them. Cooper recalled later that somehow he was not impressed and did not believe him. Once again there was a suspicion raised concerning Jacson/Mercader's bona fides but nothing was done to investigate Sylvia's husband. It also shows the lack of planning by the GPU in giving Jacson/Mercader such a poorly backstopped cover and not instructing him how to handle this kind of question about his background and acquaintances.

During the time in New York, Jacson/Mercader stayed with Sylvia at the Hotel Lexington.[630] Each day he attended meetings with Ovakimian, Rabinowitz, Eitingon and possibly others. He explained his absences to Sylvia with the wholly fictitious story the GPU gave him that he was being briefed by his Wall Street employer on the intricacies of the bull market on the stock exchange that was due to begin. Once again no one, least of all Sylvia who was so delighted to be living once again with her husband, bothered to check up on this cover story to see if he really did go to a brokerage house on Wall Street during the day.

It is easy in retrospect to blame only Sylvia for not checking into some of Jacson/Mercader's more fanciful and obvious fabrications. If she did not check them why did not some of Trotsky's security guardians do some checking on this very close friend of someone trusted and liked by Trotsky and in daily close association with him?

The Rosmers had arrived in New York from Vera Cruz and Jacson/Mercader made a point of visiting them. He embellished this association when speaking with Trotsky's guards when he returned to Coyoacan. He tried to solidify his cover story as a loyal political adherent of the Fourth International with the guards by citing his social acceptance by the Rosmers.

Apparently the GPU planners meeting in New York received the final approval from Stalin to use Jacson/Mercader as the assassin. He was given a large amount of Mexican currency that might be needed to help in his escape. The money was packaged in several cellophane wrappers, sealed with read wax. Jacson/Mercader kept these packets in the safe at the Hotel Lexington until 1 July when he left New York to return to Mexico. In preparation for the trip he secured the bundles of cash to his wrists under his shirt sleeves with rubber bands.

He told Sylvia he was returning to Mexico City by way of New Orleans, San Antonio and Nuevo Laredo. He partially explained this indirect route by saying he had always wanted to see New Orleans. The real reason was his spurious passport would probably not be subjected to any detailed examination when he crossed the border at Nuevo Laredo. He telephoned Sylvia from San Antonio to again say goodbye. He told her he was flying to Mexico City from there without offering any explanation as to why he had changed his mind about crossing at Nuevo Laredo.

Jacson/Mercader arrived in Mexico City in the early part of July 1940. He did not bother to contact Sylvia for some time. He was too preoccupied with the preparations for murdering Trotsky. Late in July he did, however, make arrangements for Sylvia to join him in Mexico City. He visited the Trotsky compound on Monday, 29 July at 2:40 p.m. and remained for about one hour and ten minutes. The ostensible purpose of the visit was to pick up his Buick sedan.

The guards asked him if he had visited the headquarters of the Trotskyite Fourth International while he was in New York, called the Socialist Workers' Party. When he told them he had not because he was too busy meeting with his bosses, the Wall Street brokers, he noted a look of quizzical surprise on their faces. He quickly realized that he had committed a blunder which might bring him under suspicion.

Every loyal member of the Fourth International considered a visit to New York without visiting the Trotskyite headquarters something akin to a devout Catholic visiting Rome without a peek in at St. Peters or the Vatican. Jacson/Mercader quickly back-tracked and told the guards a compounding lie.

He said that the reason he had not visited the Trotskyite headquarters was that he had considered it more important to spend time convincing Sylvia and her sisters that Trotsky's position was correct, i.e., that the Soviet Union must be defended against military attack even though it was under the domination of their arch-enemy, Stalin. This appeared to be a very weak rebuttal to the guards' skepticism.[631]

This issue, whether to defend the Soviet Union was a hotly contested one, that had long divided Trotsky's followers. Sylvia took the minority view that under Stalin the Soviet Union was not worthy of any defense from the Trotskyites. It was still a contentious issue and Jacson/Mercader thought it sufficiently controversial that the guards might accept his explanation.

Trotsky's own view was the prevailing one and that was opposite to that of Sylvia and her sisters. Jacson/Mercader thought he had a chance of being accepted by the guards if he appeared to subscribe to Trotsky's own views. Trotsky felt that the Soviet Union, the cradle of communism, even under the rule of a mad tyrant like Stalin, was still worth defending against the Nazis.

The guards did report Jacson/Mercader's failure to visit the New York headquarters to Trotsky. Ever the optimist, and personally convinced that the moral right of his philosophy would triumph eventually over all opposition, Trotsky mulled over Jacson/Mercader's lack of proper respect and failure to observe protocol. Trotsky said to Hansen in this regard,

> He [Jacson] is rather light-minded and will probably not become a strong member of the Fourth International, nevertheless, he can be won closer. In order to build a party we must have confidence that people can be changed.

Trotsky also told the guards that Jacson/Mercader was writing a statistical summary about the French situation which, with his help, might turn into something for a short magazine article. He also told Hansen that everyone should try harder to become more friendly with Jacson/Mercader so as to bring him into the fold of the Fourth International. This was, in essence, the sacrificial sheep opening the door for the murderous wolf to come in!

An elated Sylvia, delighted to be reunited with her husband, arrived in Mexico City during the first week in August 1940. The loving couple occupied room 113 in the Hotel Montejo. This lovely hotel was located in the central part of Mexico City on the scenic tree lined Paseo de la Reforma near the intersection of Calle Nisa and the lateral of the Reforma, specifically at Reforma 240.

In the early evening of 31 July when Jacson/Mercader dropped by the Trotsky compound to bring a box of chocolates to Natalia, he seemed "funny and odd, according to her. Later that evening when Sylvia arrived to visit Natalia, the two ladies commented on what a great change had taken place in Jacson/Mercader's appearance.

They agreed he was pale, nervous, irritable, unable to eat the way he used to and his complexion had become pallid with a greenish tinge. All in all an arresting description that should have caused an ordinarily prudently suspicious person to at least wonder what had brought about this drastic change in her lover and friend.[632]

During the first three weeks of August 1940 there was a markedly worsening physical deterioration of Jacson/Mercader. Sylvia was apparently oblivious to the inner turmoil and social deterioration of his manners. He became progressively more nervous. He seemed to age and wither. His coloring (normally tan and healthy) took on a grey pallor as if some poison from within was seeping through his system. His hands became slightly palsied. This was not apparent to a casual observer but it was very noticeable to those who knew the pride he took in his ability to control any outward manifestation of nerves. Sylvia attributed Jacson/Mercader's condition to amoebic dysentery, chronic diarrhea and the malaise of living in the altitude of Mexico City's 7000 feet above sea level.

The fact remains that no one checked up on Jacson/Mercader. No one checked at the Edificio Ermita where he shared an office with Siqueiros; no one tried to identify his alleged employer, the mysterious Peter Lubeck; no one tried to reconcile his stories of being a road engineer with the other stories of his working for a stock brokerage firm or for a British import house; nor did any one attempt to reconcile the conflicting stories he told Otto Schuessler that his business was dealing in sugar and oil with the story he told Natalia and Joe Hansen that he was in some way dealing with diamonds from Belgium; no one queried if it was not suspicious that he lost his way in Vera Cruz where he supposedly went two or three times every month and no one considered the drastic change in his health as a warning sign of anything. (Even for his personal well being, should not his friends and lover suggested that he consult a doctor?)

Many Americans with whom Jacson/Mercader came in contact, including Jim Shirley at the motel, officials at Pan American Airways, and the visa officials in the Consular Section of the American Embassy, all might have noticed something not quite right about this strange Canadian and reported him to the authorities at the Embassy, specifically to the FBI. At that time, just before World War Two, the FBI in Mexico was on an increasingly vigilant monitoring campaign of all suspicious persons in Mexico because the Nazis were using Mexico as a staging area to infiltrate espionage agents and saboteurs into the United States.

There was every reason to believe, and prudent speculation would corroborate, that if Jacson/Mercader had come to the attention of the FBI, they would have told their agent in the Trotsky entourage, Joe Hansen, that Jacson/Mercader was probably not everything he purported to be.

There was a story told about this time that Jacson/Mercader was driving a group of Trotsky's friends from the States on a sightseeing tour of Toluca, a town about fifty miles southwest of Mexico City known for its colorful native market. The road between Toluca and Mexico City in 1940 was steep, without guard rails, narrow around mountains, in poor repair, with large chuck holes that wrested the steering wheel from the driver's hands. This narrow road without any side curbing gave way, in many places, to deep canyons filled with boulders the size of a car and impenetrable jungle, where a wrecked car would probably not be found for weeks.

Jacson/ Mercader was driving his Buick sedan with the visitors when he suddenly swerved toward the unguarded edge of the road. There was a sheer drop of several hundred feet of jagged rocky canyon yawning beneath the front tires of the car as he braked to a stop at the very last moment before the car plunged over the edge.

This was obviously a self-generated crisis. A very dangerous and fool hardy maneuver. His passengers were aghast at his recklessness. Jacson/Mercader merely backed the car away from the precipice and said, "Everything would have been finished if I had gone over the edge, wouldn't it?"

ROBERT SHELDON HARTE EMERGES

We communists are all dead men on leave!

Eugene Levine[8]

There are things a man must not do, even for his country; he must never cry in public!"

John O'Leary, Irish nationalist

The movies, in their striving for the macabre, always have the discovery of a corpse occur in a grisly setting. Usually it is dark and ominous, accompanied by spooky, thunderous music with rain pelting down in wind-blown sheets. Lightening is flashing through the black sky with the ghostly scene vividly, starkly revealed in each flash. This scene is not written into every murder thriller by contract nor could it possibly be regulated by some little known code of the industry. Is it a supernatural fact, therefore, that all murdered bodies have to be discovered under these circumstances? Regardless, the scene is sufficiently descriptive for Colonel Sanchez's next episode in the Trotsky case.

At the same time, in direct contradiction to the theory they were proposing of self-assault by Trotsky in the 24 May raid, the GPU began a vilification campaign against Harte. The GPU charged that Harte was the "leader of the raid"; that he had "betrayed" his chief (Trotsky); that he was not an American but a Russian; that he was a typical "gangster type"; that Harte had to be involved with the raiders since he provided them with the keys to drive away in the two cars (the keys were always left in the ignitions of the cars to be ready for emergencies), and that Harte was very familiar with the mystery man "Felipe."[633]

When Robert Sheldon Harte drove away from Trotsky's compound with Felipe, the French Jew, and Nestor Sanchez Hernandez, he disappeared. His father, Jesse Harte, hurried to Mexico from New York

---

[8] German communist's last words before he was shot by a firing squad on 6 June 1919 in Munich, Germany.

City to help in the search for his son. Colonel Sanchez had already alerted police posts throughout the Republic to find the elusive Sheldon Harte.

Journalists who had come to Mexico from newspapers around the world to cover the investigation and aftermath of the 24 May raid on Trotsky's house were also looking for Harte. Most reporters at that time would have gladly given a year's wages for an exclusive interview with Sheldon Harte. Unfortunately, such was not to be the case. Although the search for Harte was one of the most intensive in Mexican memory, no clues to his whereabouts were discovered.

The Colonel recalled the field trip Herrera Vasquez told him he had made with Angelica Siqueiros to the shack at the Desert of the Lions. (The name has such a nicer ring to it in Spanish as, Desierto de los Leones, it is shame to use the English!)

True to a motion picture script, when the Colonel arrived at the Desierto de los Leones in the area of Rancho de Tlanilapa, night was falling. The winding path up the hill from the dirt road to the shack was slippery from the afternoon's rain. The sky was black and ominous with more rain, thunder and lightening threatening. The dark clouds obscured the sky and lighting was flashing in the distant rain clouds clinging to the surrounding mountains.[634]

When Herrera Vasquez described the place that he had visited with Angelica Siqueiros as a shack, he had glorified the structure. It was more of a hovel. It was a small lean-to, burrowed against the side of a hill. It was made of sun-dried adobe bricks, painted throughout with whitewash with a sagging roof made of palm and grass rushes. The surrounding area was wooded and the lush foliage of tress shaded the area and added to the gloom of the setting.

The shack was entered from the slope of the hill on the upper floor, or second story American style.[635] There were three rooms on the upper floor, with wooden plank flooring which was the ceiling of the rooms on the ground floor below. A visitor had to go out of the shack from the upper story and down a path at the side of the house, that was slippery from the rain, to reach the ground floor. There were no interior stairs or ladder connection between the levels. The lower level consisted of a servant's utility room that could have been used for washing clothes, or as a maid's bedroom, that had a straw sleeping mat on the floor.

The other room on the ground floor was a kitchen with a small comel and bracero (charcoal brazier) in one corner. There was no running water in any of the rooms but a bucket full of water stood just outside the kitchen, probably filled by the recent rains. All of the floors on the ground level were compacted natural earth as it had existed before the walls of the shack had been erected.

Upstairs in what would have been the sala, or living room, was a rickety table and three chairs, some old newspapers (all dated subsequent to the 24 May raid) and an open tin of pimento peppers in vinegar. In the bedroom there was a folding cot, (probably the one Herrera had helped Angelica Siqueiros purchase). The cot was strangely slashed near the head, as if a razor had been used to cut the fabric. The floor boards and the cot were covered with powdered lime.

The last of the three rooms on this upper level had a lovely view of the village of Santa Rosa. It contained an artist's easel, with an untouched canvas still in place, two paint brushes and several open pots of paints on a small table beside the easel. In one corner of the bedroom was a partially rolled up mattress that was similarly slashed near the head and several used .22 caliber shell casings lying on the floor close to the mattress.

The planks of the upstairs bedroom floor were profusely littered with American Lucky Strike cigarette butts. American cigarettes were too expensive for the average Mexican worker to afford and it was assumed that only Americans were rich enough to smoke them.

The conglomeration of furnishings and effects were mysterious in themselves. There was nothing, however, that was sufficiently pertinent to the investigation of the raid, or the disappearance of Harte, to keep the interest of Colonel Sanchez, especially since it was getting late, he was a good hour from home and the deluge of rain was about to start at any moment.

Just as the police were about to leave, one of the men spotted a soft spot in the earthen kitchen floor that obviously had recently been disturbed. The Colonel immediately called a peasant, who was working in a field nearby, to come to the shack and dig up the kitchen floor in the freshly turned earth.

As if on cue, the rain began to fall in torrents; then the thunder roared and the lightening flashed as they exhumed the body of Robert Sheldon Harte from its grave in the kitchen floor. Covered with lime, the body had a chemical iridescence that made it glow eerily in the light of the police flashlights.

It was deduced that the actual killing had probably taken place in the upper rooms where the cot and the mattress had been slashed. It was subsequently determined that the slashings had been done in a futile attempt to eradicate the blood that had poured from the bullet wounds in Harte's head onto the mattress and cot.

"The day after the sensational events [the 24 May Siqueiros raid] on 25 May, it is supposed the Arenal brothers murdered him [Harte] and gave him an improvised burial."[636]

Harte's body had been covered with lime in the bedroom, probably under the erroneous impression that lime hastened the decomposition of the body. It had then been carried down the slippery side path to the kitchen for burial in the soft earth floor. Blood from Harte's head wounds had run between the planks of the bedroom floor and the lime had been used in an attempt to obliterate the blood. (The autopsy was conducted by Dr. Arturo Baledon Gil at the Juarez Hospital.)

In the ashes of a nearby fire the police discovered the partially burned remains of a navy blue sweater, a pair of trousers, buttons from a jacket and pieces of mattress cloth. The mattress remnants matched the slashed mattress in the shack and the clothes definitely could have belonged to Harte. It was morning before the police finished the examination of the contents of the adobe shack in the Desierto de los Leones. It was the 25th of June, exactly one month and one day since Robert Sheldon Harte had last been seen alive as he drove away from Trotsky's compound with Felipe, the French Jew.

Early that same morning Otto Scheussler, Trotsky's guard who knew Harte the best, was summoned by the Colonel to the shack to identify the body. By ten o'clock Harte's body had been removed from the shack to the police station at San Angel when Trotsky came, as required by law, as the second witness to identify the body.

Trotsky peered for a long time at Harte's body with the two bullet holes in the right side of his head. The body, rigid and shining from the lime, resembled a prone statue. The mixture of lime and clay earth had given a strange sheen to the skin. The face retained a benign countenance, as a man asleep with neither fear nor anxiety. The eyes were closed, the mouth was slightly open and the arms lying in a relaxed position beside the body.

There were bruises on Harte's body indicating that he had been tortured before being killed. The police referred to this practice as killing "a la Mexicana." This meant that Harte had been beaten first in an effort to make him fight back, so his killers might justify their act with the idea they were killing in self-defense rather than cold blooded murder.[637]

Trotsky's thought was that Harte was but another in the long list of his secretaries who had been murdered as victims of Stalin's wrath — Glasman, Butov, Blumkin, Sermuks, Posnansky, Klement and Wolf.[638]

The Colonel learned that the shack was owned by one Daniel R. Benitez, a civil engineer, who lived at Calle de Empresz No. 19, Mixcoac. Benitez told the Colonel that around the first of May a smartly dressed man had come to his house driving a Packard automobile with New York license plates and asked to rent the shack.

The prospective lessee told Benitez that he was a painter. He desired to use the shack as a place of instruction for young painters who were interested in his technique of painting frescos with a mixture of cellulose to prevent them from deteriorating.

> The person who rented the shack is believed to have been Antonio Pujol. The man said he was 'the painter Cisneros'. Benitez said the rent would be 5 pesos a month but the man said he would pay 30.[639]

At the time the lease was negotiated there was a tenant family living in the shack by the name of Cruz Hernandez. The new lessee promptly evicted Cruz and his family, explaining that he wanted complete privacy for his art school. The degree of knowledge of frescos indicated to the Colonel that the lessee was probably a bona fide painter. From the sketchy description Benitez had been able to provide, the lessee had a dark complexion with a generally healthy appearance and was smartly dressed. The Colonel believed it had been one of the Arenal brothers who had rented the shack.

On 2 July the communist lawyers for Serran Andonequi and Mateo Martinez were given a hearing by the Coyoacan Court of First Instance before Judge Raoul Carranca Trujillo and Public Prosecutor Moreno Tapia. Trotsky and Natalia were represented by Antonio Franco Rigault.

As a witness Trotsky answered the provocative questions from the baiting defense counsel. He testified that he had finally completed the purchase of his house on Viena street in Coyoacan at the beginning of May, for the sum of $9800 Mexican pesos.

Trotsky answered defense counsel,

> I am not, neither by birth nor inclination, a house owner. I have never previously owned one. During my stay in Norway however I was the object of a serious attack. Those who carried out this attack tried on several occasions to buy the house I occupied. It would have been easier for them to prepare the blow. In Mexico too a similar thing happened regarding the house which I at present occupy. When my friends in New York learned about it they asked me to tell them how much the complete assurance of my safety would amount to. As an ownership of the house was important since the GPU were preparing to acquire it, I let them know and they sent me a check for $2,100 [American] dollars.
>
> It is not exact to say that I have been expecting an attack since the month of January. I've been waiting for it for two years. But since December or January I have been expecting it at any moment. During the past few months my guards and I have held frequent conferences to try and take extra precautions because the attacks in the communist press against me had become more and more violent. I have denounced the invasion of Poland and Finland by Russia as well as the Moscow alliance with Hitler and my declarations have been reproduced in all the newspapers of the world causing an enormous sensation. The last conference of the Mexican Communist Party was influenced throughout by the battle against Trotsky and Trotskyism. It's slogan was 'Death to Trotsky.'

At this hearing Trotsky denounced Colonel Sanchez for even intimating that Harte had not been completely loyal to him. Specifically, Trotksy repeated that Harte had never been in collaboration with the GPU nor in sympathy with the objectives of the 24 May raiders. Naturally the Colonel took exception to Trotsky's defense of Harte but chose not to comment to the press nor make a personal issue of Harte's alleged loyalty to Trotksy.

A short time after this hearing, on 4 July, Colonel Sanchez interrogated Mariano Herrera Vasquez. It was a dark night and Colonel

Sanchez took Herrera Vasquez back to the shack in Desierto de los Leones. He had arranged for a police officer to play the part of another participant in the 24 May raid who was unknown to Herrera Vasquez. When the police officer refused to answer the Colonel's questions, the Colonel had him "executed" on the spot by firing blank cartridges at him. For good measure the Colonel then kicked the "dead body." He then turned to Herrera Vasquez and told him, that unless he told the truth he was the next to die that night. Herrera Vasquez immediately decided to "tell all" which, in retrospect, confirmed the Colonel's initial impression that Herrera Vasquez had been telling the truth all along.[640]

This knowledgeable witness told him that on 24 May, the morning after the raid, about eight o'clock, Luis Arenal had arrived at the shack with a tall, American man, with curly red hair who spoke Spanish very badly. Arenal told him that the man was to live at the shack for a couple of days. He also told Herrera that each day an Indian girl would come to clean the shack and he did not want the maid to see the American. Herrera was to make sure the American went out of the shack while the maid was cleaning.

Herrera asked Luis Arenal if he meant that the American was free to go out alone without being escorted. Arenal replied, absolutely, the American man was free to go out alone, anywhere he wanted to go. The Colonel felt that this was further proof that Harte had never been a prisoner of Arenal's and had come to the shack on his own free will. This emphatically proved to the Colonel that Harte had been a willing accomplice of the GPU.

This evidence of Harte's complicity was further corroborated by several subsequent visits of the Arenals to the shack. Both Luis and Leopoldo Arenal spoke in English with Harte in front of Herrera. He could not understand what they said but he assured the Colonel that the conversations among the three men were cordially animated and very friendly.

During the Arenals last visit to the shack, Luis Arenal gave Herrera 50 pesos which were due him for the five day's wages and told him to leave the shack. Herrera told the Colonel that from all of the evidence he had observed, and his personal intuition, he was convinced that Harte was voluntarily staying at the shack. Herrera also said that during the short time he knew Harte at the shack he had grown to like the Americano.

Harte, for some inexplicable, reason had asked Herrera to call him "Timo" or "Timeteo."

Harte was probably murdered because he and Jacson/Mercader had become friends before the 24 May raid, either by pre-arrangement by the GPU or by the accident of mutual attraction. Harte and Mercader were frequent drinking companions at the Bar Kit Kat on Independencia street and went to local hotels with prostitutes from the Bar Kit Kat.[641]

One of the favorite places they both enjoyed was the Salon Mexico, a cavernous dance hall with two bands playing, one at either end. The girls available there for dancing, were called "ficheras" because they collected a fichera, or token, from each customer at the beginning of each paid dance. The girls would cash in their collected ficheras at the end of each dance set, usually three dances, and be paid an appropriate amount for each fichera. (In the States this arrangement of men purchasing dances was popular about the same time and was called a taxi dance.)[642]

Uppermost for the GPU was the fact that Harte might be a hinderance to have around as they moved Jacson/Mercader's assassination operation into the active operational phase. The raid operation was blown and most of the participants had been identified. The GPU therefore did not particularly care if Harte told the newspapers all he knew about Siqueiros's boondoggle.

Had Harte been allowed to live he could still give a description, and probably a positive identification, of Felipe, the French Jew, and the GPU considered this identification a matter of utmost sensitivity.[643] The major worry for the GPU at this juncture was keeping Jacson/ Mercader unencumbered and spiritually free of any past relationships, except of course, from his very important continuing relationship with Sylvia Ageloff.

The GPU wanted Jacson/Mercader's assassination operation compartmented and uncompromised by any taint of the 24 May fiasco. They could not take the chance that Harte, once freed from the shack at Desierto de los Leones, would attempt to resume his friendship with Jacson/Mercader.

Trotsky did not know any of the details of Harte's background before the time he reported for duty at Coyoacan. He had relied on the Trotskyite organization in New York to assure the loyalty of the guards they sent to him. Unfortunately some of the key people in the New York organization were not completely loyal to Trotsky but were GPU penetrations. For example, Sylvia Franklin was a GPU penetration, who could have steered Harte through the selection process at the New York Trotskyite headquarters so that he would be chosen to be one of Trotsky's trusted guards.

Even after all the evidence of Harte's duplicity was laid before Trotsky, he still refused to believe it. Trotsky maintained that Harte was basically loyal to him and was not, and could never have been, a GPU agent. In the final analysis, Trotsky had relied on Natalia's motherly subjectivism about Harte and his own emotional intuition, which was far from objective.

Regardless of what we now know about Harte's treachery, Trotsky to the end chose to depend upon Natalia's intuitive liking for the boyish Harte. Trotsky refused to entertain the thought that Harte could have been working for the enemy. To commemorate Harte's loyalty to him, Trotsky had a plaque erected on the wall of the garden at the compound which simply stated, " In Memory of Robert Sheldon Harte, 1915-1940. Murdered by Stalin."

There was one additional factor that made Trotsky want desperately to believe that Harte was not a GPU agent, and that was Trotsky's own pride. Trotsky wanted to believe, as he had believed of all his secretaries and followers who had been murdered by the GPU, that Harte too had died for Trotsky's cause. Therefore, Harte's memory must be unblemished so that, by inference, there could be no stain on the memory of his other murdered disciples.[644]

Trotsky's enormous ego was a bulwark shielding him from reality. He always remained convinced that if ever a GPU agent did infiltrate his household that his personal charm and the dynamism of his political philosophy would win over the infiltrator as a convert to Trotskyism and cause him to repudiate the GPU.

A convincing piece of evidence that Harte was a GPU penetration was the fact that after his death, a guard was cleaning out Harte's room at the Trotsky compound. The guard found a Spanish/English dictionary with a signature of Jose Alfaro Siqueiros on the fly leaf.[645]

Fanny Yanovich, Trotsky's Russian secretary, later reported to Colonel Sanchez that every night when Harte drove her home after work at the compound, he was always extremely curious about the progress of Trotsky's book on Stalin. Trotsky was working on a biography of Stalin and dictated his thoughts while Fanny Yanovich took them down in shorthand. Every night Harte asked her how far Trotsky had gotten and what was he saying about Stalin?[646]

In 1976, the Fourth International asked Leopoldo Arenal if he or his brother had killed Harte. Arenal replied in the negative. He was then asked why he did not sue for defamation all those newspapers and writers who claimed he and his brother had killed Harte? Arenal replied that bringing such a law suit for defamation was too difficult to prove in Mexican courts.[647]

TROTSKY'S LAST DAY

Whoever recalled it later remembered the exceptional peace and serenity in the house up to the fatal hour!"

Isaac Deutscher, *The Prophet Outcast,* p.500

One more and final story and my scroll is at an end!"

Pushkin, *Boris Godonov* (Natalia Trotsky's remembrance to Leon Trotsky)

After the Siqueiros raid Trotsky was heard to ruminate, "Fate has granted me a reprieve. It will be of short duration." He was convinced that another attempt on his life would soon be made. He knew Stalin had given orders to the GPU that he must be liquidated at all costs.

The last year or so of Trotsky's life was marred by interludes of depression and hypochondria. He suffered "uncontrollable anguish" over the fate of his Soviet adherents and relatives. In March of 1940 he complained of "high and rising blood pressure" and speculated he would probably die from a "brain hemorrhage."[648]

Trotsky felt himself constantly menaced and could easily have developed a siege mentality. He continued, however, to defy the immense power that Stalin and the GPU were bringing against him. Despite the omniscient threat, his health remained tolerably good, his energy shone from the twinkle in his eyes, the firmness of his features and frame and the force of his personality proclaimed to the world that he was unafraid.

A group of American Trotskyites who had visited Trotsky before the middle of June, had implored him to "go underground," assume another identity and they would smuggle him into the Untied States where he would be safe. Trotsky refused, saying he could not sulk for the rest of his life and do his work furtively.[649]

Tuesday, the 20th of August 1940, to paraphrase a historic statement, a day of villainy that will live forever in the annals of infamy, dawned cloudless, sunny and crystal clear. It was a lovely day, as only a day can dawn in the clear, sparkling mountain atmosphere of Mexico City at that time of the year — after the rainy season when everything is green and blooming and the dust settled.

Trotsky rose shortly before seven o'clock and said to Natalia, "I feel very well this morning, better than I've felt for a long time. I took a double dose of sleeping powder last night. I've noticed it does me a world of good."[650]

Natalia replied that they had noticed the same effect from the sleeping powders in Norway when he had felt more tired than he did now. She cautioned him that it was not the drug itself that made him feel better but it was the deep sleep and the complete rest that did it.

Trotsky had a humorous, self-imposed, little ritual that he went through almost every night before he and Natalia went to bed. He would close the iron shutters on their bedroom window, which their American friends had installed for them after the 24 May raid, and say to Natalia, "Now Siqueiros won't be able to get us!"

Continuing his little charade each morning upon arising and opening the shutters, Trotsky would say, "See Natalia! They did not kill us during the night and still you aren't happy!" Sometimes he would mournfully, but half-playfully add, "Yes, they've given us a reprieve!"

Natalia gave Trotsky the usual massage to help him relax his shoulders and back muscles. He dressed quickly in his blue French peasants smock and trousers, eagerly looking forward to the prospect of doing a good day's work.

He left Natalia still dressing in their bedroom and walked vigorously out onto the patio and into the garden to tend his rabbits and fowls. Trotsky had a great love of animals and pitied the rabbits in their cages. Caring for these animals was a labor of love for him because it was a distraction from his usual intellectual labors. He hated the tedium of cleaning the cages and the filth. He looked upon these chores as a drudgery but knew that both he and the animals benefitted.

On this day Trotsky felt guilty having neglected the rabbits because, on his doctor's order, he had spent Sunday resting in bed. Trotsky worked with his pets from quarter past seven to nine o'clock. He cleaned the cages, and put in fresh water, feed, grass and greens. Usually from time to time he would interrupt his work to dictate an order or an idea which had occurred to him to one of his secretaries standing nearby, but on this day he worked without interruptions.

Trotsky washed his hands at the garden spigot and walked back across the patio to have breakfast with Natalia. He ate his usual breakfast and again could not help expressing in a buoyant mood, how well he felt. Natalia smiled at his boyish elation. Trotsky was eager to get back to work on his "poor book," his biography of Stalin, which he felt he had neglected. He did not feel he had been spending enough time on this work and vowed jokingly to Natalia that he had been naughty and neglectful long enough.

The aftermath of the raid with the constant interruptions by the police and reporters had taken time. His own efforts to rebut the propaganda charges of the local communist that he had staged the raid also left him little time to concentrate on his writings.

Trotsky assured Natalia once again that his mind was now at ease. He felt the Mexican authorities were at last inclining toward the correct identity of the perpetrators of the raid.

Trotsky told Natalia enthusiastically that he had enough energy for a lot of tasks. He was going to start right in on his Stalin book as soon as he finished another article for a small periodical. This was to be, Trotsky said, not an article for the bourgeois press but for a small selective magazine that subscribed to his theories. On this relatively humorous note, Trotsky rose from the table and went into his study to work.

The windows of the study were open to the clear mountain air. The dazzling sunlight angled a window-wide beam onto the wooden floor. The pleasant freshness of the atmosphere of the room harmoniously matched Trotsky's exuberance, and pleasantly prepared for the welcome news that was in the morning mail, waiting for him on his desk.

A cable from the librarian of Harvard University in American acknowledged the safe receipt of Trotsky's archives. There had been many obstacles, real or imaginary, caused either by the GPU or the FBI, or maybe both independently. These obstacles had delayed the delivery of his life's papers from arriving at the safe haven of Harvard.

Whether Trotsky was jousting with windmills or not he was still suspicious that the FBI wanted to examine his papers for leads to communists in the United States. Trotsky had instructed his American attorney and loyal follower, Albert Goldman, to take legal steps against the FBI should they try to pry into his papers. Trotsky had explained to Goldman that he did not care if the Bureau nosed through his personal papers since his life was literally an open book, having been made so by the investigations of many of the police organizations of the world.

He was concerned for the well being of the many people, friends, followers and some casual acquaintances who were mentioned in his papers. He had done nothing to purge any names from these papers before he had them shipped to Harvard. He was deeply concerned that to inadvertently reveal these names to either the FBI or the GPU might cause them embarrassment or even some danger in the case of the GPU.

Trotsky had insisted when he had agreed to deposit his papers with Harvard, that one section remain closed until 1980. This section contained his correspondence with his party followers. He felt he owed them the courtesy and duty as a party leader of protecting them. This condition was especially pertinent when it is remembered that in 1940 the whole of Europe was controlled by either the Gestapo or the GPU, both of which would have eagerly liquidated any Trotskyite referred to by Trotsky in his personal correspondence as a party follower.

After enjoying this bit of good news from Harvard, Trotsky wrote and dictated into the Dictaphone answers to his other correspondence. There were a few short letters to Trotskyites in the United States. He thanked one couple for their gift of a dictionary of American slang, joking that he could use it to understand the dinner table talk of his American guards. He inquired about the health of a former secretary who had returned to the United States and he also sent fraternal greetings to two of his followers who were in prison for strike activities.

He then turned his attention to the article he was eager to complete but it did not come out well. For some reason, he had difficulty marshaling his thoughts in a logical sequence. His almost incoherent ramblings on the Dictaphone (the last recording of his voice that is preserved) later revealed a confused and illogical mind unable to focus on the issues he was trying to explain.

It is known that Trotsky opposed conscription in the United States as a general thesis. At one point in his rambling on this last recording, he strongly supported conscription there. This was in direct opposition to the position taken by Max Schachtman and his other loyal party followers in America. On this last recording Trotsky claimed that Trotskyites should place themselves alongside the 70% of the workers in America who, according to a recent poll he had seen, favored conscription.

About one o'clock that afternoon, Trotsky's Mexican attorney, Antonio Franco Regault, arrived. Regault strongly advised Trotsky to reply to a scurrilous article in Vicente Lombardo Toledano's communist newspaper *El Popular*. This article had accused Trotsky, among other things, of defaming Mexico's trade unions. Trotsky agreed with Regault that he must answer the charges but feared his reply would once again drag him into senseless exchanges with the local Stalinists. He put aside his notes on the article on revolutionary defeatism that he had been mumbling into the dictaphone about and promised Regualt that he would write a rebuttal charging *El Popular* with brazen lying.[651]

When Rigault left, Trotsky went into the sitting room and told Natalia of Rigault's suggestion. Trotsky told her he felt he was just getting into the article he wanted to write and he should now have to leave it to answer *El Popular's* charges. He told Natalia he had agreed with Rigault and that he would have to take the offensive against *El Popular* and accuse it of libel. Trotsky sighed in resignation as he told her he had asked his secretary to gather all the material for the article and put it on his desk.

Trotsky again mentioned to Natalia how fine he was feeling and went into the bedroom for his customary mid-day siesta. When he awoke it was almost two-thirty in the afternoon. He felt refreshed and returned to his study for more writing.

Natalia peeked into the study and later wrote:

> I saw him seated in front of his work table. He was feeling well and I was in a happy mood. Recently he had complained of a general tiredness and sometimes gave way to it. He knew that it would not last but at such moments he thought of 'them' more often than usual. That day there seemed to be an improvement in his physical condition. He looked better too. Taking care not to disturb him, I peeped in from time to time to look at him as he leaned over his papers with his pen in his hand. 'One more episode and this story will be finished', I thought. Thus spoke the old chronicler Pimen, relating the crimes of Tsar Boris in Pushkin's drama 'Boris Godounov'. The life that Lev Davidovich led resembled that of a prisoner or a hermit with

> the difference that, in his writing, he not only related events, he waged a bitter struggle against political adversaries.

Trotsky continued working in his study until about five o'clock when he and Natalia took tea. He told her he had dictated several parts of his article on conscription in America and had written about fifty pages in reply to the article in *Wl Popular*. After tea, Trotsky started back into his study but decided to make certain his rabbits and fowls had enough water and feed.[652]

About twenty minutes past five o'clock Natalia went out on the long balcony that looked over the patio and saw Trotsky by the cages feeding his pets. Natalia noticed an unfamiliar figure wearing a hat hovering ominously near Trotksy. The momentary panic she felt changed to a feeling of dislike and revulsion when the figure came closer and she recognized it as Jacson/Mercader.

The Assassin's Last Day

> The word 'assassin' is derived from hashshashin signifying 'the eaters of hashish', for those who were sent out on their mission of murder were said to be deranged beforehand with this weed. The drug itself was merely one element of a process of mental conditioning, remarkable for the psychological knowledge it developed in those who employed it.
>
> Hugo Dewar, *Assassins at Large*, p.viii

On that same Tuesday morning, 20 August 1940, in room 113 of the Hotel Montejo at 240 Paseo de la Reforma in Mexico City, Jacson/Mercader woke about 7:45 a.m.. In the small bathroom he showered and shaved. He dressed and had coffee and pan dulce sent up to the room for breakfast for himself and Sylvia.

The manager of the Hotel Montejo, Señorita Noriega, had routinely inquired about his health the night before when she saw him in the lobby with Sylvia. He had replied that he was not feeling too well; something he had eaten was his perfunctory excuse.

Señorita Noriega had become accustomed to Jacson/Mercader's odd and sullen behavior. She did not attach any particular significance to his professed ill health. He was known at the hotel as a cranky, introverted, solitary person who did not acknowledge civilities proffered by the hotel staff and other guests.

Jacson/Mercader, on this Tuesday morning, left the Montejo Hotel about nine o'clock. He told Sylvia that he was going to the American Embassy to have his visa revalidated for their planned return to New York the next day.

Jacson had not gone to the American Embassy to have his visa revalidated as he had told Sylvia. He returned about noon time to Sylvia at the hotel in a surly and truculent mood. Sylvia had been anxiously waiting for him and asked why he had been gone so long getting a visa validated? Jacson/Mercader snarled back at her that the lines at the Visa section at the Embassy were so long that he had to wait for his turn.

Sylvia accepted this excuse and suggested a quiet walk to soothe his nerves. He pouted a little bit more, but Sylvia managed to placate him. Finally she suggested they get something to eat since it was time for the mid-day meal. Sylvia hoped that the lunch would be a divertisement that would bring him out of his foul mood. They walked north on Reforma to Avenida Juarez and east on Juarez to San Juan de Letran street (now called Cardenas). Sylvia chattering about the things she saw in the shops along the way and Jacson/Mercader, apparently preoccupied, answered her chatter with an occasional grunt.

At the corner of Juarez and San Juan de Letran, in front of the Palacio de Belles Artes (Palace of Fine Arts) they met Otto Schuessler, the young German guard at the Trotsky compound, and his girl friend. Sylvia did all the talking and told Otto and his girl that she and her husband were returning to the States the following day.

Sylvia said she and Jacson/Mercader would be out to Coyoacan to see Trotsky to say goodbye about four or five that afternoon. Jacson/Mercader spoke up at this point and suggested that Otto and his girl friend join them at a restaurant for dinner about seven-thirty that evening for a farewell despidida (going-away party).

Jacson/Mercader appeared nervous and ill-at-ease during this meeting. He suddenly, without any preliminaries, excused himself with the curt comment that he had forgotten an important appointment. Sylvia was left alone, gaping after the departing figure, and not quiet knowing what she should say to Schuessler and his girlfriend. Jacson/Mercader continued walking rapidly back on Juarez toward the Reforma.

Sylvia tried to excuse her husband's rudeness to Schuessler and his girl friend by saying he had not been feeling well. She suggested his illness might have been caused by the altitude, which was always a convenient excuse. (Almost anything can be blamed on the altitude in the rarified atmosphere of Mexico City at more than 7000 feet above sea level.) She also made an attempt to blame his indisposition on the spicy, Mexican foods he had been eating. Schuessler and his girl friend mumbled politely that "they understood" and left Sylvia, promising to meet her and her husband for dinner that evening at seven-thirty.

Sylvia walked slowly back by herself to the hotel and waited in room 113 for Jacson/Mercader to return. She made nervous trips to the lobby to ask Señorita Noriega if he had come back? Apparently not trusting Señorita Noriega's vigilance, Sylvia prowled the lobby looking for him in his favorite chair where she expected to find him sullen and pouting. She scanned the tables and chairs in the garden patio to see if he was pretending to be nonchalant, pettish and although seemingly indifferent, actually begging to be forgiven for his churlishness. All to no avail!

Eventually Jacson/Mercader returned in the early afternoon, full of apologies for his rude behavior to Schuessler and his girl friend. He tried to sweet-talk Sylvia into forgiving him. She relented enough to agree to go have an aperitif and some hors d'oeuvres at the Sala Don Quixote. After this brief reconciliation, he again left Sylvia, telling her he would return shortly. He did go to the Wells Fargo office, about this time, to see if there was any mail.

Sylvia commented that she received mail at Wells Fargo when she had been in Mexico before. One of her sisters wrote to her from New York to report two Stalin agents were in Mexico City — B. Helman (possibly Bittleman) and Jack Stachel.[653]

Jacson/Mercader went someplace to get his weapons and pick up the raincoat and hat he wore to Trotsky's house. His weapons were the large .45 caliber automatic, the knife with a 13 3/4 inch blade and the ice axe with the shortened handle. The handle had been cut from its normal length of two and one-half feet to a foot in length in order to conceal it better. The two headed metal cross piece, measuring 7" across, had a point on one end and a broad, two inch chisel-like axe on the other. The pointed end had been honed to a fine needle point.

The .45 caliber pistol (No. 195-264) was to be carried in his left rear hip pocket, the knife was carried on his person and the ice axe secured to the inside of the raincoat with a piece of cord. He also carried a propaganda piece in the form of a letter purporting to explain why he was committing this heinous act. The letter was, in essence, Jacson/Mercader's confession. It had obviously been written for him. The GPU planned to release the letter to the press even if he escaped after the murder.

The GPU thought the letter would explain to the world, and be accepted as a rational explanation for, the murder. In this confession Jacson/Mercader claimed to have originally been a loyal disciple of Trotsky. He had become disillusioned when Trotsky ordered him to go to the Soviet Union and murder Stalin. The confession explained how this proposition had so revolted him that he decided he must kill Trotsky to rid the world of this menace.

The confession was typed in French which was compatible with his cover as a French Canadian. The final version of the letter had been typed before the day it was to be used and the date was left unwritten. Therefore, he used a pencil to write in the date of the 20th of August 1940.

Jacson/Mercader's ostensible reason for visiting Trotsky was for him to edit Jacson/Mercader's article for publication. The confession and the article had been written on two different typewriters. In order to explain this difference, the GPU concocted another blatantly obvious lie for Jacson/Mercader to tell. His story was that, after leaving the Wells Fargo office that afternoon, he had gone to a deserted area near a railway track, close to a new road leading into Mexico City. There, he claimed, he burned all his personal papers and passport.

In his statement to authorities after the assassination, Jacson/Mercader claimed he had written the confession on the eve of his attack on Trotsky while sitting in the woods of Chapultepec Park. (This is a lovely wooded park in the center of Mexico City that was originally the summer home of the Aztec rulers.)

> I bought the typewriter at the Monte de Piedad [Mexico's National Pawnshop] for 140 or 150 pesos. I had given it to a man named Bartolo Perez or Paris, I don't know which, to keep for me. Then, after I had written the letter, I made a present of it [the typewriter] to Bartolo whom I thought I should not see again. I had no reason to keep it, I did not want to take it to the hotel, for what explanation should I have given Sylvia.
>
> I met Bartolo at the Kit-Kat Club, a bar near the corner of Independencia and Dolores streets. He's one of those men you find in any big town, ready to do anything to earn a bit of money. I invited him to have a drink and we got to talking. I asked him if he could get me a revolver, no matter what make, model or caliber, and he replied that he could. On Tuesday, after I had a meal with Sylvia, I went to the bar in question where I found Bartolo, who had the gun

> and the typewriter. We went towards the Square of San Juan. I took a wad of notes out of my pocket and gave them to him. Her counted them and said it was all right. I made him a present, besides, of the typewriter, thinking that I should not survive the crime. I wanted the revolver and the dagger to kill myself after I killed Trotsky with the ice axe.

Jacson/Mercader had gone to the garage near the Hotel Montejo and drove off in his Buick sedan toward Coyoacan. He knew, as it had been drilled into him by the GPU, that if he struck the correct blow with the ice axe Trotsky would die instantly without being able to make a sound. His escape and future life depended upon his striking the lethal blow with sufficient force and to the exact center of Trotksy's skull.

The Crime of the Century

It was a sound such as all those who heard it said they would never forget — a long drawn-out cry between a roar and a scream — an expression both of fury and of pain... so unlike anything else that for a moment no one knew what it was.

Mosley, *The Assassination of Trotsky*

I shall hear that cry all my life!

Jacson/Mercader, aka Ramon Mercader referring to Trotsky's scream as his Alpenstock pierced his skull.

All that is fit to live is fit to perish.

Leon Trotsky

How Jacson/Mercader murdered Trotsky has been told many times by some of the finest writers of this century; practically anyone who could put pen to paper has, at one time or another, written something about the real "crime of the century." The denouement remains one of the most dramatic, unforgettable moments frozen in time, of a palladin's effectuation of a tyrant's diabolical revenge upon a hapless victim.

After tea, with some bread and butter and jam, about five twenty-five, Trotsky rose from his seat and went to the hutches at the rear of the garden to feed his pets their afternoon meal.

All around Trotsky was the hum of activity as workmen put the finishing touches on the new security construction on top of the walls and around the shutters at the windows. Joseph Hansen with Charles Cornell and the Mexican man-of-all-work around the compound, Melquiades Benitez, were on the roof of compound near the main corner guard tower. They were making the final connection of a siren to the alarm system. In the event of trouble in the compound, particularly in Trotsky's study, the siren would shrilly announce to those inside, as well as the Mexican police outside, that assistance was immediately needed.

As this trio worked on the roof and upper battlements in the hot afternoon sun, stringing wires and connecting insulators, Hansen happened to glance down into Viena street and saw Jacson/Mercader approaching in his Buick sedan. Although he thought nothing of it at the time, Hansen later recalled that this was only the second time since he had been coming to the house, Jacson/Mercader turned his car around and backed it into a parking space so the car was headed back toward Mexico City.

Usually Jacson/Mercader pulled up to the front of the house with a display of bravado in a cloud of dust, hastily applying the brakes at the last moment, he usually skidded to a halt with his front bumper inches from the wall. Naturally Hansen, at this time in the afternoon of 20 August, did not attribute any particular significance to the Jacson/Mercader's change of parking pattern. Hansen noted the time, as a matter of routine in the log book, as exactly twenty minutes past five.

As he got out of his Buick, Jacson/Mercader waved to the three men on the roof and asked if Sylvia had arrived? The two Americans on the roof looked quizzically at each other since there had not been any appointment in the "Old Man's" notebook for either Sylvia or Jacson/Mercader. Hansen showed his surprise in the tone of voice when he told Jacson/Mercader that they were not expecting either of them. Hansen recalled later that, so far there was nothing in this exchange that caused him any concern or that should have raised his level of suspicion. Many times the "Old Man" forget to tell the guards who kept the entries in the appointment book of his pending engagements. Hansen thought that Sylvia and her husband had an appointment with Trotsky but that he had merely forgotten to tell the guards.

Hansen nodded his approval to Charles Cornell to open the gate and told Jacson/Mercader to wait a moment for the gate to open. Cornell pressed the switch that opened the new electrically controlled metal gate and Jacson/Mercader stepped into the garden patio. Harold Robins, another guard from the United States, was on duty at the patio and logged

Jacson/Mercader in the visitors' book. Robins then escorted Jacson/Mercader across the garden to the hutches where Trotsky was feeding his beloved animals.

As Jacson/Mercader approached him Trotsky drew some dry hay from a nearby bale. In a manner of being polite to a guest who is waiting for the host's attention, Trotsky explained to Jacson/Mercader that if the rabbits were fed wet grass it would swell in their bellies and cause then gas and pain.

Natalia, by some inexplicable instinct, had sensed a stir of activity in the patio above the normal clatter of the workmen on the roof. She left the kitchen and went out on the balcony overlooking the patio and garden. She saw Trotsky feeding his pets and puttering around the hutches. She did not immediately recognize the man figure standing next to her husband but, as she later explained, had an instant sense of terror and foreboding.

She could never explain why she felt that quick stab of dread when she saw that man figure next to her husband. She dismissed it as hindsight when she recalled the incident after Trotsky's death. To her last day, she attributed this warning flash of intuition as supernatural, or divine revelation, although she had never been a religious person.

The man figure turned toward the balcony and, recognizing Natalia, raised his hat in salutation. Natalia recognized him instantly. She later recalled that the first flash of warning that she had initially had persisted and she thought to herself with distaste that there he was again. Why is it that he has started coming here so often? Natalia genuinely liked Sylvia but she had reservations about Sylvia's husband. After his last visit when Trotsky himself had expressed concern about his bona fides, Natalia had started to develop a genuine dislike for Sylvia's husband.

Jacson/Mercader came over to where Natalia was standing on the balcony above the patio and exchanged greetings with her. He then said that he was "frightfully thirsty" and "could he please have a glass of water?"

One thing that every investigator has known from antiquity is that a liar, or someone under great emotional stress or faced with a frightening decision, has a dry mouth. This kind of guilty dry mouth can never be sated by any number of glasses of water. It is an emotional thirst that only telling the truth or getting the deed done and over with can slake. Jacson/Mercader's mouth was dry but not from anything he had eaten or any illness. His was the guilty dry mouth caused by his dread and fear of what he was about to do!.

Natalia, not knowing the reason for his dry mouth, suggested that he might like a cup of tea instead or a glass of water. He replied, facil with the lie, "No,no, thank you. I ate late and I can still feel my meal up here" and pointed to his throat. "It is choking me!" he added. Natalia gave him the glass of water but from the way he gulped it down it did not appear to do anything to quench his thirst.

Natalia noticed, as she looked down at this strangely acting man, that his complexion was a livid gray-green and he seemed more nervous than usual. She had noted that he always seemed nervous but she had never seen him as twitchy as he was now, standing below her in the patio. Despite her subconscious warning flags and her mental reservations concerning Jacson/Mercader, Natalia deduced no particular sinister inferences from his unusual behavior that day.

Jacson/Mercader never wore a hat as far as Natalia could recall. She did, on the other hand, remember that he used to brag frequently about his hardiness that he would never wear a hat even in the worst of storms. Natalia felt compelled to ask, why he was wearing a hat and carrying a raincoat on such a sunny day?

He gave a brief glance upward at the bright sun in the cloudless sky and replied with an inanity, that the good weather would not last long. Anyway, he said, it might rain! As he said these words he clutched his raincoat, heavily laden with his arsenal, tighter against his body.[654]

Natalia shook her head in resignation and changed the subject. "How is Sylvia feeling today?" she asked. Jacson/Mercader probably still mulling over his answer to the hat question, was momentarily taken back by Natalia's new tact. He was confused and Natalia thought he had not understood her question. Jacson/Mercader dumbly, as if in a stupor and seeking time to recoup his thoughts, repeated Natalia's question. He stuttered, "Sylvia.... Sylvia..... Oh yes, Sylvia! She is always well," he finally muttered in reply.[655]

Jacson/Mercader turned away, glad for a chance to escape Natalia's interrogation, and started to walk back to the rear of the patio where Trotsky was still puttering with his pets. Before he could leave, Natalia asked him if the article he wanted Trotsky to review was ready and neatly typed?

He reached into his raincoat pocket with his left hand and Natalia again noticed that his hand was trembling. Jacson/Mercader was also using this same left hand and arm to keep his raincoat tightly pressed against his body. His weapons inside the raincoat made it heavy and his gesture was clumsy. It was an awkward motion but he managed to take out a few typewritten pieces of paper that he turned to show to Natalia with a little flourish, as if to say, "See, I have it here."

Natalia descended from the balcony walkway into the patio. Not wanting to end her exchange with Jacson/Mercader without exercising her womanly prerogative of having the last word. She wanted to put down this upstart whom she was beginning to heartily dislike. "It's good your manuscript isn't written by hand," she dryly commented. "Lev Davidovich dislikes illegible manuscripts."

Together Natalia, with an almost regal bearing having had the final word, and Jacson/Mercader, clutching his raincoat to his body, walked the short distance across the patio to where Trotsky was still tending his pets.

Trotsky looked up as they approached and said to Natalia in Russia, "You know he is expecting Sylvia to call on us. They are leaving tomorrow."

Natalia had not known that Sylvia and her husband were leaving. It caused her a momentary pang at the thought that Sylvia would be so inconsiderate as not to have told her. She quickly replied that no, she had not known it. Trotsky, sensing her hurt and genuinely apologetic for having neglected to tell her, replied that he was sorry he had forgotten to mention it to her.

Natalia said it was too bad she had not known, as she would have liked to have sent some things with them to friends in New York. Jacson/Mercader, always seeking to worm his way into Natalia's good graces, hastily replied that if she could get the things ready that evening he would pick them up the next day at one o'clock.

Natalia, ever sensitive to Trotsky's suggestions, deduced that Trotsky thought that Sylvia and her husband should be invited to tea or dinner as a despedida or small farewell party. Still speaking in Russian, Natalia explained to Trotsky that Jacson/Mercader was not feeling well and had already refused tea.

Trotsky's expression showed immediate concern and he gave Jacson/Mercader a paternal pat on the shoulder. He commented aloud, "Your health is poor again. You look ill. That is not good!"

Uncomfortable under Trotsky's paternalistic concern, Jacson/Mercader could not think of a proper answer and merely nodded. Trotsky dropped his hand and turned to the hutches once more to make certain the latches were securely in place. Trotksy was obviously reluctant to leave one of his few relaxations and pleasures with his pets and doubly dismayed at the prospect of having to read the trash he knew Jacson/Mercader had for him.

Trotsky had gone over the outline of Jacson/Mercader's article a few days before, on the 17th of August, and had been appalled by the juvenility of the theme and the poor syntax. He was slightly repulsed at the thought of having to wade through the turgid writing a second time.

The article was entitled, "The Third Camp and the Popular Front." It dealt with the Burnham-Shachtman theory of a "third camp" in World War, i.e., a comparison of the class basis for the "third camp" with that of the French Popular Front. The idea was not original with Jacson/Mercader but had been stolen from Otto Schuessler.[656]

Trotsky recalled Jacson/Mercader's last visit with repugnance. Jacson/Mercader had been rude and had not shown the proper respect due Trotsky's eminence. Trotsky felt he was due a certain homage as the former second in command, only to Lenin, of a great nation. At least some courtesy was due him by a younger man to an older gentleman. Trotsky had been affronted during Jacson/Mercader's last visit. He had kept his hat on during the interview in Trotsky's study and he had sat on the corner of Trotsky's desk while Trotsky tried to read his ill prepared article; basically he was not a gentlemen. He had no sense of the amenities Trotsky was accustomed to receive.[657]

Natalia, always perceptive to swings of Trotsky's moods, quickly sensed that he was developing a dislike for Jacson/Mercader. Her thoughts were:

> Later, when it was all over and Trotsky beyond resurrection, Natalia was to write that on that August day she believed that Trotsky was in the process of digesting the impressions he received emanating from Jacson/Mercader and evaluating these new warning signs. Natalia sincerely believed that Trotsky had wanted to reach a conclusion not to proceed with the fatal interview if he could have somehow managed to delay it just a little while longer.

Rejecting his own negativism, Trotksy turned away from his beloved pets and said, "Well, what do you say, shall we go over your article?" Natalia later described the scene in detail:

> Lev Davidovich fastened the hutches methodically and removed his working gloves. He took care of his hands, or rather his fingers, inasmuch as the slightest scratch irritated him, interfered with his writing, He always kept his pen, like his fingers, in order. He brushed off his blue blouse and slowly, silently started walking towards the house accompanied by Jacson and myself. I came with them to the door of Lev Davidovich's study.

Trotsky later recalled from his death bed, that as he ushered Jacson/Mercader into his study and closed the door behind his visitor, the thought crossed his mind, "This man could kill me!"

Trotsky, calmly and slowly walked across his study to the chair behind the large rough-hewn table he used as a desk. The top of the desk was in its usual orderly disarray, with piles of papers, newspapers and magazines all about but Trotsky knew where everything was and could put his hand on any document he wanted. His pens and pencils, his vital working instruments, neatly up right in a jar. There was his favorite goose neck lamp for use after the sun light no longer came through the study windows from the garden. There was also a large old fashioned blotting paper, an ivory letter opener and, on the left, was his dictaphone with a supply of disks.

Close by his right hand was a .25 caliber automatic, daily placed there freshly cleaned and oiled by Joe Hansen, as a defensive weapon. There was also a .38 caliber automatic in the study but Trotsky had moved this weapon off the desk as it got in the way as he shifted papers during his research.

On the left side of the desk was the alarm button that would instantly alert his guards inside the compound and the Mexican police in the street outside. Jacson/Mercader knew about this alarm button and was able to take a position on Trotsky's left where he could more easily prevent Trotsky from reaching the button. Jacson/Mercader purposefully, and almost obviously, took up his position on Trotsky's left.

Trotsky sat down at his desk and had half turned away to catch the light from the window over his left shoulder and thus had his back to Jacson/Mercader. Jacson/Mercader's testimony to the Mexican authorities after his arrest, picks up the story at this point,

> I put my raincoat on the table on purpose so that I could take out the ice axe which I had in the pocket. I decided not to lose the brilliant opportunity which was offered to me and at the exact moment when Trotsky started to read my article, which served as my pretext, I took the piolet [ice axe] out of my pocket, took it in my fist and, closing my eyes, I gave him a tremendous blow on the head..."[658]

Trotsky must have turned toward Jacson/ Mercader at that very last moment because from the angle of the entrance wound there was evidence that the blow came from the front. The ice axe had an iron head with one end sharply pointed and the other end flat, like a broad claw or chisel. With his eyes closed Jacson/Mercader had hit Trotsky with the flat end and not the fine pointed end. The twisting, or normal change in angle when an object is swung in an arc, caused the end to rotate as he struck Trotsky. Regardless, the broad claw entered Trotsky's skull a distance of two and three-quarters inches.

Trotsky screamed instantly, bounded out of his chair and came toward Jacson/Mercader, grabbing any object he could from the desk to throw at him and fend him off. Trotsky grabbed the pen holder, his books, as he struggled with Jacson/Mercader. He tore the ice axe out of his hand and Jacson/Mercader screamed as Trotsky bit into his finger. This added to the melee heard throughout the house. Jacson/Mercader was finally able to push Trotsky away from him and Trotsky stumbled toward the door.

Continuing Jacson/Mercader's testimony,

> The man screamed in such a way I will never forget it. His scream, 'AAaaaa', very long, infinitely long and it still seems to me as if that scream were piercing my brain. I saw Trotsky get up like a madman. He threw himself at me and bit my hand.... look, you can still see the marks of his teeth... Then I pushed him, so that he fell to the floor. He

> lifted himself as best he could and then, running or stumbling, I don't know, he got out of the room.

Trotsky's scream was heard throughout the compound. Each person in the house reacted in his or her own way. Joseph Hansen, still working on the alarm wiring on the roof recalled,

> A fearful cry rent the afternoon calm... a cry prolonged and agonized, half scream, half sob. It dragged me to my feet, chilled to the bone. I ran from the guard house onto the roof. An accident to one of the ten workers who were remodeling the house? Sounds of a violent struggle came from the 'Old Man's' study and Melquiades was pointing a rifle at the window below. Trotsky, in his light blue work jacket became visible there for a moment, fighting body to body with someone.
>
> "'Don't shoot!', I shouted to Melquiades, 'you might hit the 'Old Man'." Melquiades and Cornell stayed on the roof, covering the exits from the study. Switching on the general alarm, I slid down the ladder into the library. As I entered the door connecting the library with the dining room the 'Old Man' stumbled out of his study a few feet away, blood streaming from his face.

Trotsky saw Hansen through the blood now covering his eyes and weakly said, "See what they have done to me!"

Natalia heard the scream at the same time as Hansen. She dropped her sewing and ran through the door into the dining room. Harold Robins, another guard, pushed his way through the north door into the dining room in front of Natalia. Robins continued into the study to overpower and subdue the killer.

Natalia later said that after she saw Jacson/ Mercader and her husband go into the study she had taken up her sewing. A few minutes later, it could not have been more than three or four minutes, she heard a "terrible, soul-shaking cry." Although she did not realize the scream came from his husband, she intuitively sensed he was in mortal danger.

As she entered the dining room she saw Trotsky bracing himself against the door frame between the dining room and the balcony. His face was covered with blood, his hands limp at his side, his sharp blue eyes naked in the light without his glasses and his blue peasant's smock deeply stained with brownish-purple splotches of blood.

Natalia threw her arms around Trotsky but he seemed oblivious to her presence. Natalia's first thought was that some heavy object had fallen from the ceiling as a result of the construction on the roof. She led him, stumbling, doddering and limp onto the balcony where he collapsed, dragging her down with him.

She remembered that moment with exactitude, "He said to me, calmly, without any indignation, bitterness or irritation, 'Jacson'. Lev Davidovich said it as if he wished to say 'It has happened!'"

Jacson/Mercader, speechless at the moment, utterly stupefied by the enormity of his failure to have killed neatly and silently, felt betrayed by Trotsky's failure to instantly die. He stood rooted to the place where Trotsky had left him, gasping for breath, sucking his thumb that Trotsky had bitten, his other arm limp beside his body, barely grasping the automatic pistol dangling from his hand.

In the split second that all of this was happening, Hansen shouted to Robins, as he sprang into the study, to "take care of" Jacson/Mercader, while he took care of the "Old Man."

Trotsky, with Natalia's help, managed to stagger back into the dining room. He looked at Hansen as if he wanted to issue an order and take command of the crisis. Natalia spoke for Trotsky, as if to goad Hansen into more immediate action, "See what they have done!," she cried. Trotsky's stamina ebbed once more and he seemed to wilt in her arms as he sank to the floor.

A quick first look at Trotsky's head wound gave the impression that the gash may be superficial. Hansen had heard no shot and surmised that the blow had been struck with some blunt heavy object. This kind of a blow would only have split the skin on the skull and caused the profuse bleeding. Hansen knelt besides Trotsky and asked what had happened?

"Jacson shot me with a revolver, I am seriously wounded... I feel that this time it is the end." Hansen replied that he was certain it was only a surface wound. He assured Trotsky that all would be well and that he would recover. Then Hansen, anxious to get on with the indictment of Jacson/Mercader, pressed Trotsky for an answer and asked if Jacson had hit him from behind?

At this point Trotsky started to drift in and out of consciousness and probably did not hear Hansen's question because he did not, or could not, answer. Hansen persisted, desperately trying to get through to Trotksy's rapidly failing consciousness. He raised his voice to penetrate Trotsky's apparent lack of concentration not realizing Trotsky was failing and almost beyond comprehending, Hansen shouted, "No, he did not shoot you, we didn't hear any shots. He struck you with something."

Trotsky had been mumbling to Natalia in Russian but he turned his gaze to Hansen and gave him a doubtful look, as if to tell Hansen, he must be mistaken. Trotsky was continually bringing Natalia's hand to his lips for little kisses. He seemed to gather strength and said, "Natasha, I love you." Then, with a stifled gasp of pain and a choking sound he said, "No one, no one must be allowed to see you without being searched."

Not really knowing what she was doing, Natalia, still rational and in control of herself, compassionately did what she could. She carefully placed a small pillow under Trotsky's bloody head and held a piece of ice to his wound. Cradling her badly injured husband's body in her arms, she stopped the gentle rocking to and fro, as if she had a helpless baby in her arms, thinking the motion might somehow make Trotsky's condition worse.

Trotsky, intermittingly rousing himself into consciousness and slipping out of it more rapidly each time, strained to whisper to Natalia, "Seva must be taken away from this." His voice weak, trailing off, softly returning, "You know.... in there...." indicating in the dirction of the study with his eyes, "I sensed... understood what he wanted... to do. He wanted to strike me....once more... but I wouldn't let him." This last was said with a note of pride!

Natalia hugged her Lev Davidovich closer as if to let him know she understood. "What about that one, they will kill him?" she quickly asked her husband as she saw Charles Cornell charge through the room into the study with a gun in his hand and his face contorted with rage.

Robins had Jacson/Mercader in a corner clubbing him with the butt of his revolver. Simultaneously Hansen vented his rage by repeatedly hitting Jacson/Mercader on the mouth and behind the ear, breaking his hand against the assassin in his fury. Hansen vividly recalled,

> My first impulse was to kill him. But, I decided better not, I felt sure he would talk if we could keep at him a while

> longer. So I kept the police out while Robins, then the two of us, then Robins alone, worked him over. But he wouldn't talk. I think he went unconscious twice or so, and I again became afraid that, giving way to impulse, we might kill him. That was the first I realized how hard this person really was. He would sooner die than talk!

Trotsky seemed to revive and Natalia thought she felt him try to rise. He looked up at her and slowly, with great effort and sternness he said, "No... impermissible to kill. He must be... forced to talk."

As if on cue, Jacson/Mercader's voice rose above the clamor of the guards blows and their energetic grunts as they beat him down to the floor, he yelled, "They made me do it! They've got my mother!" Robins and Cornell paused for a moment in their savagery. Jacson/ Mercader looked pleadingly up at them and sobbed, "Slyvia had nothing to do with this. The GPU had nothing to do with this." Rapidly regaining his composure at this momentary respite from the blows, he looked beseechingly at the two men and, almost calmly, said, "Why don't you kill me?" As if to say, let's get it over with because I'm not going to say anymore!

Hansen ran back to the roof and shouted to the Mexican police in the street. They were alert from the noise and shouts of confusion from the compound that something was terribly wrong. Hansen told them to call an ambulance and get the nearest doctor to come at once.

Charles Cornell also ran into the street in search of a doctor. He tried to use the Buick that Jacson/Mercader had so carefully parked facing away from the wall of the compound but Cornell could not find the ignition keys. Cornell had to take one of Trotsky's cars to get Dr. Dutren who lived nearby and had attended the family in the past.[659]

Trotsky, in a voice so soft that Natalia had to strain to hear, asked if the doctor had come? Natalia replied that one had been summoned and would be there soon.

Within minutes, Dr. Dutren, out of breath and carrying the ubiquitous black bag, burst into the room. He examined Trotsky's bleeding head and announced that the wound was not serious. Trotsky, with a smirk of resignation, knew the doctor was lying. He turned his head to Hansen and, speaking in English, pointing to his heart, said, "I feel there... that it is the end... This time... they've succeeded." Trotsky smiled, a benign look of resignation which quickly changed to a frown of warning as he implored Hansen with his eyes, he said, " Take care of Natalia. She has been with me many, many years."

Then consternation spread across his face and he pulled Hansen's head down close to his lips and whispered, "He was a political assassin. Jacson was a member of the GPU or a fascist. Most likely GPU."

Trotsky's left arm soon lost all feeling and was paralysed. His right hand moved continuously in small circles touching Natalia's hands. He spoke with increasing difficulty, words coming slowly, raspy with rushing gasps for air. Natalia shushing him to conserve his strength and not talk, asked him how he was feeling? Pretending to be encouraged by the doctor's lies, Trotsky replied that he was feeling better.

The ambulance arrived at the gate surrounded by mounted police. A crowd of the curious was milling around the front of the compound. Natalia tried to stop Trotsky from being taken to the hospital. She remembered how her son had been murdered in the hospital in Paris by the GPU.

Trotsky's loyal guards protested his removal and would not let the ambulance attendants take Trotsky out of the house. First, they had a natural reverence and respect for Natalia's wishes. Secondly, there was their own reluctance to allow the still living Trotsky to go beyond the protective walls of the compound where they would have a much more difficult time protecting him. As far as they knew at the time, the lone assassin may be only one of a larger group poised to complete the killing. Thirdly, and probably the most important, was that Hansen had not given his permission to allow Trotsky to be moved.

Hansen was chief of the guards and they were waiting for his authoritative word. Hansen wanted to prolong the departure as long as possible. He knew that as soon as the excitement of moving Trotsky was over, the Mexican police would want to question the suspect. Hansen wanted all the time he could get alone with Jacson/Mercader to question him before he had to turn him over to the Mexican police.

Trotsky drifted back into consciousness. With much difficulty he told his guards they must do something but he would leave the decision to them.

When they arrived at the hospital Natalia's worries were intensified when she saw the crowds that had gathered to await the arrival of the Great Trotsky. The news of the assault on Trotsky had spread rapidly among his many friends, followers and enemies in Mexico City. Natalia ordered Hansen that only Trotsky's loyal followers must surround the stretcher as it moved between the ambulance and the door of the hospital. She implored Hansen, "There may be enemies among them, as there always are in such cases."

Meanwhile, as pre-arranged, exactly at seven-thirty, Otto Schuessler and his fiance arrived at the place in downtown Mexico City where they were to meet Jacson/ Mercader and Sylvia. Sylvia arrived about fifteen minutes late, and was surprised that Jacson/ Mercader was not already there. Sylvia telephoned several times to the Hotel Montejo seeking Jacson/Mercader but he was not to be found.

About ten minutes past eight, Schuessler telephoned Trotsky's compound and was told to return at once because "Jacson has just tried to kill the Old Man!" Sylvia accompanied Schuessler and his fiance to Trotsky's house. When she heard the details, Sylvia cried despondently and asked to be taken to the police delegation. When she arrived there she was arrested.

When Sylvia arrived Colonel Sanchez described her as wearing gold rimmed spectacles, with thick lenses, she was extremely nervous and assured Colonel Sanchez that Jacson "had used her as a means of getting into Trotsky's house to assassinate him."[660]

Back at the hospital, Trotsky was taken to a private examining room. The doctors, aware of the historical significance of the case, examined the wound closely. They rapidly recognized that there was not much that could be done; that death seemed inevitable. The ice axe had penetrated deeply enough and had done enormous damage even though the blow had not been struck as efficiently as the assassin had planned. The ice axe had entered the brain and done irreparable harm to the soft tissues. The doctors agreed that if the ice axe had penetrated another fraction of an inch, death would have been instantaneous.

In order to better see the gaping gash and the surrounding scalp, the doctors ordered the nurses to cut away some of Trotsky's hair. Trotsky turned to Natalia, and with a wry smile was able to weakly say, "You see, the barber also has come." Trotsky's attempt at humor, no doubt offered primarily to ease Natalia's concern, referred to the fact that just that day they had discussed having a barber come to the house to cut Trotsky's hair.[661]

While the nurse was snipping his locks, Trotsky asked Hansen to take some notes for he had something to say. Ever an efficient secretary, Hansen quickly produced a notebook and waited, bent over Trotsky, with pen poised. Trotsky motioned Hansen closer and said, "I am close to death from the blow of a political assassin. I was struck down in my room. I struggled with him. We had entered... talked about French statistics. He struck me. Please say to our friends.... I am sure of the victory... of the Fourth International. Go forward!"[662]

Natalia stood at the head of Trotsky's bed talking soothingly to him in Russian, patting and gently stroking him and holding an ice compress on his wound. They started to undress him. First cutting his blue peasant's jacket with a pair of scissors. A nurse took off his watch and his undergarments without cutting them. Trotsky turned his head slightly back so he could better see Natalia at the head of the bed and softly said to her, "I don't want them to undress me. You do it."

For posterity to treasure, these were the last words Trotsky spoke. Maybe they do not encapsulate enough of an historically important legacy from a man who had been one of the most proficient users of beautiful prose in his speeches and writings — one who had used persuasive, philosophically compelling rhetoric. But commonplace though these last words may be they somehow bequeath to future generations the simple goodness of the man, his sincerity and thoughtfulness. We can only imagine that somewhere in their meaning there was a secret, sweetly shared intimacy, that was known and appreciated by Natalia.

As Natalia finished undressing Trotsky she bent over and kissed him on the lips. Through his receding awareness Trotsky must have recognized the love in this tender touching and Natalia felt his lips respond. She kissed him again, and again he responded. This was the last conscious response to life from Leon Davidovich Bronstein, known to most of the world simply as Leon Trotsky. This last conscious moment fittingly shared with one who had loved him best and who knew him simply as her beloved Lev Davidovich.

The doctors operated on Trotsky but it was a futile attempt.[663] The assassin and the GPU had carried out Stalin's orders too well. Stalin had triumphed at last!.

Natalia stayed with him throughout the night hoping for a miracle. His breathing varied from the heavy and labored to the sonorous and peaceful. His eyelids never flickered and, except for the erratic breathing, he showed no signs of life.

Around the middle of the following day, 21 August 1940, the doctors thought they detected some strengthening of his vital signs but their optimism only raised false hopes. In late afternoon, Trotsky's breathing became faster and more shallow. The assembled doctors and hospital attendants tried to appear calm for Natalia's sake. They knew the end was fast approaching and hovered closely by the bed. They tried to lift him a little to ease his breathing but his head merely drooped closer onto his shoulders and his arm fell listlessly to his side.

Natalia described Trotsky's posture in this position as similar to that of Christ in Titan's "Descent from the Cross," except that Trotsky's head had a crown of bandages instead of Christ's crown of thorns. Trotsky died at 7:25 p.m. on the 21st of August 1940, a Wednesday.

Shortly thereafter, Colonel Sanchez announced Trotsky's death to the waiting newspapermen and the crowds of praying followers outside the hospital. There were some who were only curious who had come to the hospital for the excitement, there were the genuine mourners who had stood silent vigil since Trotsky's arrival at the Green Cross Hospital, and lastly, far in the majority, there was the common people of Mexico who revered Trotsky as the embodiment of a saviour-father figure of revolutionary prominence. A few of these were under the mistaken impression that somehow Trotsky had been involved in their own Mexican revolution but, even after they were corrected, they revered his reputation as a revolutionary all the same.

Trotsky's body lay in state for five days at the Alcazar Funeral Parlor on Puente de Alvarado street across from the Alameda park. His bier was guarded, around-the-clock, by at least one of his American guards or one of his faithful Mexican followers at each corner at the catafalque at all times. There was a large banner strung above the casket which simply stated, "Fourth International." No exact count was made of the mourners but the official estimate was that more than 300,000 people filed past the casket in homage to Trotsky's last public appearance.[664]

Newspapers around the world carried banner front page headlines announcing the death of Trotsky despite the important war bulletins from Europe that would ordinarily have dominated the news. Even Stalin's *Pravda* carried a short squib that Trotsky had been murdered by a disillusioned follower.[665]

The Daily Worker, the newspaper of the CPUSA gave the official Party line:

> Leon Trotsky died at the hands of one of the small gang of dubious social elements and provocateurs who alone remained for him to lead after he had long been exposed as an enemy of the working class.
>
> Trotsky's activities attracted this kind of irresponsible, criminal element. The tiny circle of adherents lived in an atmosphere of mutual hatred, suspicion and violence, utterly without principles. It appears now that one of his own followers, one of the members of his degenerated circle, which Trotsky had created for his own designs, struck him down."[666]

The official voice of the CPUSSR reported with equal rancor:

> Trotsky having gone to the limits of human debasement, became entangled in his own net and was killed by one of his own disciples. Thus, a hated man came to his inglorious end, going to his grave with the stamp of murderer and international spy on his forehead.[667]

Stories in other newspapers varied; in some the assassin was described as a disgruntled Trotskyite and in others he was a Stalinist GPU assassin. The PCM media expressed regret for the assassination having occurred on Mexican soil but, with a straight face, claimed the assassination was the work of a paid Trotskyite agent provocateur![668]

Siqueiros, still a fugitive in hiding because of the 24 May raid, sent a cable from his hiding place expressing regret. He explained that he was only waiting for confirmation of Trotsky's death before he could come out of hiding.

An autopsy was performed by Mexican authorities. Trotsky's brain was exceptionally large, weighing three and one-half pounds. The brain was not replaced in the body. The Mexican police said it would be returned to Mrs. Trotsky or preserved in a museum.[669]

On 27 August Trotsky's body was cremated.[670] The ashes were buried in his beloved garden close to where his rabbit hutches had been in the compound at Coyoacan.[671]

Without fanfare or publicity, and only Natalia, Seva, his loyal American guards, the servants from the house, some of the Mexican police who had served in the guard unit plus a few other personal friends present, the last homage was paid to Trotksy in a simple garden ceremony.

There is a plain, square stone oblisk about four feet high erected in the garden of his home in Coyoacan. It has a hammer and sickle emblazoned on the front with "Leon Trotsky" in simple letters.[672] Natalia lived on in the house for another twenty years. Every morning as she arose, her eyes turned to the white stone in the garden.[673]

In March of 1953, Stalin died a horrible choking death. His daughter who was present, described his last terrible glance as, "full of fear of death." Years later Soviet Premier Nikita Khruschev, in a speech before the 20th Communist Party Congress, denounced Stalin. He enumerated Stalin's many crimnes. In the speech Khruschev stated that he was starting a process "which Trotsky had predicted." This was the first time in many years that Trotsky had been given credit for anything in the Soviet Union which he helped create.

AFTER THE CRIME OF THE CENTURY

> This crime [Trotsky's assassination] will be censured throughout all time by history as dishonorable for those who inspired it and foul for those who actually perpetrated it.
>
> Lazaro Cardenas, President of Mexico, 30 August 1940

> Whatever history's verdict upon him [Trotsky] may be, it will not fail to record that he helped fill some of its most colorful pages."
>
> *New York Times,* 22 August 1940

There were two unfortunate aspects inherent in the Mexican government's investigation of Trotsky's murder. President Cardenas was incensed at the blatant insult to Mexico's policy of guaranteeing safe haven for all political exiles. This outrageous act of unprovoked violence by a foreign intelligence organization on Mexican soil was an unforgivable transgression to the sovereignty of Mexico.

President Cardenas's sincere vow to uncover the whole truth regarding the murder and punish the wrongdoers was quickly and quietly frustrated and stalled by the PCM members in the Mexican government. Certainly the confessed killer went to jail but the complicity of the GPU, the PCM, the Soviet Union and Stalin himself was quietly ignored and swept under the carpet.

The second unfortunate circumstance was that Cardenas's term as president of Mexico was completed in a few months after the murder. Although in Mexico the outgoing president hand picks his successor, the successor (in this instance Miguel Aleman) is not bound to follow his predecessor's policies. Aleman was a staunch anti-communist and did nothing to hinder the investigation of Trotsky's murder, but World War Two intervened and there were other matters that demanded his attention. Trotsky's murder was not investigated as fully as it could have been.

Approximately thirty minutes after the crime had been committed, General Jose Manuel Nunez, chief of the Mexican police, arrived at Trotsky's compound to take charge. He took Jacson/Mercader to the Sixth Delegation which was located in the same building as the Green Cross hospital where Trotsky was taken.

Nunez searched Jacson/Mercader and found a letter in his jacket breast pocket.[674] It was described as a "confession" by the press but that was only a part of it. It was in reality a propaganda ploy, an alleged justification for his act, which the GPU wanted the world to see. If Jacson/Mercader had been successful and been able to leave the study after the murder without being detained, he would have left this justification behind at the scene to be found and exploited by the media.

The killer's alleged justification/confession was typewritten but signed and dated in pencil as, " JAC 20 August 1940." The salient features are translated from the original French:

> Gentlemen: In writing this letter I have no other object, [except that] in the event an accident comes to me, than to explain to public opinion, the motives which induce me to execute the act of justice which I propose.
>
> I am from an old Belgian family..." Jacson/Mercader then proceeds to tell of his study of journalism and his recruitment into Trotsky's Fourth International in Belgium. He was induced, he claimed, to go to Mexico where he would be required do a special job for Trotsky. "I remain chilled by his [Trotsky's] skill in sewing discord in his own party."
>
> It was proposed [by Trotsky] that I go to Russia in order to organize there a series of attempts against different persons and in the first place against Stalin," i.e., to sabotage industry in Russia and to murder Stalin. (This was a man supposedly being ordered to go to Russia by one of the world's most proficient undercover operatives, where he would commit sabotage and murder the nation's leader, who did not speak the language and did not know anything about the area — farcical!)
>
> Trotsky had no other object in his life than to utilize his followers in order to satisfy his own personal ends and

> paltriness."... " I [Jacson/Mercader] was astonished to see with what contempt he spoke of the Mexican revolution and everything that was Mexican."
>
> Jacson/ Mercader asks rhetorically, "... from whence came the money.." for the construction work on [Trotsky's] compound, and answers with, "Perhaps the consul of a great foreign nation [the United States] who often visited him could answer that question for us."
>
> The reference that Trotsky could count on the backing "not merely the support of a great nation but also the backing of a certain parliamentary committee " was obviously as reference to the Dies Committee of the United States Congress.[675] Jacson/Mercader cited his progressive disenchantment with Trotsky, until he could bear no more and decided to rid the world of Trotsky. (He elaborated on this theme extensively later in his oral confession.)
>
> He claimed that he wanted to marry Sylvia Ageloff and take her with him on his mission to Russia but Trotsky forbade him to marry her. "It is probable that after my act she may not wish to know me any more, nevertheless it was also for her sake that I decided to sacrifice myself entirely, removing the chief of the workers' movement who did not do more than prejudice it, and I am sure that later, not only the party, but also the entire history will know how to see me in the right light, when they will see that the bloody enemy of the working class is gone."
>
> "In case anything unfortunate occurs to me, I ask the publication of this letter."[676]

Although Colonel Sanchez did not outrank General Nunez in this case, he remained nominally in charge of the overall Trotsky investigation. Of immediate concern was to verify the killer's identity which Jacson/Mercader now claimed was Jacques Mornard Vanderdreschel, the son of a wealthy Belgian diplomat. He claimed he was born in Tehran in 1904 when his father was the Belgium ambassador to Iran.

Mornard/Mercader claimed long discussions with Trotsky in his home but the logs maintained by the guards revealed a different story. The time Mornard/Mercader spent alone with Trotsky in his study was a total of twenty minutes during the ten visits to the compound.[677] He also declared that he had never been in Russia and did not speak the language. With absolutely nothing to recommend him for a sensitive assignment, the master of espionage craftsmanship was supposed to have chosen him to infiltrate Russia, sabotage the industries and murder the country's leader.

Ten days after the murder, the Belgian Charge d'Affaires in Mexico, Walter Loridan, reported that everything in Mornard/Mercader's biographic claim was false. Further, the genuine Jacques Mornard reported to the Mexican police that he was in Belgium working on a newspaper called *Le Nation Belge*, and had nothing to do with Trotsky's murder and had never been to Mexico.[678]

Colonel Sanchez decided that a confrontation between Mornard/Mercader and Sylvia Ageloff might provide a wedge in what he considered to be a well-rehearsed but completely false story. Mornard/Mercader, his head swathed in bandages from the wounds suffered from the pummeling by Cornell's and Robin's pistols, was brought into the room where Sylvia was sobbing on a bed.

As soon as Mornard/Mercader saw Sylvia he cried out, "Colonel what have you done? Take me out of here," as he attempted to break free from the two detectives holding him.

At the noise, Sylvia turned on the bed and realizing what was happening shouted, "Take that murderer away! Kill him! Kill him!" but Colonel Sanchez tried to persuade her to talk to Mornard/Mercader.

She turned toward the wall, sobbing, "A thousand times no. He is a hypocrite and an assassin. I want to see him killed the way Trotsky died. He is a canaille."

Squirming to try to free himself from the detectives and leave, Mornard/Mercader cried out again, "Colonel why did you bring me here?"

Colonel Sanchez could not terminate the confrontation without another try. He asked Sylvia if her husband had been disillusioned with Trotsky. Sylvia cried out, "Rubbish!" Then turning toward Jacson/Mercader she spoke directly to him, "Tell the truth, traitor! Tell the truth, even if you pay with your life!"

"Were you sure of his affection?" asked a fatherly Colonel Sanchez. "I was sure of it but when he killed Trotsky I saw that he had been deceiving me!," Sylvia sobbed.

The Colonel plumbed deeper and asked Sylvia about the points in Mornard/Mercader's confession? Sylvia, by now exhausted by the confrontation, weakly replied, "Lies! Lies! He is a liar! A canaille!"

The formal judicial proceeding began on 3 September 1940 when Judge Raoul Carranca Trujillo, of the Coyoacan Court of First Instance, appointed psychologists to examine the defendant.[679] After six months of intensive interviewing and testing (972 hours) they issued a report of 1,359 pages entitled, *Organic-Functional and Social Study of the Assassin of Leon Trotsky*.

Mornard/Mercader displayed an attitude of amused but polite toleration for the psychologists' efforts; a disdainful acceptance that they had a job to do. The final result, in addition to reams of physical and psychological findings was that Mornard/Mercader was essentially a weak person, easily led and a latent homosexual.

Mornard/Mercader never showed any remorse for his crime but acknowledged that the law required some punishment. He agreed with the legal requirement that he must pay the widow of his victim a cash indemnity. Unfortunately, he never mentioned Sylvia during the time of the examinations nor during his entire period of confinement.[680]

Before Mornard/Mercader was brought to trial, Colonel Sanchez had captured Siqueiros. The Mexican courts, dominated by PCM members and thoroughly sympathetic to their famous Mexican artist of international repute, reduced the charges against Siqueiros and the Arenal brothers to inconsequential misdemeanors. Siqueiros was allowed to leave Mexico for Chile to paint murals.

Mornard/Mercader attempted to change his defense to self-defense but his confession refuted this contention. On 17 April 1943, two years and eight months after the murder, Mornard/Mercader was sentenced to nineteen years and six months for pre-meditated murder and six months for illegal possession of firearms.[681]

Mornard/Mercader served the initial portion of his sentence in the Black Palace of Lecumberri in Mexico City. In the summer of 1958 he was transferred to the new federal prison located about ten miles from Mexico City on the Puebla highway. He was permitted, as are all Mexican prisoners, connubial visits with his girl friend, Roquelia Mendoza.

In September of 1950 the assassin was positively identified as Ramon Mercader del Rio Hernandez.[682]

On 6 May 1960, some three months before the expiration of his sentence, Ramon Mercader was released from custody and taken to the International airport. Two officials of the Czechoslovakian Embassy in Mexico City joined him and the three flew to Havana on a Cubana de Aviacion plane.[683]

ENDNOTES

1. Trotsky's archives can be found at the Houghton Library at Harvard University in Boston, Massachusetts. There is also an extensive collection of Trotskyite and Fourth International material at the Hoover Institution, Stanford, California.

2. For a modest chronology of Trotsky's life see, Warth, Robert D., *Leon Trotsky*, Twayne Publishers, Boston, (1977), p.9 and 10.

3. Stalin was born on 24 December 1879, when Trotsky was then a month old. Lenin was a boy of nine at that time. (Wolfe, *Three Who Made a Revolution*, Loc. Cit., p. 408.)

4. Payne, Robert, *The Life and Death of Lenin*, Simon and Schuster, New York, (1964), p.162.

5. There is an old saying that exemplifies the lack of harmony and cohesiveness in the Trotskyite movement that augurs against it ever being the nucleus of a militant organization, "Wherever there are two Trotskyites these are always three opinions!" (Andrew,,*KGB*, Loc. Cit., p.156.)

6. Deutscher, Isaac, *The Prophet Outcast*, Oxford University Press, London, (1963), p.125.

7. Barron, John, *KGB*. Readers Digest Press, New York, (1974), p. 2.

8. Medvedev, Roy A., *Let History Judge*, Alfred A. Knopf, New York, (1972), p.140n. (The Soviet secret service has changed names many times over the years; it has been known as the GPU, NKVD and KGB among others.)

9. The preceding three paragraphs are taken from, Sterling, Claire, *The Time of the Assassin*, Holt, Rinehart and Winston, New York, (1983), p. 125.

10. *CIA Review*, "Leon Trotsky: Dupe of the NKVD", Loc. Cit., p. 2.

11. Cookridge, E.H., *Soviet Spy Net*, Frederick Muller, Ltd., London, (1954), p. 162.

12. Kirk, Betty, *The Mexican Front*, University of Oklahoma, Norman, Oklahoma, (1941), p.214.

13. Segal, Ronald, *Leon Trotsky*, Pantheon Books, New York, (1979), p. 11.

14. Nedova, Joseph, *Trotsky and the Jews*, Jewish Publications Society of America, Philadelphia, (1971), p.22.

15. Segal, *Leon Trotsky*, Loc. Cit., p.12.

16. Segal, *Leon Trotsky*, Loc. Cit., p.13.

17. The members of Peoples' Liberty (Narodnaya Volya) were called Narodniks.

18. For a comprehensive chronology of Trotsky's life see, Smith, Irving H., *Trotsky*, Prentice Hall Inc., Engleside Cliffs, New Jersey, (1973), p.7.

19. The Russian calendar was twelve days behind that of Western Europe in the nineteenth century and thirteen days behind in the twentieth century. On 1 February 1918, the Soviet government adapted the Western or "new style" calendar. The Western calendar is used throughout this book. The Bolshevik seizure of power took place on Trotsky's birthday, i.e., 25 October according to the old calendar and 7 November according to the new calendar. The Bolshevik Revolution is referred to as the October Revolution (old style) but is now always celebrated on 7 November according to the new style, (Wolfe, Bertram D. *Three Who Made a Revolution*, Dell Publications, Inc., New York (1948), p. 171.)

20. Trotsky's father David Bronstein died in 1922 when his son was at the peak of his powers. At that time David was the director of a government owned grinding mill near Moscow. A position obtained, no doubt, through his son's political clout. (Wolfe, *Three Who Made a Revolution*, Loc. Cit., p. 185.)

21. David, the father, was fortunate to have purchased this extra land when he did. Two years later a Czarist edict prohibited Jews from acquiring any additional land.

22. Wolfe, *Three Who Made a Revolution*, Loc. Cit., p.173.

23. Trotsky's younger sister, Olga, became a revolutionary like her brother. She became the wife of the Bolshevik leader, Leo Kamenov. (Wolfe, *Three Who Made a Revolution*, Loc. Cit. p.189.)

24. Skousen, W. Cleon, *The Naked Communist*, Ensign Publishing Company, Salt Lake City, Utah, (1961) p.97.

25. Smith, *Trotsky*, Loc. Cit., p.2.

26. Wolfe, *Three Who Made a Revolution*, Loc. Cit., p.201.

27. This piece of seditious literature, "Our Cause", deserves a place in history only because it was Trotsky's first propaganda challenge to the government.

28. In order to follow the many changes in the names of the Russian security and counterintelligence organizations, consult the following chart. In almost all cases it is the same organization with the same charter and personnel but the name has been changed. The author has tried to conform to these name changes but verification can be made of the proper name by consulting the date of the incident against this chart:

Czarist Okhrana gave way to communist Cheka - 1917 -1922
Cheka became GPU - 1922-1923.
GPU became OGPU (but still popularly known as GPU) 1923-1934
OGPU became NKVD - 1934-1941
NKVD became NKGB - 1941-1946
NKGB became MGB - 1946-1953
MGB became KGB - 1954-1991
KGB became FSK - 1991-1996

(Source: DeGramont, Sanchez, *The Secret War*, Andre Deutsch, London, (1962), p.42.)

29. The religious ceremony was offensive to Trotsky's beliefs, but there was no other way to make the marriage legal (Segal, *Leon Trotsky*, Loc. Cit., p.30). The marriage had no force of law and was a "party marriage of convenience". (Wolfe, *Three Who Made a Revolution*, Loc. Cit., p.210.)

30. Vladimir Ilyich Ulyanov used the name Lenin and was known as N. Lenin. The N did not stand for Nikolai as supposed for his father's middle name of Nikolai. The N probably stood for the Russian nyet, a controversial symbol for anonymity. Among others Lenin also used alias of K. Tulin, Petrov, Fray, Karpov, Meyer, Starik, Vilyin and Vladimir Ilyin. (Burnham, Tom, *The Dictionary of Mis-Information*, Thomas Y. Crowell Company, New York, (1975), p.137.)

31. Shapiro, Leonard, *Russian Studies*, Elizabeth Sifton Books-Viking, New York, (1987), p.310

32. Payne, *The Life and Death of Lenin*, Loc. Cit., p.171.

33. Wolfe, *Three Who Made a Revolution*, Loc. Cit., p.171.

34. Colonel Abel (William Fisher) died on 16 November 1971. His grave in Moscow is inscribed with the birth date of "11 July 1903" and his true name as "William Genrykovich Fisher". British birth records reveal that William August Fisher was born at 140 Clara Street, Benwell, in the Westgate district of Newcastle Upon Tyne to an engine fitter named, Henry Matthew Fisher and Lubow Gidova (West, Nigel, *Games of Intelligence*, Crown Publishers, Inc., New York, (1989), p. 224.)

35. "Illegal" is used her to denote an outside unofficial cover agent in the accepted KGB terminology, as opposed to the "legal residentura" of the KGB which is usually housed inside the Soviet Embassy or other official installation.

36. The difference in ages disparages any proof of closeness between Trotsky and Abel, plus the difference in their positions of lofty editor (Trotsky) and menial office boy (Abel), the fact remains that, from a counterintelligence point of view, Trotsky, and a deep cover Soviet agent of later years, did have an association. This does have relevance if only to illustrate past contacts and associations can be the basis for a deeper investigation of a suspect.

Conspiratorial alliances which are tied together over half-a-century again emphasizes the faulty operational methods the KGB uses when it resurrects and re-uses agents. Long forgotten relationships that may have been active for the GPU years ago have a way of dove-tailing with operations again sometime in the future. Sleeper agents must be activated as operational requirements dictate. Once the ready reserve of sleeper agents is depleted, the re-treads may have to be resurrected and this automatically breeds a security factor into the operation.

37. Trotsky was attractive to women. He had little time for the wooing, but a number of affairs, temporary and not engaging his spirit very deeply, came his way without much expenditure of effort on his part. (Wolfe, *Three Who Made a Revolution*, Loc. Cit., p.217.)

38. Rue Gassendi 46, (Segal, *Leon Trotsky*, Loc. Cit., p.43.)

39. Trotsky's idyllic, revolutionary bred, relationship with Natalia was apparently not resented by Trotsky's first wife, Alexandra, since all three remained on good terms until their deaths.

40. Sayers, *The Great Conspiracy*, Loc. Cit., p. 188.

41. Segal, *Leon Trotsky*, Loc. Cit., p.53.

42. Segal, *Leon Trotsky*, Loc. Cit., p.114.

43. Wolfe, *Three Who Made a Revolution*, Loc. Cit., p.289.

44. Wolfe, *Three Who Made a Revolution*, Loc. Cit., p.300.

45. Taylor, Edmond, *The Fall of the Dynasties*, Doubleday and Company, Inc., Garden City, New York, (1963), p.285.

46. Floyd, David, *Russia in Revolt*, MacDonald Publishers, London, (1969), p.64.

47. Wolfe, *Three Who Made a Revolution*, Loc. Cit., p.286.

48. Wolfe, *Three Who Made a Revolution*, Loc. Cit., p.301.

49. "Soviet" means "council" as a common noun.

50. Segal, *Leon Trotsky*, Loc. Cit., p. 76.

51. Segal, *Leon Trotsky*, Loc. Cit., p.79.

52. Wolfe, *Three Who Made a Revolution*, Loc. Cit., p.335.

53. In those days Trotsky disliked Lenin. He spoke of Lenin, as a man who always looked at other men as if he wanted to give them orders. (Payne, *Lenin*, Loc. Cit., p. 201.)

54. Wolfe, *Three Who Made a Revolution*, Loc. Cit., p.336.

55. See *Life*, 23 April 1956, pages 47 to 51.

56. For a more comprehensive treatment on the subject see; Levine, Isaac Don, *Stalin's Great Secret*, Coward-McCann, New York, (1956).

57. In Vienna Trotsky and Natalya had found refuge with Dr. Victor Adler, noted protegee of Sigmund Freud. (*New York Times*, 22 August 1940). (Dr. Adler's son had assassinated Count Stuerkh in 1916. Ibid.)

58. Trotsky's slogan in 1914 was "peace without indemnities and annexation, peace without victor or vanquished". (Shub, David, *Lenin: A Biography*, Doubleday and Company, Inc., Garden City, New York, (1948). p. 132.)

59. Another source reports that the family moved into a three room flat on Vyse street in the Bronx. (*New York Times*, 22 August 1940.)

60. Trotsky worked on *Novy Mir* with his old friend and Lenin's opponent, Nicolas Bukharin. (Sayers, *The Great Conspiracy*, Loc Cit., p.189.)

61. The development of the CPUSA and how it relates to the Trotsky operation is the subject of another chapter. The alliance between Trotsky and Fraina at this time in New York is noted here for future reference and to provide continuity of the underlying theme that relationship between Soviet agents of one era can, and profoundly do, have an effect on clandestine relationships of another time and place.

62. Heller, Mikahil and Aleksander Nekrich, *Utopia Power*, Summit Books, New York, (1982), p.25.

63. The so-called "February revolution" that caused Czar Nicholas II to abdicate actually took place, according to the western calendar, in March of 1917. The Czar's abdication permitted the Provisional Government of Alexander Kerensky to rule Russia for eight month's.

The "October revolution" in which the Bolsheviks ousted the Provisional Government actually took place according to the western calendar in November of 1917.

World War One ended in November of 1918, just one year after the Bolsheviks seized power in Russia. By way of historical perspective, America entered World War One in April of 1917, less than one month after the first revolution in Russia although the latter was not brought about or influenced by the former.

64. Bruce Lockhart. British intelligence officer in Russia at the time, said that the British Government made a serious mistake in the way they handled Trotsky in Canada. (Sayers, *The Great Conspiracy*, Loc. Cit., p.190.)

65. Lieutenant Colonel John Bayne MacLean, of the Royal Canadian Intelligence Service during World War One, claimed personal knowledge that Trotsky was not a Russian but a German whose real name was not Bronstein, as it would be if it were Russian, but Braunstein, as it would be in German. He claimed Trotsky spoke better German than he did Russian and, before the war, Trotsky was supposed to have been the leader of the German Black Band in Berlin, a notorious political organization. MacLean charged that in August of 1913, Trotsky with the connivance of German intelligence was deported in a mock proceeding, so that he could go to the United States under cover as a Russian refugee. Trotsky, was to organize the Germans and Austrians in the United States and then go to Russia to support the Bolshevik revolution. Once the Bolsheviks were in power, with the support of the Germans and Austrians in the Untied States, Trotsky would pledge a peace with Germany without the Allies. MacLean also reported that not only Lenin and Trotsky were German agents but so was Alexander Kerensky, the leader of the Russian Provisional government.

66. As a matter of historical interest it might be noted that Ambassador William Dodd's daughter, Martha Dodd Stern, was a Soviet agent who was active in the United State. She later fled American prosecution to live in Mexico but this was long after the Trotsky assassination.

67. Byrant, Louise, *Six Red Months in Russia*, George H. Doren, New York, (1918), pps. 66-67.

68. Trotsky was moving at such a violent rate of speed to assure the success of the Bolshevik revolution that he projected himself outside history altogether and entered the world of legends. (Payne, *Lenin*, Loc. Cit.,p.406.)

69. Alexandra Kollontai and the talented agitator, Anatoly Lunacharsky, were released from prison at the same time as Trotsky. (Gardener, Virginia, *"Friends and Lovers": The Life of Louise Bryant*, Horizon Press, New York, (1982), p.86.

70. Cowles, Virginia, *The Romanovs*, Harper and Row Publishers, New York, (1971), p.254.

71. Shub, *Lenin*, Loc. Cit., p.263.

72. Lincoln, W. Bruce, *Passage Through Armageddon*, Simon and Schuster, Inc., New York, (1986), p.432.

73. Rapoport, Vitaly and Yuri Alexee, *High Treason*, Duke University Press, Durham, (1985), p.108.

74. Crankshaw, Edward, *The Shadow of the Winter Palace*, The Viking Press, New York, (1976), p.318.

75. Dukes, Paul, *Red Dusk and the Morrow*, Doubleday, Page and Company, Garden City, New York, (1922), p.vii.

76. Crankshaw, Edward, *Putting Up With the Russians*, Elizbaeth Siffon Books/Viking, New York, (1984), p.266

77. Dallin, David J., "Mark Zborowski, Soviet Agent", *The New Leader*, New York, 19 March 1956.

78. Friedin, Seymour and George Bailey, *The Ex-perts*, The Macmilllan Company, New York, (1968), p.41.

79. Mazour, Anatole B., *Tsarist and Communist Russia*, P. Van Nostrand Company, Princeton, New Jersey, (1962), p.588.

80. Thayer, Charles W., *Diplomat*, Harper and Brothers Publishers, New York, (1959) p.184.

81. Shub, *Lenin*, Loc. Cit., p. 298.

82. Browder, Robert Paul, *The Origins of Soviet-American Diplomacy*, Princeton University Press, Princeton, New Jersey, (1953), p. 5.

83. Clarkson, Jesse D., *A History of Russia*, Random House, New York, (1968), p.522.

84. Dziak, John J., *Chekisty: A History of the KGB*, Lexington Books, Lexington, Massachusetts, (1988), p.4.

85. De Gramont, *The Silent War*, Loc. Cit., p.47.

86. Richards, Guy, *The Rescue of the Romanovs*, Devon-Adair Company, Old Greenwich, Connecticut, (1947).

87. Payne, *Lenin*, Loc. Cit., p.516/17.

88. Payne, *Lenin*, Loc. Cit., p. 514.

89. Sheridan, Clare, *Mayfair to Moscow*, Boni and Liveright, New York, (1921), p.152.

90. *New York Times*, 21 November 1920.

91. *New York Times*, 27 November 1920.

92. *New York Times*, 25 November 1920.

93. Payne, *Lenin*, Loc. Cit., p.535.

94. Payne, *Lenin*, Loc. Cit., p.537.

95. Payne, *Lenin*, Lco. Cit., p. 510.

96. Payne, *Lenin*, Loc. Cit., p. 548/549

97. Levytsky, Boris, *The Uses of Terror, "The Soviet Secret Police 1917-1940"*, Coward, McCann and Geoghegan, Inc., New York, (1972), pps. 8 and 54.

98. Sayer, *The Great Conspiracy*, Loc. Cit., p.200. (The entire thrust of Sayer's book is anti-Trotsky and, with the exception of the quoted *Daily Worker* article, is without attribution. This one article offers no proof that it was written by Trotsky and is without doubt merely another example of some Stalin directed misinformation.)

99. Payne, *Lenin*, Loc. Cit., p. 601. (See also, Trotsky, Leon, "Did Stalin Poison Lenin?", *Liberty*, August 1940.)

100. Liversidge, Douglas, *Lenin; Genius of Revolution*, Franklin Watts, Inc, New York, (1964), p.171.

101. Daniels, Robert V., *Red October*, Beacon Press, Boston, (1967), p.220.

102. Essad-Bey, *OGPU*, The Viking Press, New York, (1933), p.165

103. Payne, *Lenin*, Loc. Cit.,, p.590.

104. Bazhanov, Boris, *Bazhanov and the Damnation of Stalin*, Ohio University Press, Athens, Ohio, (1990), p.104.

105. Bailey, Geoffrey, *The Conspirators*, Harper and Brothers Publishers, New York, (1960), p.279/80.

106. Moorehead, *The Russian Revolution*, Loc. Cit., p.285.

107. For a general description of Stalin's demeanor at this time see; Orlov, Alexander, "The Ghastly Secrets of Stalin's Power", *Life*, Nos. I to IV, 6 to 27 April 1953.

108. De Jonge, Alex, *Stalin*, William Morrow and Company, Inc., New York, (1986), p.208.

109. Finder, *Red Carpet*, Loc. Cit., p.59.

110. Clemenceau had charged the French government with inefficiency and virtual defeatism before he himself took power and waged the war against Germany to a victorious conclusion.

111. Deutscher, Isaac, *Stalin; A Political Biography*, Oxford University Press, New York, (1949), p.310.

112. De Jonge, *Stalin*, Loc. Cit., p. 209.

113. Carr, *Socialism in One Country*, Loc. Cit., pp.139-152.

114. Carr, *Socialism in One Country*, Loc. Cit., pp.139-152.

115. Rapoport, *High Treason*, Loc. Cit., p.110

116. Gardner, Virginia, *"Friend and Lover': The Life of Louise Bryant*, Horizon Press, New York, (1982), p.113.

117. Generally speaking *black propaganda* is propaganda ostensibly emanating from some third party. *Black propaganda* could denigrate the target but without implicating the real sponsoring party as the originator. *Grey propaganda* is propaganda that is non-attributable to the originator but is obviously issued by some one siding with the originator. *Grey propaganda*,for example, would be a pro-American article published in any newspaper praising the America government, but written by an ostensibly impartial non-government reporter. *Black propaganda* would be an article praising the American government published in a Japanese newspaper ostensibly written by a German politician. *White propaganda* would be the same as for grey propaganda but published by an American government source such as USIA.

118. Kominski, Morris, *The Hoaxers*, Brandon Press, Boston, (1970), p.252.

119. Sayers, *The Great Conspiracy*, Loc. Cit., p. 24.

120. Mahoney, M.H., "Dmitri Kalamatiano", *Journal of Intelligence and Counterintelligence*, Winter 1995, and Fogelson, David S., *America's Secret War Against Bolshevism*, University of North Carolina Press, Chapel Hill, (1995).

121. Emanuel Voska was the principal agent of an extensive network of agents in Russia on a joint Anglo-American mission. (See Miller, Nathan, *Spying for America*, Paragon House, New York, (1989), p.207. Also see, Voska, Emanuel and Will Irwin, *Spy and Counterspy*, Doubleday, Doren and Co., Inc., New York, (1940).

122. See Mahoney, M.H., *American Prisoners of the Bolsheviks*, Wheaton Press, Illinois, (1996). Also see, Churchill, General Marlborough, "The Military Intelligence Division, General Staff", *United States Artillery Journal*, April 1920, p.1.

123. Smith, *America Enters the World*, Loc. Cit., p.542.

124. Sayers, *The Great Conspiracy*, Loc. Cit., p.3.

125. Sayers, *The Great Conspiracy*, Loc. Cit., p.27.

126. The salaries for Foreign Service Officers were too low to attract outstanding men. Many FSOs were unprepared by experience, language proficiency or education for the requirements thrust upon them. Many of the deficiencies that plagued the Department and Foreign Service during World War One can be attributed to the incompetent political appointments made by former Secretary of State William Jennings Bryan prior to his resignation. The situation was not much improved by his successor's (Lansing's) lack of influence with President Wilson. (Estes, Thomas S., and E. Allan Lightner, *The Department of State*, Praeger Publishers, New York, (1976). p.34.)

127. The Foreign Service Officers in the Political Section in the American Embassy, if they were doing their job, should have warned Ambassador Francis of the possible pitfalls in a continuing relationship with Matilda This function is as much a part of the Political Section's responsibility as are its duties as a passive reporting agency, merely commenting on the overt political events of the host country's government

128. There is an estimate that one in every four city dwellers was an informant for the Cheka. (Freemantle, Bruce, *KGB*, Holt, Rinehart and Winston, New York, (1982), p.23.)

129. Major Raymond Robins was "permitted" by Ambassador Francis to maintain an informal liaison with Trotsky.

130. Sisson, *One Hundred Days*, Loc. CVit., p, 155.

131. Kennan, George F., *Russia Leaves the War*, Princeton University Press, Princeton, New Jersey, (1956), p.50.

132. Kominsky, *The Hoaxers*, Loc. Cit., p. 252.

133. Volkman, *Secret Intelligence*, Loc. Cit., p.17.

134. Sayers, *The Great Conspiracy*, Loc. Cit., p.32.

135. The countries were Great Britain, France, Japan, Germany, Italy, United States, Czechoslovakia, Serbia, China, Finland, Greece, Poland, Rumnia and Turkey.

136. Sayers, *The Great Conspiracy*, Loc. Cit., p.79.

137. Andrew, *Secret Service*, Loc. Cit., p.203.

138. It is said that the British wartime intelligence organization in pre-revolutionary Russia was also almost as confused as the Czarist regime. (Andrew, *Secret Service*, Loc. Cit., p,203)

139. Hill, George A., *Go Spy the Land*, Bernhard Tauchnitz, Leipzig, (1935), p.254.

140. Sayers, *The Great Conspiracy*, Loc. Cit., p.10.

141. Maugham, W. Somerset, *Ashenden: The British Agent*, Doubleday and Company, Inc., New York, (1927).

142. Andrew, *Secret Service*, Loc. Cit., p. 209.

143. West, Nigel, *MI-6*, Random House, New York, (1983),p.10.

144. Andrew, Christopher, *Her Majesty's Secret Service*, Viking Press, Inc., New York, (1986), p.209.

145. Hill, *Go Spy the Land*, Loc. Cit., p.166. (Hill, as a personal view, was repelled by the entire Bolshevilk doctrine. He disliked them for their tyranny and ruthlessness and for their "hatred of the British Empire."

146. Smith, Page, *America Enters the World*, Volume 7, McGraw-Hill Book Company, New York, (1985), p.534.

147. Sayers, *The Great Conspiracy*, Loc. Cit., p.25.

148. Lockhart, Bruce, *British Agent*, Garden City Publishers, New York, (1933), p.34.

149. Andrew, Christopher, *Her Majesty's Secret Service*, Viking Books, Inc., New York, (1986), p.212.

150. Volkov, Fydor, *Secrets from Whitehall and Downing Street*, Progress Publishing, Moscow, (1986), p.21.

151. Dukes, *The Story of ST-25*, Loc. Cit., p.272.

152. Ashford, Kent, "Memorandum on the Work of the Section of Civil Affairs and Security," British General Staff, 12 December 1929.

153. Hill's code designation was "I.K. 8." Like Reilly and Trotsky Hill was predisposed toward the opposite sex. Apart from collecting a series of mistresses he was married four times. (Lockhart, *Reilly: The First Man*, Loc. Cit., p.61.)

154. Hill, *Go Spy the Land*, Loc. Cit., p.174. (Other sources are skeptical regarding some of Hill's claims. In this regard see; Andrews, *Secret Service*. Loc. Cit., p.216.)

155. According to Kim Philby, Hill was one of the few living Englishmen who had actually put sand in axle-boxes -- the school solution to sabotaging railroad engines. (Andrew, *Secret Service*, Loc. Cit., p.216.

156. Hill, *Go Spy the Land*, Loc. Cit., p. 177.

157. West, *MI-6*, Loc. Cit., p.17.

158. Some of Reilly's biographers write that "Reilly was recruited into SIS" but at the time SIS did not exist. (Andrew, Christopher, *Secret Service*, Heinemann, London, (1985), p.83.)

159. Hill, *Go Spy the Land*, Loc. Cit., p. 164.

160. The American deep cover agent of the State Department, Dmitri Kalamatiano, who worked with Reilly, expressed doubts as to Reilly's loyalties to the British. (Lockhart, *Reilly: The First Man*, Loc. Cit., p. 102. See also, Mahoney, *Kalamatiano*, Loc. Cit.)

161. Andrew, *Secret Service*, Loc. Cit., p.84.

162. Dukes, Paul, *The Story of "ST-25,"* Cassell and Company, Ltd., London, (1938), p.195.

163. Sayers, *The Great Conspiracy*, Loc. Cit., p.87 n.

164. Andrews, *Her Majesty's Secret Service*, Loc. Cit., p. 223.

165. West, *MI-6*, Loc. Cit., p.22.

166. Dukes, *The Story of "ST-25,"* Loc. Cit., p. 238.

167. Andrew, *Secret Service*, Loc. Cit., p. 261.

168. Some code pads are glued all around and stitched with twine to make breaking into the pad extremely difficult. Extreme care must be taken when the pad is initially violated that it can be re-stitched and re-glued *exactly* as it was originally or the code clerk with know the pad has been opened and will not use it for encoding.

169. For a general over-view of this Soviet operation see; "The Trust" *CIA Historical Intelligence Collection*, dated March 1969, (Approved for release on 1 March 1980.)

170. Savinkov is alleged to have tried to convince Dzerzhinsky to permit him to work for the Cheka. Dzerzhinsky refused and Savinkov was left with no alternative. He allegedly threw himself from a prison window to his death in the courtyard below in May of 1925. (Andrew, *Her Majesty's Secret Service*, Loc. Cit., p. 314.)

171. Other reports claim that Reilly was shot in early November 1925. (Andrew, *Her Majesty's Secret Service*, Loc. Cit,., p. 315.)

172. West, *MI-6*, Loc. Cit., p.26.

173. Cookridge, *Soviet Spy Net*, Loc. Cit., p.77.

174. CI Staff report of the Central Intelligence Agency, "Leon Trotsky, Dupe of the NKVD," released on 3 March 1985, page 5. (Preceeding three paragraphs.)

175. CI Staff report of Central Intelligence Agency, "Leon Trotsky, Dupe of the NKVD," released on 3 March 1985, p.3.

176. Sayers. *The Great Conspiracy*, Loc. Cit., p.200/201.

177. Sayers, *The Great Conspiracy*, Loc. Cit., p.225

178. Sayers, *The Great Conspiracy*, Loc. Cit., p.198. (On page 202 of this source, Kristinsky is noted as being the official Soviet liaison officer with the German High Command, which makes his contact with the Germans routine and legal. It is the sinister implications the trial testimony tried to adduce that makes the relationship compromising, if true.

179. Bryant, Louise, *Six Red Months in Russia*, George H. Doran Company, New York, (1918), p.146.

180. Essad-Bey, *OGPU*, Loc. Cit., p. 77.

181. Sayers, *The Great Conspiracy*, Loc. Cit., p.27.

182. Pipes, Richard, *The Russian Revolution*, Alfred A. Knopf, New York, (1990), p.589.

183. Sisson, *One Hundred Red Days*, Loc. Cit., p.445 et seq.

184. Gardiner, Virginia, *"Friend and Lover"; The Life of Louise Bryant*, Horizon Press, New York, (1962), p.243. (They were married in Paris on 10 December 1923. This was two days after her birthday, she was then thirty-eight years of age. Bullitt believed her to be twenty-nine. He was thirty-two.)

185. *New York Times*, 9 July 1921, "Trotsky and Lenin Patch Up Quarrel."

186. Rositzke, Harry, *The KGB: The Eyes of Russia*, Doubleday and Company, Inc., Garden City, New York, (1981), p.72.

187. Cookridge, E.H., *Soviet Spy Net*, Frederick Muller, Ltd., London, (1954), p.27.

188. Sayer, *The Great Conspiracy*, Loc. Cit., p.206.

189. Churchill, Winston, *Great Contemporaries*, G.P. Putnam's Sons, New York, (1937), p.170.

190. Andrew, *KGB*, Loc. Cit., p.115.

191. Agabekov, George, *GPU*, Brentwood, New York, (1931), p.224.

192. Trotsky was to be in exile for eleven and one-half years:

Turkey from the beginning of 1929 to the summer of 1933.
France from the summer of 1933 to the summer of 1935
Norway from the summer of 1935 to December of 1936.
Mexico from January 1937 to 21 August 1940.

Andrew, Christopher and Oleg Gordievsky, *KGB — The Inside Story*, Harper Collins Publishers, New York, (1990), p.154.

193. Sergie, Trotsky's older son, and his wife, Ana, decided to remain in Russia rather than share Trotsky's exile.

194. Sayers, *The Great Conspiracy*, Loc. Cit., p.211

195. Van Heijenoort, Jean, *With Trotsky in Exile; From Prinkipo to Coyoacan*, Harvard University Press, Cambridge, Massachusctts, (1978), p.18.

196. This house should not be confused with the Villa Izzet Pasha on the island of Prinkipo where the Trotskys later settled even though the names are the same.

197. In Turkish the island of Prinkipo is called "Buyuk Ada" which means the "Great Island." "Prinkipo," the word itself, means "princes'" island. It was there the Byzantine emperors banished princes of their realm who were out of favor after having them blinded by tearing out their eyeballs.

198. Van Heijenoort, *With Trotsky in Exile*, Loc. Cit., p. 22.

199. They called the house "Sea Spray" because the waves that crashed on the rocks below sent a spray of salt water into the garden. (Trotsky, Natalia Sedova, *How It Happened*, Merit Publishers, New York, (1969), p.35.

200. Sayers, *The Great Conspiracy* has been cited in this book on several occasions for anti-Trotsky misinformation. An indication of the lack of valid research in *The Great Conspiracy* is evidenced by the claim that Sheldon Harte was a bodyguard of Trotsky's on Prinkipo Island. (See page 212.) Harte did not join Trotsky's entourage until many years later when Trotsky was in Mexico in the summer of 1940.

201. Deutscher, Isaac, *Stalin: A Political Biography*, Oxford University Press, New York, (1949), p.348.

202. The fire was set by the GPU and during the ensuing melee they stole many of Trotsky's papers. (CI Staff report of Central Intelligence Agency, "Leon Trotsky, Dupe of the NKVD," released on 3 March 1985, p.9.)

203. Van Heijenoort, *With Trotsky in Exile*, Loc. Cit., p.43.

204. Eastman, Max, *Great Companions*, Farrar, Straus and Giroux, New York, (1959), p.53 - 69.

205. Valentine Olberg in the first Moscow Purge Trials in 1936 was prosecuted by Stalin.

206. For three years, from the spring of 1929, both Sobolevicius brothers were among Trotsky's closest confidants. They had access to his codes, secret inks and cover addresses used by him to correspond with his supporters in the Soviet Union. Trotsky even entrusted much of his correspondence to them. (Andrew, *KGB*, Loc. Cit., p. 155.)

207. A source used in this book with extreme caution, supposedly objective but often cited as severely biased as dispensing Soviet disinformation is Sayers, *The Great Conspiracy*, It should be noted that Sayers, in relating Trotsky's residence on Prinkipo, makes no mention of any of the GPU provocations who attempted to penetrate Trotsky's entourage during his stay on the island.

208. In an article in the New York *Journal American*, in 1957, Jack Soble, the former Senin, claimed his wife Myra had been held hostage by the GPU in order to force his compliance to spy on Trotsky at Prinkipo. (CI Staff of the Central Intelligence Agency study, "Leon Trotsky, The Dupe of the NKVD," released 3 March 1985, p. 156.)

209. Thirty years after the Prinkipo incident, Abraham Senin, then known as Jack Soble, was convicted in the United States as a Soviet espionage agent He was a periferal member of the Rosenberg atom spy network. During his interrogation by the FBI, after he became a witness for the prosecution, he confessed to the FBI of having spied on Trotsky on Prinkipo for the GPU.

210. Leonid Eitingon was the "legal" GPU resident in Istanbul and was the case officer who directed Lisa Gorskaya to seduce Bylumkin The difference between "illegal" and "legal" residents is, that the former is usually under nonofficial cover while the latter is usually under official cover in the embassy. Eitingon was later in charge of the Trotsky assassination in Mexico (Andrew *KGB*, Loc. Cit., p. 155.)

211. Lisa Gorskaya later married Vasili Zarubin, who was at the time the GPU resident in Berlin Later Zarubin, as Zubilin, was the chief of the NKVD in Washington, D.C. and one of the Soviet master-minds in the theft of the atomic secrets from the United States through the Rosenberg spy ring. Liza (or Lisa) was known to Elizabeth Bently and Hede Massig as "Helen" when she acted as their case officer in the United States.

212. Wyndham, *Trotsky*, Loc. Cit., p.126.

213. Van Heijenoort, *With Trotsky in Exile*, Loc. Cit., p. 22.

214. Nine years before Kirov had replaced Zinoviev on the Politburo and had since served the Soviet government honorably and with distinction.

215. Glotzer, Albert, *Trotsky; Memoir and Critique*, Prometheus Books, Buffalo, New York, (1989), p.245.

216. Sixty years after Stalin accused Trotsky of Sergie Kirov's murder, the papers of the former Soviet Premier, Nikita Khrushchev were released. Khrushchev reported that Stalin had arranged Kirov's murder saying, "... I have no doubt Stalin was behind the plot." *(Washington Post*, 24 September 1990)

217. Deutscher, *Stalin*, Loc. Cit., p.373.

218. Van Heijenoort, *With Trotsky in Exile*, Loc. Cit., p.35.

219. Klehr, Harvey, *The Heyday of American Communism*, Basic Books, New York, (1984), p. 83.

220. Van Heijenoort, *With Trotsky in Exile*, Loc. Cit., p.77/78.

221. Sayers, *The Great Conspiracy*, Loc. Cit., p.276. (These alleged secret deals made by Trotsky are uncorroborated.)

222. In addition to the supporters of Zinoviev, Lev Kamenov, Nikolai Bukharin, Trotsky and others, Stalin ordered the murder of everyone who had ever opposed him. Also, nearly everyone in the Party establishment over thirty-five and much of the Red army, were arrested, charged with espionage and liquidated. (Beschloss, Michael R., *May-Day*, Harper and Row, Publishers, New York, (1986), p. 166.)

223. There is another version of Kirov's murder, this one hewing closer to the Party line and not blaming the GPU for botching the job. According to this version, Kirov was shot by Leonid Nikolayeu, the estranged husband of Milda Draule, who worked as a waitress in Kirov's secretariat at the Smolny Institute. It is alleged that Kirov forced his attention on this young lady. This sordid sex scandal was suppressed by Stalin who did not want to blemish the reputation of one of his henchmen. (Sudaplotov, *Special Tasks*, Loc. Cit. p. 51.)

224. Written by Leon Blume in 1931. Blume later became the Socialist Premier of France. (CI Staff report of the Central Intelligence Agency, "Leon Trotsky, Dupe of the NKVD," released on 3 March 1985.)

225. Petrov, Vladirmir and Evdokia, *Empire of Fear*, Andre Deutsch, Publishers, London, (1956), p.67.

226. The Soviet government was threatening to stop the importation of large amounts of Norwegian herring unless Trotsky was muzzled. (Van Heijenoort, *With Trotsky in Exile*, Loc. Cit., p. 88.

227. *Committee on Un-American Activities*, "The American Aspects of the Assassination of Leon Trotsky," Eighty-First Congress, Washington, D.C. (1950), p.IX.

228. CI Staff of Central Intelligence Agency report "Leon Trotsky, Dupe of the NKVD," released 3 March 1985.

229. Churchill, Winston, *Great Contemporaries*, G.P. Putnam's Sons, New York, (1937), p.167.

230. The Republicans represented the remaining legally and duly elected government of Spain. In addition to being called "Republicans" they were variously known as "Loyalists," "Reds" or "Leftists." They stood against Fascism, against the Roman Catholic Church (especially against the priests), against the Spanish military caste and against all privileged classes of Spain. This would include, but not be limited to, wealthy absentee landlords, entrenched nobility and all commercial exploiters who were called "capitalists."

The Nationalists opposed the Republicans. General Francisco Franco was the titular leader of the Nationalists. They were aligned against Marxism, leftist labor unions, squatter land-hungry peasants and in general, they were against anyone who opposed them. They were especially against anyone who claimed to be a Republican or a communist. The Nationalists were variously known as the "Insurgents," "Fascists," "Rebels," "Rightists" and "Falangists." Sometimes the Nationalists were also known by their leader's name "Franquistas."

231. Fischer, Louis, *The Nation*, New York, 13 March 1939.

232. Lyons, *The Red Decade*, Loc. Cit., p. 269.

233. Van de Rhoer, *The Shadow Network*, Loc. Cit., p.74.

234. Lockhart, *Reilly: The First Man*, Loc. Cit., p.36.

235. Lockhart, Robin Bruce, *Reilly: The First Man*, Penguin Books, New York, (1987), p.37/38.

236. See generally, Donovan, James R., *Strangers on a Bridge: The Case of Colonel Abel*, Atheneum, New York, (1964).

237. Abel subsequently surfaced in the Soviet Union teaching prospective Soviet espionage agents destined for deep cover assignments in the United States how best to survive.

238. For a detailed account by the Soviet case officer who handled these British agents see, Modin, Yuri, *My Cambridge Friends*, Farrar, Straus and Girous, (1994).

239. Glees, Anthony, *The Secrets of the Services*, Carroll and Graf Publishers, New York, (1987), p.367.

240. Burgess was recruited into Soviet espionage by Semyon Nikolayevich Rostovsky, under the operational direction of Orlov. Rostoveky used the alias of Ernest Henri. Burgess then went on to recruit Anthony Blunt and Blunt recruited an American Michael Straight. (Lockhart, *Reilly: The First Man*, Loc. Cit., p.45.)

241. Wyden, Peter, *The Passionate Way*, Simon and Schuster, Inc., New York, (1983), p.406.

242. *Chicago Tribune*, 12 May 1988, page 12.

243. Fraser, Ronald, *Blood of Spain*, Pantheon Books, New York, (1979), p.12.

244. Andrew, *KGB*, Loc. Cit., p.160.

245. The communist newspaper in Lerda, Spain procliamed, "The POUM and its leader, Nin, must be exterminated in the interests of the freedom of Spain. (Cookridge, *Soviet Spy Net*, Loc. Cit., p.177.)

246. Sayers. *The Grand Conspiracy*, Loc. Cit., p.283.

247. Thomas, Hugh, *The Spanish Civil War*, Harper and Row, New York, (1977), p.708.

248. Eby, Cecil D., *Between the Bullet and the Lie*, Holt, Rinehart, New York, (1969), p.xxi.

249. Eby, *Between the Bullets and the Lie*, Loc. Cit., p.316.

250. GRU is the military intelligence department of the Soviet government.

251. Lyons, *The Red Decade*, Loc. Cit., p.273.

252. In 1957 Orlov testified before the United States Senate Sub-Committee on Internal Security He reported that Eitingon, under General Sudoplatov, had been assigned by Stalin to organize the Trotsky assassination

253. Dallin, David J., "Mark Zborowski, Soviet Agent," *The New Leader*, New York, (19 March 1956).

254. Medvedev, *Let History Judge*, Loc. Cit., p.40.

255. CI Staff of the Central Intelligence Agency report, "Leon Trotsky, Dupe of the NKVD," released 3 March 1985, page 22.

256. It might be noted that in view of the subsequent attendance of these two "doctors" to Sedov when he became ill, that neither of these two were licensed to practice medicine in France.

257. CI Staff of Central Intelligence Agency report, "Leon Trotsky, Dupe of the NKVD," released on 3 March 1985, page 27.

258. Later these brothers were known in Soviet espionage in the United States as the spy master Jack Soble (Martha Dodd Stern's case officer) and his brother, Dr. Robert Soblen.

259. CI Staff of Central Intelligence Agency report, "Leon Trotsky, Dupe of the NKVD," released on 3 March 1985, p. 27.

260. Poretsky, Elizabeth, *Our Own People*, Loc. Cit., p.239.

261. For a complete report on the murder of Ignace Reiss by the NKVD see, Poretsky, Elizabeth, *Our Own People*, The University of Michigan Press, Ann Arbor, (1970). (Poretsky was Reiss wife and although her writing may at times be subjective she does have all the available facts.)

262. See Dallin, David, "Mark Zborowski, Soviet Agent," *The New Leader*, New York, (19 March 1956).

263. CI Staff of the Central Intelligence Agency report, "Leon Trotsky, Dupe of the NKVD," released 3 March 1985.

264. Orlov actually learned about Zborowski from one of Smirnov's Soviet case officers, Alexee, who used to work for Orlov. Alexee was showing off his prize agent to his former boss. Alexee used only the agent's code name, which also happened to be his true name of Mark but Orlov did not know that.

265. *Washington Star*, 25 November 1962.

266. Dallin, David J., "Mark Zborowski, Soviet Agent," *The New Leader*, New York, (19 March 1956).

267. For his immigration into the United States, Zborowski obtained an affidavit from Harry Liverman, allegedly a native of Uman, who claimed to have known the Zborowski (sic) family. It was discovered that Liverman's affidavit contained many lies similar to those he made in affidavits for others, including Lilia Ginsburg's brother, Rafael, and her subsequent husband, David J., Dallin. When these affidavits were submitted to the immigration authorities, it was found that Liverman was in correspondence with Maurice Stern, a suspected Soviet agent. (CI Staff of the Central Intelligence Agency report, "Leon Trotsky, Dupe of the NKVD," released 3 March 1985, page 22.)

268. For what is obviously a contrary and biased point of view on this issue see, Dallin, David, "Mark Zborowski, Soviet Agent," *The New Leader*, New York, (19 March 1956). See also a much more skeptical report; United States Senate, 93rd Congress, "The Legacy of Alexander Orlov," *Government Printing Office*, August 1973, Washington, D.C..

269. Kasson, Henry, "The Zborowski Case," *The New Leader*, New York, 21 November 1955.

270. Zborowski denied that he had met Jack Soble between the years 1944 and 1947. During those three years the FBI proved that the pair had had more than fifty meetings. (Morris, *My Ten Years As a Counterspy*, Loc. Cit., p. 247.)

271. FBI file on Fraina, 100-118061 (264 pages of a total of 502) which is all that could be released to the author under the Freedom of Information Act. No.351.960, dated 2 February 1992.

272. Carr, Edward Hallett, *Socialism in One Country*, The Macmillan Company, New York, (1958), pp.139-152.

273. Calomiris, Angela, *Red Masquerade*, J.B. Lippencott Company, New York, (1950), p.246.

274. For general information on Borodin see, Jacobs, Dan Norman, *Borodin: Stalin's Man in China*, Harvard University Press, Cambridge, Massachusetts, (1981).

275. Haynes, John E., and Harvey Klehr, "'Moscow Gold', Confirmed at Last," *Labor History*, Volume 33:2, (Spring 1992), p.279.

276. Lyons, Eugene, *The Red Decade*, The Bobbs-Merrill Publishers, New York, (1941), p.20.

277. Kahn, David, *The Code Breakers*, The Macmillan Company, New York, (1967), p.634.

278. Budenz, Louis, *This Is My Story*, McGraw-Hill by Whittlesey House, New York, (1947), p.239.

279. Dunham, Donald, *Kremlin Target: USA*, Ives Washburn, Inc., New York, (1966), p.6.

280. For purposes of clarity the Bureau of Investigation was the predecessor of the Federal Bureau of Investigation,

281. Smith, *America Enters the World*, Loc. Cit., p. 771 and 772.

282. Lyons, *The Red Decade*, Loc. Cit., p. 38.

283. Kahn, *The Code Breakers*, Loc. Cit., p.634.

284. Sayer, *The Great Conspiracy*, Loc Cit, p.353.

285. Also spelled Grosenberg by his alma mater Valparaiso University (Personal letter from Alice Koly, Archivist at Valparaiso University, Valparaiso, Indiana, dated 21 September 1990)

286. Haynes, "Moscow Gold," *Labor History*, Loc Cit, p. 281.

287. Vonsiasky was arrested by the FBI after the Japanese attack on Pearl Harbor. He was accused of giving military information to the Japanese. He admitted this but claimed he gave the information to the Japanese before the attack on Pearl Harbor and only so that they could use the

information to attack the Soviet Union. (Sayer, *The Great Conspiracy*, Loc. Cit., p. 348.)

288. Draper, Theodore, *American Communism and Soviet Russia*, The Viking Press, New York, (1960), p. 368.

289. GRU is the Soviet military intelligence service.

290. Baker had replaced J Peters as being in charge of the CPUSA's "special work", i.e., espionage.

291. Klehr, Harvey, John Earl Haynes and Fridrikh I Firsov, *The Secret World of American Communism*, Yale University Press, New Haven, Connecticut, (1995), p.89.

292. Sudaplotov, *Special Tasks*, Loc Cit., p. 101.

293. There is a story that before World War One Caridad did have among her lovers a rich Belgian named Mornard, the name and country that Ramon used as his alias when he met Sylvia Ageloff. (Mosley, *The Assassination of Trotsky*, Loc. Cit., p. 174.)

294. For the record Ramon's older brother, Jorge, was born in 14 December 1911. Two younger brothers, Pablo, was born on 25 November 1915 and Luis in June of 1923. A daughter, Montserrat, was born in Barcelona in June of 1922.

295. Please see the chapter on Zborowski for a full explanation about the person suspected of being Klement's assassin

296. Retrospectively this letter bears a striking similarity in sophistry, syntax and form to the letter of explanation for his murder of Trotsky supposedly written by Ramon Mercader that was found in his possession after his arrest by the Mexican police

297. Sudoplatov, in what may be the quintessential attempt at misinformation, suggests that during this time in Paris Caridad earned her living by knitting. (Sudoplatov, *Special Tasks*, Loc. Cit.,p.70.)

298. Leonid Eitingon was in Paris and formally recruited Caridad into Soviet intelligence. (Sudoplatov, *Special Tasks*, Loc. Cit., p. 70.)

299. Many strange threads ran from the Spanish Civil War into the history decades later. None is more curious than Ramon Mercader's transformation from a Catholic youth into Trotsky's assassin. (Matthew, Herbert, *Half of Spain Died*, Charles Scribner's Sons, New York, (1973), p.196.)

300.. According to her own confidential admissions, Caridad had personally carried out some twenty executions of Trotskyites during the Spanish Civil War. (Levine, *Mind of an Assassin*, Loc. Cit., p.51.)

301. There are no other references to Lena Imbert and it is believed that the name may be a pseudonym for another communist woman from Catalonia who accompanied Caridad Mercader on this journey, Carmen Brufau Civit.

302. Caridad's very attractive daughter, Montserrat Mercader, was also in the party. Diego Rivera became so entranced with Montserrat's beauty that he insisted on painting her portrait. (Fourth International, *Trotsky's Assassin At Large*, Labor Publications, New York, (1977). p.3.)

303. Using forged Mexican passports to enter Mexico where they would be subject to minute scrutiny appears to be the height of operational folly and very poor planning.

304. For a subjective pro-Siqueiros assessment, his participation in Soviet Russian espionage and the Trotsky assassination see Stein, Philip, *Siqueiros; His Life and Works*, International Publishers, Co. Inc., New York, (1994).

305. Siqueiros is considered an irresponsible, exhibitionist yearning for theatrical adventure. (Kirk, *Covering the Mexican Front*, Loc. Cit., p. 221.)

306. It is alleged that Siqueiros had two Mexican passports under the fictitious names of Miguel Gonzalez and Alfonso Rojas. He used the alias of Marcario Sierra while his wife, Angelica, used Dona Eusebita as an alias during their flight after the raid on Trotsky's home on 24 May 1940.

307. Hooks, *Modotti*, Loc Cit., p.242.

308. There is considerable evidence that Eitingon accompanied Siqueiros on the 24 May raid and was known to the raiders as "Felipe, the French Jew"

309. Stein, *Siqueiros*, Loc Cit., p.117 (No where in this apologia for Siqueiros's criminal acts is the accepted genuine reason for the raid cited, i.e., that Siqueiros was ordered to make the raid by the GPU in order to murder Trotsky.)

310. Stein, *Siqueiros*, Loc Cit, p.117.

311. It is interesting to note that the principal painters that revolve around the Trotsky case in Mexico were all muralists with the exception of Frida Kahlo; they were Diego Rivera, Javier Guerrero and Siqueiros, all muralists of international renown.

312. The concept of a GPU-lead revolution in Mexico to turn it into a base for Soviet operations against the Untied States was true. The same idea was attempted in 1971. (See Barron, John, *KGB*, Reader's Digest Press, New York, (1974), p.312 et. seq.)

313. Vidali/Contreras was a Comintern agent, a tough organizer and disciplinarian who had many a coward and deserter shot. (Stein, *Siqueiros*, Loc. Cit., p.103.)

314. Angelica Siqueiros allegedly carried a classified pouch from the Mexican government to the government of the Spanish Republic that supposedly contained a code the two governments would use to communicate with each other. (Stein, *Siqueiros*, Loc. Cit., p.105)

315. Mina was a Spaniard who had fought in Mexico for the country's independence.

316. Richardson, *Comintern Army*, Loc. Cit., p. 184,n.44.

317. Oswald, J. Gregory and Anthony J. Strover, *The Soviet Union and Latin America*, "Soviet Relations with Latin America," by Stephen Clissold, p.16., Praeger Publishers, New York, (1970)

318. Levine, *Mind of an Assassin*, Loc. Cit., p. 70.

319. Richardson, R. Dan, *Comintern Army*, The University of Kentucky Press, Lexington, (1982), p.21.

320. Draper, Theodore, *American Communism and Soviet Russia*, The Viking Press, New York, (1960), p.179 (Two preceeding paragraphs.)

321. Although this report is mostly factual there is some obvious errors. (Rexroth, Kenneth, *An Autobiographical Novel*, Doubleday and Company, Garden City, New York, (1966), p.344.

322. There were long standing rumors in Mexico that Vittorio Vidali murdered Mella. (Hooks, *Modotti*, Loc. Cit., p. 247.)

323. Rivera continued to claim the killer had been a gunman hired by the Cuban government. He insisted that the Cuban government had breached and insulted Mexican sovereignty. As a result the Cuban ambassador to Mexico was recalled to Cuba. Mexico never did go so far as to break diplomatic relations with Cuba but the Cuban government was without any diplomatic representation in Mexico for over a year.

324. Vidali was one of the killers of Andres Nin, the leader of the Trotskyite POUM. Vidali took part in the grotesque torture and mutilation of Nin before he died. The top Spanish communist, Jesus Hernandez, reported that Vidali directed a group of Germans from the International Brigade, posing as Gestapo officers, to stage a raid on the jail ostensibly to free Nin but on Vidali's orders murdered him instead. (Richardson, *Comintern Army*, Loc. Cit., p.164, n.24.)

325. There is a prevalent rumor that Ernest Hemingway used Modotti as his model for the character Maria in his novel "For Whom the Bell Tolls."

326. Martinez Carton brought Vidali a list of political opponents slated for assassination including the name of Leon Trotsky on the list. (Hooks, Margaret, *Tina Modotti*, Pandora, London, (1993), p.241.)

327. Salot was the street name of the Comintern agent in charge of the International Communist Youth whose true identity was known only to officials of the Comintern and probably her own husband.

328. Hooks, *Modotti*, Loc. Cit., p.242.

329. Hooks, *Modotti*, Loc Cite, p.247.

330. Hooks, *Modotti*, Loc. Cite., p.251.

331. Vidali was detained and questioned by the police regarding the Tresca murder but released for lack of evidence. (Dewar, *Assassins at Large*, Loc. Cit., p. 104.)

332. Levine, Isaac Don, *Mind of An Assassin*, Farrar, Straus and Cudahy, New York, (1959), p.70.

333. Mink is cited as one of three "master executioners" of the GPU together with Andre Simone and Ernest Wollweber by the Soviet expert E.H. Cookridge in his book *Soviet Spy Net*, Frederick Muller, Ltd., London, (1956), p.28. (Most Soviet experts would also include Vittorio Vidali in this group of "master executioners.")

334. Valtin, Jan, *Out of the Night*, Alliance Book Corp., New York, (1941), p.310.

335. There is a possibility that Lozovsky was a distant relative of Mink's. Solomon Lozovsky was the director of the Soviet International Trade Council in Moscow, better known as the Profitern.

336. Klehr, Harvey, *The Heyday of American Communism*, Basic Books, Inc., New York, (1984),p.160.

337. Jan Valtin's true name was Herman Krebs. He died on 1 January 1957 at the age of 45 years. He wrote about Mink in the account of his (Krebs's) life as a GPU agent under the title *Out of the Night*, Loc. Cit., p.310. He wrote "Wissinger was murdered by Hugo Marx and George Mink, He was shot to death in his bed in his apartment on the Muehlenstrasse on 22 May 1932." The communist press, in an attempt to avoid a scandal, attributed the killing to Nazi terrorists.

338. Leon Josephson was the legal advisor to the International Labor Defense. Josephson worked with Cyril Lambkin and George Mink to develop an ingenious Comintern espionage organization in the United States. Lambkin was an executive of the Amtorg Trading Company, a Soviet cover company for GPU agents. The Amtorg officials would insist that an American industrial company that wished to trade with the Soviet

Union must permit their factories and research facilities to be examined by Soviet technicians. (Spolansky, Jacob, *The Communist Trail in America*, The Macmillan Company, New York, (1951), p.146.)

339. Spolansky, *The Communist Trail in America*, Loc. Cit., p.147.

340. Richardson, *Comintern Army*, Loc. Cit., p.164.

341. Valtin, *Out of the Night*, Loc. Cit., p.313.

342. Dziak, John J., *Chekisty; A History of the KGB*, Lexington Books, Lexington, (1988), p.99.

343. Dallin, David J., *Soviet Espionage*, Yale University Press, New Haven, Connecticut, (1955), p.409.

344. Kirk, *Covering the Mexican Front*, Loc. Cit., p.218.

345. Hansen, *Trotsky; The Man and His Works*, Loc. Cit., p.7.

346. Dallin, *Soviet Espionage*, Loc. Cit., p.410.

347. Chambers, Whittaker, *Witness*, Random House, New York, (1952), p.303, n..

348. *The Gelfand Case*, Volume I, Labor Publications, New York, p.50. (There is a crater, actually a deep hole that has never been plumbed, in this area located on the road from Mexico City to the Pyramids of Tenochtitlan. It is claimed the Aztecs threw maidens down this hole as sacrifices to the gods. Rumor has it that the Mexican police, on occasions, have used this so-called "bottomless pit" to dispose of bodies they do not know otherwise what to do with them.)

349. Lockhart, *Reilly: The First Man*, Loc Cit., p.75.

350. Peters real name was Alexander Goldberg. He also used the alias of Alexander Stevens.

351. Lamphere, *The FBI-KGB War*, Loc. Cit., p.37.

352. Lockhart, *Reilly: The First Man*, Loc. Cit., p.89.

353. Levine, Isaac Don, "Why Soviets Take Back Trotsky's Murderer." *U.S. News and World Report*, 33 May 1960, p.85.

354. Pincher, Chapman, *Too Secret Too Long*, St Martin's Press, New York, (1984), p. 417.

355. Levine, Isaac Don, *Mind of an Assassin*, Farrar, Strauss and Cudahy, New York, (1959), p44

356. Levine, *Mind of an Assassin*, Loc Cit., p.45.

357. Dallin, *Soviet Espionage*, Loc Cit., p.406.

358. Andrew, Cristopher and Oleg Gordievski, *KGB: The Inside Story*, Harper Collins, New York, (1990), p. 237.

359. Budenz, Louis, *This Is My Story*, Whittlesey House, London, (1947), p. 255.

360. Fourth International, *Trotsky's Assassin at Large*, Loc. Cit., p.15.

361. Levine, *Mind of an Assassin*, Loc. Cit., p.46.

362. *Committee on Un-American Activities*, "American Aspects of the Trotsky Assassination," 81st Congress, Washington, D.C., (1950), p.ix.

363. Haberman recruited Bertram D. Wolfe, one of the founders of the Mexican Communist Party and Soviet agent before his defection to the FBI. Haberman was also closely associated through marriage to Agnes Smedley, a Soviet agent in the Far East involved with the Richard Sorge network. Haberman worked as an official in the Mexican Ministry of Education.

364. See *The Report of the Royal Commission*, 27 June 1946, Ottawa, Canada.

365. Dallin, *Soviet Espionage*, Loc. Cit., (1955), p.405.

366. Andrew, *The KGB: The Inside Story*, Loc. Cit., p. 281.

367. Sudoplatov, *Special Tasks*, Loc. Cit., p.76.)

368. Andrew, *KGB: The Inside Story*, Loc. Cit., (1990), p. 681, footnote 64.

369. Gaik Ovakimian was the chief of the Jacob Golos network and, as such, the overall superior of Louis Budenz and Elizabeth Bentley among others. (Levine, *Mind of an Assassin*, Loc. Cit., p.102.)

370. Davidson, Bill, "The People Who Stole It From Us," *Look*, 29 October 1957

371. Lamphere, *The FBI-KGB War*, Loc Cit., p. 172.

372. Lamphere, *The FBI-KGB War*, Loc Cit., p.175.

373. Kahn, David, *The Code Breakers*, The Macmillan Company, New York, (1967), p635.

374. Dallin, *Soviet Espionage*, Loc Cit., p.423.

375. The Soviet Union never fully honored their promise to release the six Americans (Levine, *Mind of an Assassin*, Loc. Cite., p.103.) Three of the six Americans never reached the United States. Before they could leave Russia, the Germans overran the area where two were detained and they were never heard of again. One was kept in jail in Russia despite Soviet promises to the contrary. The other three came to the United States, two promptly started a pro-Soviet propaganda campaign; and only one, Dr. Michael Devenis, seemed to have been a genuine American hostage of the Soviet Union. (Dallin, *Soviet Espionage*, Loc. Cit., p. 424.) (For additional details on these six see, *Committee on Un-American Activities*, "The Shameful Years," Washington, (30 December 1951), p.17.)

376. Lamphere, *The FBI-KGB War*, Loc. Cit., p.37.

377. Lamphere, *The FBI-KGB War*, Loc. Cit., p. 218.

378. Sudoplatov, *Special Tasks*, Loc. Cit., p.406.

379. Pincher, *Too Secret Too Long*, Loc. Cit., p.21. Smedley became part of the notorious Sorge espionage network in the Far East and also served as Sorge's mistress. (See also Mahoney, M.H., *Women in Espionage*, ABC Clio Publishers, Santa Barbara, (1995).

380. Ewert and his wife returned to the States to be GPU agents in Michigan before joining Louis Carlos Prestes and Olga Benario in the abortive GPU attempt to overthrow the Brazilian government of President Getulio Vargas. Ewert was tortured into insanity by the Brazilian police and, after World War Two, was repatriated to East Germany where he died in an insane asylum. After repatriation to Nazi Germany during World War Two, Elsie Ewert was executed together with Olga Benario by the German Gestapo for being Soviet agents. (See Mahoney, *Women in Espionage*, Loc. Cit. section on Benario.)

381. Budenz, *This Is My Story*, Loc. Cit., p.241.

382. Lampere, *The FBI-KGB War*, Loc. Cit., p.45/46.

383. Josephson was an intimate of George Mink.

384. Lamphere, *The FBI-KGB War*, Loc. Cit., p.49.

385. Glazer was Poyntz's mother's name and is sometimes cited as her matronymic.

386. Cookridge, *Soviet Spy Net*, Loc. Cit., p. 183.

387. Gallagher, Dorothy, *All The Right Enemies: The Life and Murder of Carlo Tresca*, Rutgers University Press, New Brunswick, New Jersey, (1988).

388. Gitlow, Benjamin, *The Whole of Their Lives*, Charles Scribner's Sons, New York, (1948).

389. According to another source, Poyntz's body was buried behind a brick wall in a Greenwich Village house. (Andrew, *KGB - The Inside Story*, Loc. Cit., p.161.

390. Gitlow, *The Whole of Their Lives*, Loc. Cit., p.335.

391. Dewar, *Assassins At Large*, Loc. Cit., p.202.

392. Klehr, *The Heyday of American Communism*, Loc. Cit., p.161, n.21.

393. Hingley, Ronald, *The Russian Secret Police*, Simon and Schuster, Inc., (1970), p.183.

394. Leonid Eitingon should not be confused with one Max Eitingon. Some scholars claim Max Eitingon, a psychiatrist, is the same as Leonid Eitingon the GPU officer, or at least that Max is closely related to Leonid Eitingon. Max Eitingon was a close collaborator of Sigmund Freud and a psychiatrist in his own right. (See Dziak, John J., *Chekisty-A History of the KGB*, Lexington Books, Lexington, (1988), p.100, especially footnote #79.) There are several diametrically opposed views questioning whether Leonid Eitingon is Max Eitingon or is related to him. See:

1. *New York Review of Books*, 24 January 1988, "Intellectuals and Assassins - Annals of Stalin's Killerati" by Stephen Schwartz.
2. *New York Review of Books*, 6 March 1988, Letters to the Editor
3. *New York Review of Books*, 14 April 1988, "The Mystery of Max Eitingon" by Theodore Draper.
4. *New York Review of Books*, 16 June 1988, "The Eitingon Mystery Solved; An Exchange" by Theodore Draper and others.

The matter was unequivocally resolved when Pavel Sudoplatov, Leonid Eitingon's superior in the Department of Special Tasks and close personal friend, published his book. He wrote that Leonid did use the alias of Leonid Naumov, his wife's maiden name, when he was in Turkey planning the assassination of Franz von Papen. Sudoplatov categorically denies that any of the alleged relatives mentioned in the *New York Review of Books* articles are listed in Leonid Eitingon's NKVD personnel permanent file. Leonid Eitingon never contacted any of them or mentioned having a brother named Max. (Sudoplatov, Pavel and Anatoli Sudoplatov, *Special Tasks*, Little Brown and Company, New York, (1994), p.37.)

Even authoritative texts persist in some of the erroneous details of the life of Leoind Eitingon. See in this regard, Rapoport, Vitaly, and Yuri Alexeev, *High Treason*, Duke University Press, Durham, North Carolina, (1985), p.391/392.

395. Levytsky, *The Uses of Terror*, Loc. Cit., p.218.

396. Levine, *Mind of and Assassin*, Loc. Cit., p.32.

397. Dziak, *Chekisty*, Loc. Cit., p.102.

398. The scar was the result of an automobile accident. Eitingon enjoyed letting people think it came from being wounded in some adventurous escapade.

399. Ramon Mercader may have been on a mission for the regular Republican army and not on an assignment for Eitingon's Department of Special Tasks. Dziak, an authority on the KGB and predecessor organizations, writes that "Spain during the Civil War was a veritable NKVD killing field --- but *behind* (emphasis Dziak's) Republican lines, not against the Nationalists"> (Dziak, John, J., *Chekisty - A History of the KGB*, Lexington Books, Lexington, (1988), p.100.)

400. Levine, *Mind of an Assassin*, Loc. Cit., p. 36.

401. Eitingon was an incurable romantic. His first wife died after giving birth to their son. In China he fell in love with the wife of Vassili Zarubin, Olga Naumova and married her. Without divorcing Naumova, he took to Spain with him the wife of an officer from the Third Section named Aleksandra Kochergina. Eitingon never registered his second and third marriages and, for a while, lived alternately in two households. (Sudoplatov, *Special Tasks*, Loc. Cit., p.34.)

When Kochergina died in 1941 Eitingon took a champion parachutist named Muza Malinovskaya with him to Turkey as his female companion. In 1968 he married another wife, Yevgenia Puzirova. (Sudoplatov, *Special Tasks*, Loc. Cit., p.34.)

402. Conquest, *The Great Terror*, Loc. Cit., p. 446.

403. Levytsky, Boris, *The Uses of Terror*, Coward, McCann and Geoghegan, Inc., New York, (1972),p.147.

404. It was rumored that William Fisher (Colonel Rudolph Ivanovich Abel) an old friend of Eitingon's since they served together in the military during World War Two, pleaded with the Soviet hierarchy for Eitingon's release. Abel was a KGB spy in the United States for many years. (Dziak, *Chekisty*, Loc. Cit., p.101.)

405. Sudoplatov, *Special Tasks*, Loc. Cit., p.350.

406. When Lie visited Moscow in 1923 he was identified as being with the Comintern, according to Trotsky. (Levine, Isaac Don, *Mind of an Assassin*, Farrar, Straus and Cudahy, New York, (1959), p.8.) This is the same Trygiv Lie who later became Secretary General of the United Nations.

407. Warth, *Leon Trotsky*, Loc. Cit., p.172.

408. Wolfe, Bertram D., *The Fabulous Life of Diego Rivera*, Stein and Day Publishers, New York, (1963), pp. 6 and 7.

409. For a general report on the formation of the PCM see the report of an interview with Manny Gomez (real name Charles Phillips), "From Mexico to Moscow," *Survey*, October 1964.

410. Alexander, Robert J., *Trotskyism in Latin America*, Hoover Institution Press, Stanford University, Stanford, California, (1973), p. 179/180.

411. Arquin, Florence, *Diego Rivera*, University of Oklahoma Press, Norman, (1971), p.4.

412. Herrera, Hayden, *Frida; A Biography of Frida Kahlo*, Harper and Row, Publishers, New York, (1983), p.201.

413. Angela Beloff, Rivera's wife referred to him as, "He is just a child, a great big child." (Wolfe, *The Fabulous Life of Diego Rivera*. Loc. Cit., p.251.)

414. *New York Times*, 7 June 1940.

415. Herrera, *Frida*, Loc. Cit., p.202.

416. Herrera, *Frida*, Loc. Cit., p.198, 370 and 400.

417. Herrera, *Frida*, Loc. Cit., p. 196.

418. Herrera, *Frida*, Loc. Cit., p.197.

419. Herrera, *Frida*, Loc. Cit., p.79. (Rivera left his first wife, Lupe Marin, because of his affair with Modotti.)

420. Frida habitually referred to Rivera as "carasapo" (frog face) and "panzas" (fat belly). (Wolfe, *The Fabulous Life of Diego Rivera*, Loc. Cit., p.244.)

421. Herrera, *Frida*, Loc. Cit., p.100.

422. There were thousands of Irishmen hired as diggers when a canal was attempted across the Isthmus of Tehuantepec before the Panama Canal was built. The canal project was abandoned but many of these handsome red and blond haired Irishmen cohabited with the already beautiful Tehuana women. The result was a race of light brown women, tall and stately, with blue eyes and blonde or red hair.

423. Frida also had an affair with Jean van Heijenoort, Trotsky's bodyguard/secretary, who was a notorious womanizer. Many years later Heijenoort was killed in Mexico by his wife, whom he had divorced and remarried, in a jealous rage over his infidelities.

424. Herrera, *Frida*, Loc. Cit., p.203.

425. Although still popularly referred to as the Mexican Trotskyite Party it had formally changed its name in 1934 to Liga Comunista Internacionalista and published, as their official magazine, *Nueva Internacional*, (Alexander, *Trotskyism in Latin America*, Loc. Cit., p. 182.) In 1936 Diego Rievra was elected to the Political Bureau of the LCI and the periodical was renamed *IV Internacional.*

426. Herrera, *Frida*, Loc. Cit., p.204.

427. Fourth International, *Trotsky's Assassin at Large*, Labor Publications, New York, (1977), p. 3.

428. Frida's father was a Jew born in Germany and her mother was of Spanish-Indian lineage.

429. Mosley, Nickolas, *The Assassination of Trotsky*, Michael Joseph, Ltd., London, (1972), p.29.

430. Jean van Heijenoort, Trotsky's principal secretary/bodyguard/ translator and administrator arrived at Coyoacan from Laredo about an hour after the Trotskys. (Van Heijenoort, Jean, *With Trotsky in Exile; From*

Prinkipo to Coyoacan, Harvard University Press, Cambridge, Massachusetts, (1978), p.104.) (Heijenoort remained with Trotsky in Mexico until 5 February 1939 when he returned to the United States.)

431. Trotsky, Leon, *My Life*, "Introduction by Joseph Hansen," Pathfinder Press, Inc., New York, (1970). p.viii.

432. Kirk, Betty, *Covering the Mexican Front*, University of Oklahoma Press, Norman, Oklahoma, (1941), p.212.

433. Heijenoort, *With Trotsky in Exile*, Loc. Cit., p. 108.

434. Payne, Robert, *The Life and Death of Leon Trotsky*, McGraw-Hill Company, New York, (1947), p.412.

435. Levine, *Mind of an Assassin*, Loc. Cit., p.37.

436. Alexander, *Trotskyism in Latin America*, Loc. Cit., p.190.

437. *Time*, 25 January 1937.

438. Glotzer, Albert, *Trotsky: Memoir and Critique*, Prometheus Books, New York, (1989), p.244.

439. Segal, Ronald, *Leon Trotsky*, Pantheon Books, New York, (1979), p.367.

440. Segal, *Leon Trotsky*, Loc. Cit., p.370.

441. Anderson, *Trotskyism in latin America*, Loc. Cit., p.183.

442. Cornell, Charles, *With Trotsky in Mexico*, Merit Publishers, New York, (1967), p.64.

443. The Mexican Stalinists did what they could to whip up mass hysteria against "the traitor cowering at Coyoacan." (Deutscher, Isaac, *The Prophet Outcast*, Oxford University Press, New York, (1963), p.482.

444. Cornell, *With Trotsky in Mexico*, Loc. Cit., p. 66.

445. "Introduction to Collection of Raya Dunayevskaya," Wayne State University Archives, Detroit, Michigan, April 1985.

446. Glotzer, *Trotsky*, Loc. Cit., p.256.

447. Lyons, *The Red Decade*, Loc. Cit., p. 252.

448. This counter-commission included such notable Americans as Nathaniel West, Dorothy Parker, Theodore Dreiser and others who urged a boycott of the Dewey Commission. (Segal, *Leon Trotsky*, Loc. Cit., p.370.)

449. Segal, *Trotsky*, Loc. Cit., p.371.

450. Herrera, *Frida*, Loc. Cit., p.207.

451. Harry Black, an intimate among Stalinist circles in Mexico, was the editor of a mimeographed news clip-sheet distributed free of charge in the United States by the Stalinist Workers' University. Block was the liaison between Vicente Lombardo Toledano and the Soviet Ambassador to the United States, Constantine Oumansky.

452. Black married the daughter of Vicente Lombardo Toledano, a Stalinist, a probable GPU agent and leading pro-communist labor leader in Mexico. (Glotzer, *Trotsky*, Loc. Cit., p.267.)

453. Glotzer, *Trotsky*, Loc. Cit., p.270.

454. Segal, *Trotsky*, Loc. Cit., p.371.

455. Glotzer, *Trotsky*, Loc. Cit., p.237.

456. Carmichael, Joel, *Trotsky*, St. Martin's Press, New York, (1975), p.438.

457. Kirk, *Covering the Mexican Front*, Loc. Cit., p.216/217.

458. Sayer, *The Great Conspiracy*, Loc. Cit., p. 305/306.

459. Heijenoort, *With Trotsky in Exile*, Loc. Cit., p.131.

460. Herrera, *Frida*, Loc. Cit., p.209.

461. Herrera, *Frida*, Loc. Cit., p.209.

462. At Frida's suggestion Cristina posed nude for Rivera on many occasions. (Wolfe, *The Fabulous Diego Rivera*, Loc. Cit., p.246.)

463. Herrera, *Frida*, Loc. Cit., p.209. (For a thinly disguised version of the same incident see, Heijenoort, *With Trotsky in Exile*, Loc. Cit., p.118/119.

464. Heijenoort, *With Trotsky in Exile*, Loc. Ciut., p.114.

465. Herrera, *Frida*, Loc. Cit., p.212.

466. The breakup between Trotsky and Rivera had an element of mutuality. Wolfe claimed that Trotsky could no longer stand Rivera's "fabulous fantasies" or lies. (Wolfe, *The Fabulous Life of Diego Rivera*, Loc. Cit., p.238.)

467. Curtiss was also trying to issue a Trotskyite periodical *El Bolchevique*. (Anderson, *Trotskyism in Latin America*, Loc. Cit., p. 196.

468. There were two other stories regarding the source of the money that Trotsky used to purchase the house. One has a Mexican book seller and publisher of Russian origin who was a descendent of Russian revolutionaries giving Trotsky the money. The other report comes from the editor of an American magazine who told Deutscher that Trotsky withdrew the money from a bank account he and Lenin had in the States. Supposedly he and Lenin had opened this account in their names as a fund to use to carry on the revolution from America if they had been defeated in Russia. (Deutscher, Isaac, *The Prophet Outcast*, Oxford University Press, New York, (1963), p. 445.)

469. Fourth International, *The Case for a Commission of Inquiry*, Labor Publications, New York (1976), p.3.

470. The house belonged to a family of shopkeepers named Turati who used it as a country retreat. The owners were happy to rent it to Trotsky. It had high walls and surroundings that were easy to keep under observation, because the neighborhood was not yet built up. (Van Heinjenoot, *With Trotsky in Coyoacan*, Loc. Cit., p.135. (The neighborhood today is a densely packed suburban area of Mexico City.)

471. Payne, Robert, *The Life and Death of Trotsky*, McGraw-Hill Book Company, New York (1977), p.416.

472. Deutscher, *The Prophet Outcast*, Loc. Cit., p.478.

473. In 1946 Claire Booth Luce, former United States Ambassador to Italy and widow of Henry Luce, the founder of *Time* magazine, purchased this painting. Frida's self-portrait that is dedicated to communist Trotsky is hanging in the home of capitalist Luce in Honolulu, Hawaii.

474. Herrera, *Frida*, Loc. Cit., p.250.

475. Herrera, *Frida*, Loc. Cit., p. 272.

476. Wolfe, *The Fabulous Life of Diego Rivera*, Loc. Cit., p.363.

477. *New York Times*, 6 June 1940.

478. Wolfe, *The Fabulous Life of Diego Rivera*, Loc. Cit., p.361.

479. Herrera, *Frida*, Loc. Cit., pp.276 and 355.

480. In late 1953 Frida had the fourteenth operation in sixteen years. Her leg became gangrenous and was amputated below the knee. (Wolfe, *The Fabulous Life of Diego Rivera*, Loc. Cit., p. 394.)

481. Herrera, *Frida*, Loc. Cit., p.393.

482. Herrera, *Frida*, Loc. Cit., p.402.

483. "In 1954, shortly before his death he [Rivera] permitted himself to be reinstated in the party [PCM]" Arguit, Florence, *Diego Rivera*, Loc. Cit., p.5. (This is a gross misstatement since Rivera had not "permitted himself to be reinstated" but had begged and pleaded for many years for readmission . This is a pitfall for authors who become too subjectively involved in glorifying their subject.)

484. Herrera, *Frida*, Loc. Cit., p.438.

485. *Committee on Un-American Activities*, " American Aspects of the Assassination of Leon Trotsky," Eighty-First Congress, Washington, D.C., (1950), p.V

486. Later in his expose' of Soviet intelligence operations in the United States, Budenz commented, that during the course of his more than ten years of underground party work, many of the Soviet intelligence officers he met used Anglo first names such as Michaels and Richards as street alias.

487. Franklin was later married again to a man called Doxsee and, in retirement, she was known as Sylvia Doxsee. She lived in a suburb west of Chicago.

488. International Committee of the Fourth International, *Sylvia Franklin Dossier*, Labor Publications, New York, (1977), p.2.

489. During 1935-36 Sylvia Ageloff worked as a propaganda director for James P. Cannon, head of the Socialist Workers' Party in his New York office. (*Brooklyn Daily Eagel*, 27 August 1940.)

490. International Committee of the Fourth International, *Sylvia Franklin Dossier*, Labor Publications, New York, (1977), p.21.

491. Pierre Broue, of the Organisation Communiste Internationaliste, and Jean van Heijenoort, Trotsky's close friend and International Secretary of the Fourth International headquarters in New York, both subsequently admitted that they had strong suspicions that Franklin was a GPU agent while she worked for Cannon.

492. Dallin, David J., *Soviet Espionage*, Yale University Press, New Haven, (1955), p.417/419.

493. The Socialist Workers' Party was, in essence, the Trotskyite party in the United States.

494. Bernard Schuster is identified on the VENONA transcripts with the party name of Chester and the code name of ECHO. (VENONA is the code name for enciphered messages between the CPUSA and GPU in the United States with their headquarters in Moscow which were intercepted and decoded by the United States authorities.) According to VENONA, Schuster served as liaison between the CPUSA and the NKVD (GPU),

recruiting and checking the reliability of potential spies. For example, a 21 September 1944, NKVD communication from New York to Moscow reads,

> In the FELLOW COUNTRYMAN [CPUSA] line, LIBERAL [Julius Rosenberg] is in touch with CHESTER [Bernard Schuster]. They meet once a month for the payment of dues. CHESTER is interested in whether we are satisfied with the collaboration. In as much as CHESTER knows about the role of LIBERAL's group we beg consent to ask CHESTER through LIBERAL about leads from among the people who are working on ENORMOUS [atom bomb project] and in other technical fields. (Schneir, Walter and Miriam Schneir, "Cryptic Answers," *The Nation*, 14/21 August 1995.)

495. In 1931 Hilda Ageloff took a three and one-half month tour in the Soviet Union to study the Russia methods of progressive education in kindergartens and the care of children on the communal farms. She claimed to have had an interview with Lenin's widow. (*Brooklyn Daily Eagle*, 21 August 1940.)

496. Both Hilda and Ruth worked in the real estate office of their father, Samuel Ageloff, at 191 Jeralemon street in Brooklyn. (*Brooklyn Daily Eagle*, 21 August 1940.)

497. Wyndham, Francis and David King, *Trotsky*, Praeger Publishers, New York, (1972), p.176.

498. Heijenoort, *With Trotsky in Exile*, Loc. Cit., p.146.

499. Before she worked as a social worker for the Department of Welfare, Sylvia was employed at the Jewish Children's Clearing Bureau in Manhattan. (*Brooklyn Daily Eagle*, 23 August 1940.) She was employed as a child psychologist at the New York Board of Education from July 1936 to June 1938,(*New York Times*, 27 August 1940.)

500. It was Eitingon's idea for the Trotsky operation to use as agents in Europe and Latin America, people who had never been involved before

in operations against Trotsky and Sylvia Ageloff fitted this requirement. (Sudaplatov, *Special Tasks*, Loc. Cit., p.70.)

501. Sylvia later testified that she had not even known of the conference in advance and had decided to go abroad on her own, solely for pleasure. (Levine, *Mind of an Assassin*, Loc. Cit., p.47.)

502. Weil was under the original impression that Gertrude's last name was Sauzea.

503. For additional background and aftermath of Gertrude Allison de Machado and her husband Eduardo Machado, please see Garcia Ponce, Guillermo, *Memorias de un General de la Utopia*, Cooperativa de Trabajadores Graficos, Caracas, Venezuela, (1992). This author is deeply indebted to Dr. Hernan Gonzalez for bringing this source to his attention. (See also, Mahoney, M.H., *Women in Espionage*, ABC CLIO, Santa Barbara, California, (1993),p.145.)

504. After she introduced Weil and Sylvia to Mornard/Mercader in Paris , Allison and her husband went to Mexico where they no doubt assisted in the Trotsky operation. They later returned to Venezuela where Eduardo and his brother, Gustavo, established the Communist Party of Venezuela.

505. *Committee on Un-American Activities*, "Testimony of Sylvia Ageloff," Eighty-First Congress, (December 1950), Washington, D.C.

506. For clarity's sake the name that Mercader was using at the time in the narrative will be used except where a double name might be needed for positive identification.

507. Sylvia was 28 years of age, moderately good-looking, whose appearance was spoilt by small, near-sighted eyes behind thick spectacles. She had had few, if any, love affairs but was intelligence with a pleasant personality. (Wyndham, *Trotsky*, Loc. Cit., p.176.)

508. There is a difference of opinion regarding Sylvia's personal beauty or attractiveness. One eminent Trotsky historian claims Sylvia was " a lonely spinster of a rather unattractive appearance." (Deutscher, Isaac, *The Prophet Outcast*, Oxford University Press, New York, (1963), p.483. (What is believed to be an objective appraisal by this author, based upon a study of many photographs and interviews with people who

knew Sylvia, she was a better than average good looking lady with a slim attractive figure.)

509. Ramon Mercader was left-handed, a very heavy smoker, abnormally little body hair, his skin was unusually sensitive, he suffered from stomach swelling with gas, had an umbilical hernia and had been circumcised when he was twenty-one. (Mosley, *The Assassination of Leon Trotsky*, Loc. Cit., p. 149.)

510. Mornard/ Mercader also impressed Sylvia with sizeable contributions to the impoverished treasury of the Trotskyite organization. (Levine, *Eyewitness to History*, Loc. Cit., p. 101.)

511. Deutscher, *The Prophet Outcast*, Loc. Cit., p.483.

512. Sylvia later testified that Mornard/Mercader displayed a boredom with politics so acute that "he never even read ordinary news articles." (Levine, *Mind of an Assassin*, Loc. Cit., p.50.)

513. This meeting was held on 3 September 1938 at the home of Alfred and Marguerite Rosmer in a Paris suburb. (Wyndham, *Trotsky*, Loc. Cit., p.175.) Max Schactman chaired this organizational meeting. There were twenty-eight Trotsky adherents from eleven countries. The Russian Section was represented by Mark Zborowski. (Carmichael, Joel, *Trotsky*, St, Martin's Press, New York, (1975), p.460.)

514. Dallin, David, "Mark Zborowski, Soviet Agent," *The New Leader*, New York, (19 March 1952).

515. There is no implication that the Rosmers lack of suspicion about Sylvia's charming lover had any direct connection with Trotsky's death. The Rosmers were just as taken in by Mornard as was Sylvia.

516. This may be better phrased as " shortly after Sylvia became Mornard's mistress." Better still, as one European writer phrased it, "shortly after the lovers consummated their bliss."

517. Jacson, aka Jacques Mornard, whose true name was Ramon Mercader, also used the alias of Leon Jacome and Leon Haikys. (Sayers, *The Great Compromise*, Loc. Cit., p.319.)

518. Sylvia's parents, her natural father, Samuel, and her step-mother, Fannie, were of the Jewish faith and disapproved of their daughter living with a man without being married in a religious ceremony.

519. Sylvia later admitted that she never doubted that Mercader's love for her was sincere right up to the time he murdered Trotsky. (Andrew, Cristopher and Oleg Gordievski, *KGB: The Inside Story*, Harper Collins, Publishers, New York, (1990), p. 169.)

520. Sylvia did not know that Jacson/Mercader's real mother, Caridad Mercader, was in Mexico with her lover who was the principal director in Mexico for Operation Trotsky, Leonid Eitingon, (Levine, *Mind of an Assassin*, Loc. Cit., p.75.)

521. Rosmer emphatically denied that he had met Jacson/Mercader in Paris but did admit he met him in Mexico. According to Deutscher, Jacson/Mercader confirms this. (Deutscher, *The Prophet Outcast*, Loc. Cit., p.451.)

522. Levine, *Mind of an Assassin*, Loc. Cit., p.77.

523. Sylvia specifically warned Trotsky that Jacson/Mercader not be invited to the house. (Glotzer, *Trotsky*, Loc. Cit., p.193.)

524. Soble, Jack and Jack Lotto, "How I Spied On The United States For The Reds," *Journal American*, New York, 10 through 20 November 1957.

525. Ruvin Sobolevicius, under the name of Dr. Robert Soblen, was also convicted of espionage for the Soviets. He fled from the United States to Israel in 1962. He was denied refuge in Israel. When he was being returned to the United States via England, he attempted suicide two times and died in London. (*United States Senate*, "Hearing of Subcommittee to Investigate the Administration of the Internal Security Act, dated 21 November 1957, pp. 4875-6," Washington, D.C. This reference also refers to Jack Soble.)

526. Andrew, Cristopher and Oleg Gordievski, *KGB: The Inside Story*, Harper Collins, Publishers, New York, (1990), p.155.

527. Heijenoort, *With Trotsky in Exile*, Loc. Cit., p.98.

528. Zborowski's espionage activities in the States were revealed to the FBI by Alexander Orlov who defected after being in charge of GPU operations in Europe. Once Zborowski was arrested the FBI soon rolled up Soble and his espionage network.

529. Levin, *Mind of an Assassin*, Loc. Cit., p.41.

530. Zublin's wife, Elizabeth, was his co-worker and equal in Soviet espionage. She was also a contact for Elizabeth Bentley's Soviet spy ring under the name of Helen. As Helen she was also the case officer for Hede Massig, a Soviet espionage agent of considerable importance in the United States. Massig later defected to the FBI. Even after Massig's break with the NKVD, Helen repeatedly came to her exhorting her to return to Soviet espionage. (Unpublished debriefing of Hede Massig.) See also, Massig, Hede, *This Deception*, Duell, Sloan and Pearce, New York, (1951). (It should be noted that Hede Massig had been previously married to Gerhardt Eisler before her marriage to Massig and in her book reveals information about Eisler's espionage operations.)

531. The exact timing is not known but the GPU is alleged to have introduced Sylvia Franklin to Jack Soble and Robert Soblen. (Klehr, Harvey, John Earl Haynes and Fridrikh Igorevich Firsov, *The Secret World of American Communism*, Yale University Press, New Haven, (1995), p. 142.

532. Morris, *My Ten Years as a Counterspy*, Loc. Cit., p. 203.

533. Bentley, Elizabeth, *Out of Bondage*, with "Afterward" by Hayden Peake, Ivy Books, New York, (1988), P. 328.

534. The scientist was identified as Joseph W. Weinberg. (*Committee on Un-American Activities*, "The Shameful Years," United States House of Representatives, Washington, D.C. (30 December 1951), p.31.)

535. Nelson's true name was Steve Mesarosh, born in Chaglich, Yugoslavia on 1 January 1903. He entered the United States with his mother and two sisters. He used Anthony Gerlack, the organizational secretary of District 7 of the CPUSA as his character witness for his naturalization. He was assigned to China in 1933 by the Comintern to assist Arthur Ewert, a known GPU agent. Nelson was supposedly elevated to the Central

Executive Committee of the CPUSA. (Spolansky, Jacob, *The Communist Trail in America*, The Macmilllan Company, New York, (1951), p.158.)

536. Some of the unwitting recording stars were Francis Langford, Bob Crosby and Hoagy Carmichael. Some of the hit records managed by this company were "Rum and Coca Cola" and "Chattanooga Choo Choo."

537. Morris, Boris, *My Ten Years as a Counterspy*, The Viking Press, New York, (195(), p. 17.)

538. In August of 1944 Zubilin introduced Soble to Stephan Choundenko, another NKVD case officer, who took Zubilin's place. Choundenko ordered Soble to "straighten out the difficulties between Stern and Morris." (Soble, "How I Spied On The United States For The Reds" Loc. Cit.)

539. Stern was born in 1897, attended Phillips Exter Academy and Harvard University. He left college to serve in the United States Army. He served as director of the Rosenwald Foundation for ten years. His marriage to Marion Rosenwald brought him social prominence and considerable wealth.

540. For Soble's character sketch of Martha see, Soble, "How I Spied On The United States For The Reds," Loc. Cit.,

541. When Zubilin first introduced Morris to the Sterns, Martha Stern and Zubilin engaged in passionate kissing and fondled each other. This implied to Morris that they were probably having an affair. Morris refers to Alfred Stern later as a cuckold. (Morris, *My Ten Years as a Counterspy*, Loc. Cit., p.47 and 55.)

542. According to Soble, "Jane Foster Zlatovski spelled Sex appeal." (Soble, "How I Spied On The United States For The Reds," Loc. Cit..

543. Jane Foster had gathered some of this information on Indonesia when she lived there with her first husband, an officer in the Netherlands Foreign Service. (Morris, *My Ten Years as a Counterspy*, Loc. Cit., P.96.)

544. Soble, "How I Spied On The United States For The Reds," Loc. Cit..

545. Morris, *My Ten Years as a Counterspy*, Loc. Cit., p.245.

546. Soble, "How I Spied On The United States For The Reds," Loc. Cit.. Also see *United States Senate*, "Subcommittee to Investigate the Administration of the Internal Security Act," (21 November 1957), p.4886.

547. Soble, "How I Spied On The United States For The Reds," Loc. Cit.,

548. The spy ring that Bentley handled was associated with the Whittaker Chambers and Alger Hiss cases.

549. Dallin, *Soviet Espionage*, Loc. Cit., p. 419.

550. Luis Arenal was well known to the American authorities for his former connection with the communist magazine *New Masses*, He was also a regular contributor to the PCM dominated *Futuro*, a magazine published by Vicente Lombardo Toledano. Many of the sketches published in Mexico attacking Trotsky are unmistakenly from the pen of Luis Arenal. (Hansen, et.al., *Leon Trotsky; The Man and His Work*, Loc. Cit., p.7.)

551. Morris, *My Ten Years as a Counterspy*, Loc. Cit., p.237.

552. Quint, Burt, "US Reds Have Haven in Mexico; Draft Strategy, Live Amid Luxury," *Herald Tribune*, New York, (3 August 1937), p.1.

553. The nucleus of the ACGM had been several American communists who had been sent to Mexico by the GPU to work on the Trotsky case and remained after the assassination. Jacob Epstein, an American who lived at Dublin 12 in Mexico City, was the principal agent for a letter drop for Soviet espionage on the Trotsky case. His contact in the Soviet Embassy in Mexico City was Pavel Klarin a close collaborator with Vassili Zubilin. The ACGM. as a group of ex-patriot Americans with leftist leanings, stems from Epstein's initial residence in Mexico City.(Committee on Un-American Activities, *The Shameful Years*, United States House of Representatives, Washington, D.C., (30 December 1951.) (Epstein was a veteran of the Abraham Lincoln Battalion from the Spanish Civil War.)

554. United States of America vs. Alfred K. Stern and Martha Dodd Stern, Defendants, 9 September 1957. "....to deliver, communicate and transmit information relative to the national defense of the Untied States to the Union of Soviet Socialist Republics."

555. O'Dwyer was himself going through a sensational divorce case at about this time. The case had been brought against him by his former wife, the lovely but anoretic, ex-fashion model, Sloane Simpson O'Dwyer.

556. In police jargon this procedure is known as an "unofficial extradition." This was the procedure used to return Morton Sobell from Mexico to the United States. Sobell was pushed across the International Bridge at Nuevo Laredo, Tamaulipas, Mexico by the Mexican Federal police into the waiting arms of the FBI on the other side of the bridge in Laredo, Texas.

557. United States vs. Alfred K. Stern and Martha Dodd Stern, Exhibit No.540-E. The defendants claimed the service of subpoenas was improper in that the grand jury proceeding is not a "criminal proceeding" that justifies criminal subpoenas.

558. The Sterns had enjoyed the good life of Mexico for three years, six months and twenty-six days. They lived in a luxury apartment in the exclusive Polanco area, had two offices on the Paseo de la Reforma and were building a palatial vacation home in Cuernavaca with a large swimming pool and other amenities. They were rumored to have cashed in securities and retained the money in Mexico of over fifteen millions of dollars.

559. The Sterns managed to reach Prague, Czechoslovakia on these illegal Paraguayan passports. The Paraguayan government in Asuncion then discovered the blatantly fraudulent misuse of their passports They declared that Paraguayan passports Nos. 30 and 31 as "void and without value" and punished the Paraguayan official who sold the passports to Kolomyakov.

560. *Information Digest*, 17 February 1989.

561. The return address on Martha Stern's last letters to friends in Mexico was Jinonicka 18, Prague, Czechoslovakia. Letters to that address before Alfred or Martha died were never acknowledged or returned.

562. Hansen, et.al., *Leon Trotsky; The Man and His Works*, Loc. Cit., p.7.

563. Alexander, *Trotskyism in Latin America*, Loc. Cit., p.196.

564. CI Staff of Central Intelligence Agency report, entitled, "Leon Trotsky, Dupe of the NKVD," released on 3 March 1985, p.41.

565. Deutscher, *The Prophet Outcast*, Loc. Cit., p.481.

566. Hansen, et.al.,*Leon Trotsky; The Man and His Works*. Loc. Cit., p. 8. (Oscar Greydt Abelenda, a professor at the PCM controlled Workers' University and a contributor to *La Voz de Mexico* wrote virulent articles defaming Trotsky. {Ibid.})

567. Alexander, *Trotskyism in Latin America*, Loc. Cit., p.193.

568. Siqueiros's wife, Angelica Arenal, is the sister of the brothers Luis and Lepoldo Arenal, subsequently accused of murdering Robert Sheldon Harte.

569. Alamazan was an opposition candidate to the then current President Lazaro Cardenas's party in power, the PRI, whose candidate was General Avila Camacho.

570. For the record, Serrano himself was about thirty-six, dark complexion, slim build, with an oval shaped face and sharp Castillian features.

571. Hansen, et.al., *Leon Trotsky; The Man and His Works*, Loc. Cit., p.7.

572. The person(s) at the Majestic Hotel were probably Eitingon, Mink or Contreras or all three. Some veterans from the Spanish Civil War may also have been at the Majestic, such as Codovilla.

573. Antonio Pujol and Luis Arenal were old friends of Siqueiros. In 1939 they went to New York together to an artists' convention. (Helm, MacKinley, *Modern Mexican Painters*, Books for Libraries Press, Freeport, New York, (1946), p.94.)

574. Jacson/Mercader took several large suitcases from storage at Shirley's Courts the day before the raid. It is not known definitely if these suitcase are the same ones cited here.

575. The house had originally been built as a summer home away from the heat of the valley of Mexico City. It had been built around the end of the last century, about 1895.

576. Brandenburg, Frank, *The Making of Modern Mexico*, Prentice-Hall, Inc., Englewood Cliffs, New Jersey, (1964),p.84.)

577. To a native Spanish speaker, accents are extremely revealing of the speaker's place of origin, i.e., insular Spanish as spoken in Cuba, Spanish with a preponderance of Indian intonations and the Iberian Spanish with the pronounced "th" sound, are all examples.

578. Jacson/ Mercader and Harte had become close friends. It is suggested by some experts on the Trotsky case that Jacson/Mercader was with Siqueiros and had Harte open the gate. It is doubted that Eitingon would have allowed Jacson/Mercader to be exposed to the Trotsky guards even for this brief moment. Other experts categorically state that one Joseph Grigulevich, sent to Mexico by Lavrenti Beria for the Trotsky operation, had made friends with Harte and that he was in the Siqueiros team and had Harte open the gate. (Sudoplatov, *Special Tasks*, Loc. Cit., p. 74.) (This Grigulevich, according to Sudoplatov, later went to Santa Fe, New Mexico where he opened a drug store (also described as a barber shop) that was used as a way station and message drop for Soviet agents in the Soviet atom bomb espionage case.)

579. Deutscher, *The Prophet Outcast*, Loc. Cit., p.488.

580. Harte had been in Mexico only eight weeks at the time of the raid. (Hansen, et.al., *Leon Trotsky; The Man and His Works*, Loc. Cit., p.5.)

581. The raiders were:

    David Alfaro Siqueiros
    Nestor Sanchez Hernandez
    Antonio Pujol
    Juan Zuniga Camacho
    Felipe the "French Jew"
    Jesus Alfaro Siqueiros
    Mariano Herrera Vasquez (?)
    David Serrano Andonequi
    Luis Mateo Martinez
    Luis Arenal
    Leopoldo Arenal
    Jose Alvarez Lopez
    Genero Casillas
    Narciso Padilla

Suggested and probably true:

George Mink — according to J. Edgar Hoover of the FBI

FNU Luna — Town clerk from Cinco Minas

Felix Guerrero Mejia — (Source: Hansen, *Leon Trotsky; the Man and His Works Man*

FNU Blanco — (Source: Trotsky, *Stalin's Gangsters.*

Pedro Martinez Carton — Sent by El Campesino from Spain to join Vittorio Vidali in Mexico as part of the assault team.

Suggested but doubtful:

Leon Haykis (former Soviet official in Mexico)

582. *New York Times*, 25 May 1940.

583. "A police agent in Mexico City said the information they had been able to obtain [about the 24 May raid on Trotsky's house] indicated there were about twenty men in the raiding party and they were [mostly] foreigners [non-Mexicans]. The agent said that only three men did the actual firing and that it [the raid] lasted about three minutes." (*New York Times*, 27 May 1940.)

584. There are as many estimates of the number of shots fired as there as stories about the raid. The most authoritative figure is Colonel Sanchez's count of the bullet holes in the plaster of Trotsky's bedroom, which is 73. This, of course, would not count shots fired that did not strike the plaster but went through the windows. (*New York Times*, 25 May 1940, reported that 300 shots were fired.)

585. For Trotsky's full description of the raid and his rebuttal that the raid was instigated by him see, Trotsky, *Stalin's Gangsters*, New Park Publications, London, (1977).

586. When Siqueiros was in Spain he wore the fanciest uniform ever seen in Spain. When he was asked where it came from, he said, "What is the good of being an artist if I can't design my own uniform?" This uniform, adorned with a long cape, was the one he wore on the right of the raid, along with a pair of theatrical false whiskers. (Kirk, *Covering the Mexican Front*, Loc. Cit., p. 221.)

587. Mosley, *The Assassination of Trotsky*, Loc. Cit., p.40.

588. Sanchez, *Murder in Mexico*, Loc. Cit., p.14.

589. *New York Times*, 25 May 1940.

590. Deutscher, *The Prophet Outcast*, Loc. Cit., p.489. (Abel Gonzalez Ortiz, one of the police officers stationed in front of Trotsky's house who saw Harte leave with the raiders stated, "They were not ill-treating him. He seemed to be going very quietly and of his own free will." Also, Jesus Rodriguez Casas, chief of Trotsky's police guard force asked, "Why did he [Harte] open the gate to the assailants when it was not even opened to the police officers on duty?" (Sanchez, *Murder in Mexico*, Loc. Cit., p.17/18.)

591. There is a persistent rumor that Trotsky had been tipped off to the raid and was not in his bedroom at the time the shots were fired. (Mosley, *The Assassination of Leon Trotrsky*, Loc. Cit., p.42.)

592. *New York Times*, 19 June 1940.

593. *New York Times*, 19 June 1940.

594. Javier Guerrero Saucedo, a Mexican muralist and GPU agent, was the president of The Friends of the USSR. His wife Clara Porcet de Guerrero was a militant member of the PCM and a leading organizer of the Friends of the USSR. Before his marriage to Porcet, Guerrero had been a lover of Tina Modotti, a GPU agent.

595. Sanchez, *Murder in Mexico*, Loc. Cit., p.30.

596. Payne, *The Life and Death of Trotsky*, Loc. Cit., p.422/424.

597. Trotsky, Natalia Sedova, *How It Happened*, Merit Publishers, New York, (1967), p.36.

598. Hansen, et.al., *With Trotsky to the End*, Merit Publishers, New York, (1967), p.19.

599. Sanchez Salazar, General Leandro, *Murder in Mexico*, Secker and Warburg, London, (1950). This book, written in collaboration with Julian Gorkin, a former leader of the Spanish Trotskyites, is deprecated as too much of a "mystery novel" by Alfred Rosmer. (Rosmer, Alfred, *A Fictionalized Version of Murder*, Merit Publishers, New York (1967), p.77) On balance, however, it does present facts not found in other sources and is rich with local color not otherwise available.

600. *New York Times*, 21 August 1940.

601. Sanchez, *Murder in Mexico*, Loc. Cit., p.4.

602. Sanchez, *Murder in Mexico*, Loc. Cit., p.10.

603. Hansen, et.al., *Leon Trotsky; The Man and His Works*, Loc. Cit., p.9.

604. They were Carmen Palma, cook; Belen Estrada, parlor maid and Melquiades Benitez, handy man and gardener. (Sanchez, *Murder in Mexico*, Loc. Cit., p.11.)

605. A fellow newspaper reporter of Black's was Frank Jellinek, the representative for communist-oriented *PM* magazine. He also wrote for Stalin's Federated Press. Jellinek was a GPU agent. (Hansen, et.al., *Leon Trotsky; The Man and His Works*, Loc. Cit., p.20.) Jellinek worked in close collaboration with Frank Kluckhohn, a Stalinist journalist who wrote according to GPU dictates. (Ibid., p.23.)

606. Dispatch # 162, dated 25 July 1940 from American Embassy in Mexico to Secretary of State in Washington.

607. Sanchez, *Murder in Mexico*, Loc. Cit., p.35.

608. Levine, *Mind of an Assassin*, Loc. Cit., p. 95.

609. The French Jew was reported to have handled the extensive financing involved in the raid and paid hirelings. There was an effort to identify the French Jew ("un tipo judio frances") as Leon Haykis, whom Diego Rivera had reported two months before as being in Mexico. Haykis had been the first secretary of the Soviet Embassy in Mexico in 1923 and was later Soviet Ambassador to Republican Spain. (Kirk, *Covering the Mexican Front*, Loc. Cit., p.222.)

610. Hansen and several other members of the Fourth International believe that Jacson/Mercader was Felipe, the French Jew. (Hansen, et.al., *Leon Trotsky; The Man and His Works*, Loc. Cit., p.22.) Conversely, there is a considerable body of opinion, including the author, that supports the thesis that Leonid Eitingon was Felipe, the French Jew.

After the assassination when Jacson/Mercader was in police custody, Colonel Sanchez had him brought before two of the assailants who had taken part in the 24 May raid on Trotsky's house. Jacson/Mercader was ordered to speak so the assailants could judge if his voice was at all similar to Felipe's. The two men looked Jacson/Mercader over. listened to his voice and categorically denied that he was close to Felipe, the French Jew. They both said that Felipe was a larger man, older and did not wear glasses. (Sanchez, *Murder in Mexico*, Loc. Cit., p.126.)

611. *New York Times*, 27 May 1940.

612. Jesse Samuel Harte listed his address as 1148 Fifth Avenue, New York City. (*New York Times*, 26 May 1940.)

613. Sanchez *Murder in Mexico*, Loc. Cit., p.49.

614. Evidence was produced years later from a woman who claimed to have been Harte's girlfriend, that Harte was, in fact, a committed communist who sympathized with Stalin's views. (Mosley, *The Assassination of Trotsky*, Loc. Cit., p. 43.

615. Segal, Ronald, *Leon Trotsky*, Pantheon Books, New York., (1979), p.394.

616. Colonel Sanchez was personally convinced that Harte had been in collusion with the raiders and had left with them of his own free will. (Deutscher, *The Propeht Outcast*, Loc. Cit., p.491.)

617. Although this statement of Trotsky's may be true and Harte could have killed him many times in the compound but could he have ever escaped from Trotsky's loyal followers in the compound who would have torn him apart.

618. Trotsky, Natalia Sedova, *Father and Son*, Merit Publishers, New York, (1967), p.42.

619. Hansen, et.al., *Leon Trotsky; The Man and His Works*, Loc. Cit., p. 7.

620. *Ahora*, 9 October 1972 (A Dominican magazine.)

621. Hansen, et.al., *Leon Trotksy; The Man and His Works*, Loc. Cit., p.,10.

622. Volkogonov, Dmitri A., "Unlocking Russia's Unknown History" *Prologue*, (Winter 1992), p.355.

623. Trotsky is really stating that there must be some program in the Soviet Union to defend the people against Stalin.

624. Segal, *Leon Trotsky*, Loc. Cit., p.397.

625. Jacson/Mercader visited the house a total of four times: August 8th, 10th, 17th and 20th. (Payne, *The Life and Death of Trotsky*, Loc. Cit., p. 453.)

626. Segal, *Leon Trotsky*, Loc. Cit., p.396.

627. Petrov, Vladimir and Evdokia, *Empire of Fear*, Andre Deutsch, London, (1956), p.222.

628. Shirley's Courts has been replaced by a large Telefonos de Mexico building.

629. The female caller was probably Caridad Mercader, his mother, who was in Mexico City with her lover Leonid Eitingon.

630. Other sources report they stayed at the Hotel Pierpont in New York. (FBI Report 65-43302, Subject "Caridad Mercader, et.al.," dated 31 January 1952.

631. Deutscher,*The Prophet Outcast*, Loc. Cit., p.496.

632. Deutscher, *The Prophet Outcast*, Loc. Cit., p. 497

633. Hansen, et.al., *Leon Trotsky; The Man and His Works*, Loc. Cit., p.9.

634. Deutscher, *The Prophet Outcast*, Loc. Cit., p.492.

635. American style differs from the Mexican style of floor numbering that begins with a planta baja or ground floor. The American style begins with the first floor without reference to a ground floor. The American second story, for example, would be the first story Mexican style.

636. *Excelsior*, 27 June 1940. (According to one report, the Arenal brothers had confessed to killing Harte, *New York Times*,26 June 1940.)

637. *New York Times*, 26 June 1940.

638. Deutscher, *The Prophet Outcast*, Loc. Cit., p.493.

639. *Excelsior*, 26 June 1940.

640. Sanchez, *Murder in Mexico*, Loc. Cit., p.85.)

641. Fourth International, *Trotsky's Assassin at Large*, Labor Publications, New York, (1979), p.3.

642. Jean van Hcijenoort used to take Frida Kahlo and her sister, Cristina, to the Salon Mexico when Trotsky and his entourage were living at the Blue House.

643. If Eitingon was Felipe the French Jew and was identified as a participant in the 24 May raid, it would have severely implicated the Soviet Union and, ultimately, Stalin.

644. Colonel Sanchez remained convinced that Harte was a GPU agent. (Deutscher, *The Prophet Outcast*, Loc. Cit., p.493.

645. Robins, *Security at Coyoacan*, Loc. Cit., p. 7.

646. Fourth International, *Trotsky's Assassin at Large*, Loc. Cit., p.12.

647. Fourth International, *Trotsky's Assassin at Large*, Loc. Cit., p. 18.

648. Warth, *Leon Trotsky*, Loc. Cit., p. 179.

649. Deutscher, *The Prophet Outcast*, Loc. Cit., p.494.

650. Segal, *Leon Trotsky*, Loc. Cit., p. 399.

651. Deutscher, *The Prophet Outcast*, Loc. Cit., p.502.

652. Deutscher, *The Prophet Outcast*, Loc. Cit., p.503.

653. Sanchez, *Murder in Mexico*, Loc. Cit., p. 146.

654. The label inside the raincoat read, "The Clifton Raincoat - Made in England."

655. Segal, *Leon Trotsky*, Loc. Cit., p.400.

656. Hansen, et.al., *Leon Trotsky; The Man and His Work*, Loc. Cit., p.18.

657. There is no intention to create Trotsky's thoughts but only to indicate his state of mind from previous conversations he had with Natalia on the subject.

658. Sanchez, *Murder in Mexico*, Loc. Cit., p.135.

659. Hansen, et.al., *Leon Trotsky; The Man and His Works*, Loc. Cit., p.17.

660. For posterity the Colonel noted that Sylvia was wearing a white pique dress and "a slightly used brown coat." (Sanchez, *Murder in Mexico*, Loc. Cit., p. 107.)

661. Segal, *Leon Trotsky*, Loc. Cit., p.401.

662. Deutscher, *The Prophet Outcast*, Loc. Cit., p. 507.

663. Five doctors under the direction of Gustavo Baz, rector of the University of Mexico, undertook the delicate operation of trepanning Trotsky's skull. (Sanchez, *Murder in Mexico*,Loc. Cit., p.103.)

664. Hansen, et.al.,*Leon Trotsky; The Man and His Works*, Loc. Cit., p.24.

665. Bernard Wolfe, who was a guard at Trotsky's in Mexico for eight months wrote a fictionalized psychoanalysis of Trotsky that was not objective and very derogatory. (Hansen, *Trotsky "Psychoanalyzed,"* Merit Publishers, New York, (1967).

666. *The Daily Worker*, 23 August 1940.

667. *Pravda*, 24 August 1940.

668. In the pro-Soviet biased book *The Great Conspiracy*, the authors repeat these assertions that Jacson/Mercader was hired by Trotskyites to murder Trotsky. These authors also claim the murderer was merely another disgruntled Trotskyite and still give his name as Jacques Moranrd van der Dresche.

669. *New York Times*, 23 August 1940.

670. Segal, *Leon Trotsky*, Loc. Cit., p.402.

671. Trotsky had originally been opposed to the mummifying of Lenin's body in the Kremlin. He had expressed the desire to Natalia that his remains should be cremated. (Hansen, et.al., *Leon Trotsky;The Man and His Works*, Loc. Cit., p. 24.

672. Personal knowledge of the author. For details see Deutscher, *The Prophet Outcast*, Loc. Cit., p. 509.

673. Deutscher, *The Prophet Outcast*, Loc. Cit., p.509.

674. Jacson/Mercader also had $890 in his pocket. (Hansen, et.al., *Leon Trotsky; The Man and His Works*, Loc. Cit., p.18.

675. The Dies Committee of the United States Congress was also referred to as the Un-American Activities Committee which was investigating communist influence in the United States. (Also see: *New York Times*, 8 December 1939, in which Trotsky is quoted as saying it is his political duty to testify before the Dies Committee.)

676. Goldman, Albert, *The Assassination of Leon Trotsky*,Pioneer Publishers, New York, 1947, p.8.

677. Hansen, et.al., *Leon Trotsky; The Man and His Works*, Loc. Cit. p.20.

678. Levine, *Mind of an Assassin*, Loc. Cit., p.142.

679. The two psychologists were Dr. Alfonso Quiroz Cuaron, professor of law, University of Mexico, and Dr. Juan Gomez Robledo, director of the Department of Social Medicine at the National University of Mexico.

680. Of interest to students of behavioral sciences is Mornard's adoption of a phlegmatic stoicism of the legendary Kamo (true name Ter-Petrosian), a Bolshevik hero. (See Levine, *Mind of an Assassin*, Loc. Cit., p.180 et.seq.)

681. His lawyer filed an appeal but on 19 May 1944, the Court of Appeals upheld the sentence of the lower court.

682. The identification was done largely through the efforts of Dr. Quiroz who traveled to Spain to verify the records.

683. Warth, *Leon Trotsky*, Loc. Cit., p. 185.

BIBLIOGRAPHY

Abraham, Richard, *Alexander Kerensky*, Columbia University Press, New York, (1972).

Agabeeov, Georges, *OGPU*, Brentano's, New York, (1931).

Alexander, Robert J., *Trotskyism in Latin America*, Hoover Institution, Stanford University, Stanford, California, (1973).

Andrew, Christopher, *Her Majesty's Secret Service*, Viking Books, Inc., New York, (1986).

Andrew, Christopher, *Secret Service*, William Heineman, Ltd., London, (1985).

Andrew, Christopher and Oleg Gordievsky, *KGB: The Inside Story*, Harper Collins, Publishers, New York, (1990).

Arquin, Florence, *Diego Rivera; The Shaping of an Artist,1889-1921*, University of Oklahoma, Norman, (1971).

Bailey, Geoffrey, *The Conspirators*, Harper and Brothers Publishing, New York, (1960).

Barron, John, *KGB; The Secret Work of Soviet Secret Agents*, Bantam Books, New York, (1974).

Barron, John, *KGB Today;The Hidden Hand*, Berkley Books, Inc., New York, (1985).

Bazhanov, Boris, *Bazhanov and the Damnation of Stalin*, Ohio University Press, Athens, Ohio, (1990).

Beatty, Bessie, *The Red Heart of Russia*, The Century Company, New York, (1918).

Bentley, Elizabeth, *Out of Bondage*, with an "Afterward" by Hayden Peake, Ivy Books, New York, (1988).

Beschloss, Michael R., *May Day*, Harper and Row, New York, (1986).

Blackstock, Paul W., *The Secret Road to World War Two*, Quadrangle Books, Chicago, (1969).

Bolloten, Burnett, *The Grand Conspiracy*, Hollis and Carter, London, (1961).

Bower, Tom, *The Perfect English Spy- Dick White*,St. Martin's Press, New York, (1995).

Brandenburg, Frank, *The Making of Modern Mexico*, Prentice-Hall, Inc., Englewood Cliffs, New Jersey, (1964).

Brenner, Anita, *The Wind That Swept Mexico*, Harper and Brothers, New York, (1950).

Brook-Shepherd, Gordon, *The Storm Birds*, Henry Holt and Company, New York, (1989).

Brown, Anthony Cave, *On A Field of Red*, G.P. Putnam's Sons, New York, (1981).

Bryant, Louise, *Six Red Months in Russia*, George H. Doran Co., New York, (1918).

Carmichael, Joel, *Trotsky*, St. Martin's Press, New York, (1975).

Charques, R.D., *A Short History of Russia*, E.P. Dutton and Company, Inc., New York, (1956).

Clarkson, Jesse D., *A History of Russia*, Random House, Inc., New York, (1968).

Cockridge, E.H., *Soviet Spy Net*, Frederick Muller, Ltd., London, (1954).

Conquest, Robert, *Inside Stalin's Secret Police*, Macmillan Press, Ltd., London, (1985).

Conquest, Robert, *The Great Terror*, The Macmillan Company, New York, (1968).

Corson, William R. and Robert T. Crowley, *The New KGB*, William Morrow and Co., Inc., New York, (1985).

Costello, John and Oleg Tsarev, *Deadly Illusions*, Crown Publishers, Inc., New York. (1993).

Cowles, Virginia, *The Romanovs*, Harper and Row, Publishers, New York, (1971).

Crankshaw, Edward, *The Shadow of the Winter Palace,* The Viking Press, New York, (1976).

Dallin, David J., *Soviet Espionage*, Yale University Press, New Haven, (1955).

Daniels, Robert V.,*Red October; The Bolshevik Revolution of 1917*, Beacon Press, Boston, Massachusetts, (1967).

Daniels, Robert V.,*Russia; Roots of Confrontation*, Harvard University Press, Cambridge, Massachusetts, (1985).

De Gramont, Sanchez, *The Secret War: The Story of International Espionage Since 1945*, Andre Deutsch, London, (1962).

DeJonge, Alex, *Stalin*, William Morrow Company, New York, (1986).

Deriabin, Peter and T.H. Bagley, *The KGB: Masters of the Soviet Union*, Hippocrene Books, New York, (1990).

Deutscher, Isaac, *The Prophet Outcast*, Oxford University Press, New York, (1963).

Deutscher, Isaac, *Stalin; A Political Biography*, Oxford University Press, New York, (1949).

Dewar, Hugo, *Assassins at Large*, Wingate Publishers, New York, (1951).
Dodd, Martha, *Through Embassy Eyes*, Harcourt, Brace and Company, New York (1939).
Draper, Theodore, *The Roots of American Communism*, The Viking Press, New York, (1957).
Draper, Theodore, *American Communisn and Soviet Russia*, The Viking Press, New York, (1960).
Dukes, Paul, *Red Dusk and the Morrow*, Doubleday, Page and Company, Garden City, New York, (1922).
Dulles, Allen, *The Craft of Intelligence*, Harper and Row, Publishers, New York, (1963).
Dunham, Donald, *Kremlin Target: USA*, Ives Washburn, Inc., New York, (1961).
Dzhirkvelov, Ilya, *Secret Service; My Life with the KGB and the Soviet Elite*, Harper and Row, Publishers, New York, (1987).
Dziak, John J., *Chekisty; A History of the KGB*; Lexington Books, Lexington, Massachusetts, (1988).
Ebon, Martin, *KGB: Death and Rebirith*, Praeger Publishers, Westport, Connecticut, (1994).
Eby, Cecil D., *Between the Bullets and the Lie*, Holt, Rinehart, New York, (1964).
Essad-Bey, *OGPU*, The Viking Press, New York, (1933).
Estes, Thomas, S. and E. Allen Lightner, Jr., *The Department of State*, Praeger Publishing, New York, (1976).
Finder, Joseph, *Red Carpet*, Holt, Rinehart and Winston, New York, (1983).
Fogelson, David S., *America's Secret War Against Bolshevism*, University of North Carolina Press, Chapel Hill, (1995).
Fourth International, *The Case for a Commission of Inquiry*, Labor Publications, New York, (1976).
Fourth International, *Sylvia Franklin: Dossier*, Labor Publications, New York, (1977).
Fourth International, *Dossier of a Double Agent*, Labor Publications, New York, (1978).
Fourth International, *Trotsky's Assassin at Large*, Labor Publications, New York, (1977).

Freemantle, Brian, *KGB*, Holt, Rinehart and Winston, New York, (1982).

Freidin, Seymour and George Bailey, *The Ex-Perts*, The Macmillan Company, New York, (1968).

Gardner, Virginia, *"Friends and Lovers": The Life of Louise Bryant*,Horizon Press, New York, (1982).

Gitlow, Benjamin, *I Confess*, E.P. Dutton and Company, Inc., New York, (1948).

Glees, Anthony, *The Secrets of the Services*, Carroll and Graf Publishers, Inc., New York, (1987).

Glotzin, Albert, *Trotsky; Memoir and Critique*, Prometheus Books, Buffalo, New York, (1989).

Goldman, Albert, *The Assassination of Leon Trotsky*, Pioneer Publishers, New York, (1947).

Goldston, Robert, *The Russian Revolution*, Fawcett Publications, Greenwich, Connecticut, (1966).

Golitsyn, Anatoliy, *New Lies for Old*, Dodd, Mead and Company, New York, (1984).

Grass, Peter, *Gentleman Spy: The Life of Allen Dulles*, Houghton Mifflin Company, New York (1994).

Hansen, Joseph and Others, *Leon Trotsky: The Man and His Work*, Merit Publishers, New York, (1969).

Haynes, John E. and Harvey Kllehr, "'Moscow Gold'; Confirmed at Last", *Labor History*, Volume 33:2, (Spring 1992).

Heald, Edward T., *Witness to Revolution*, The Kent State University Press, Kent, Ohio, (1972).

Heilbrunn, Otto, *The Soviet Secret Services*, George Allen and Univen, Ltd., London, (1956).

Heller, Mikhail and Aleksandr Nekrich, *Utopia in Power*, Summit Books, New York, (1982).

Helm, MacKinley, *Modern Mexican Painters*, Books For Librarians Press, Freeport, New York, (1941).

Herrera, Hayden, *Frida: A Biography of Frida Kahlo*, Harper and Rowe, Publishers, New York, (1983).

Hill, Christopher, *Lenin and the Russian Revolution*, Penguin Books, New York, (1971).

Hingley, Ronald, *The Russian Secret Police*, Simon and Schuster, New York, (1970).
Jacobs, Dan Norman, *Borodin; Stalin's Man in China*, Harvard University Press, Cambridge, Massachusetts,(1981).
Jamis, Rauda, *Frida Kahlo*, Compania Editorial, S.A., Mexico City, (1987).
Jeffreys-Jones, Rhodri, *American Intelligence*, The Free Press, New York, (1977).
Jones, Mervyn, *The Antagonists*, Clarkson N. Potter, Inc., New York, (1962).
Kahn, David, *The Code Breakers*, The Macmillan Company, New York, (1967).
Kaplan, Martin A.,(Editor), *The Many Faces of Communism*, The Free Press, New York, (1978).
Kirk, Betty, *Covering the Mexican Front*, University of Oklahoma Press, Norman, Oklahoma, (1941).
Kirk, Irina, *Profiles in Russian Resistance*, The New York Times Book Company, New York, (1975).
Klehr, Harvey, *The Heyday of American Communism*, Basic Books, Inc., New York, (1984).
Klehr, Harvey, John Earl Haynes and Fridrikh Igorevich Firsov, *The Secret World of American Communism*, Yale University Press, New Haven, Connecticut, (1995).
Kochan, Lionel, *Russia in Revolution*, Granada Books, London, (1966).
Lamphere, Robert J. and Tom Shachtman, *The FBI-KGB War*, Random House, New York, (1986).
Lazitch, Branko and Milorad M. Drachkovitch, *Biographical Dictionary of the Comintern*, Hoover Institution Press, Stanford, California, (1973).
Leacacos, John P., *Fires in the In-Basket*, World Publishing Co., Cleveland, (1968).
Leggett, George, *Cheka*, Clarendon Press, Oxford, (1981).
Leonhard, Wolfgang, *Child of the Revolution*, Henry Regnery Company, Chicago, Illinois, (1958).
Levchenko, Stanislau, *On The Wrong Side; My Life with the KGB*, Pergamon-Brassey's, New York, (1988).
Levine, Isaac Don, *Mind of an Assassin*, Farrar, Straus and Cudahy, New York, (1959).

Levine, Isaac Don, *Eyewitness to History*, Hawthorn Books, Inc., New York, (1973).

Levytsky, Boris, *The Uses of Terror*. Coward, McCann and Geoghegan,Inc., New York (1972) List of purge trials.

Lincoln, W. Bruce, *Passage Through Armageddon*, Simon and Schuster, New York, (1986).

Liversidge, Douglas, *Lenin; Genius of Revolution*, Franklin Watts, Inc., New York, (1969).

Lockhart, Robin Bruce, *Reilly: The First Man*, Penguin Books, New York, (1987).

Lyons, Eugene, *The Red Decade*, The Bobbs-Merrill Company, New York, (1941).

Mahoney, H.M., *Women in Espionage*, ABC Clio Publishers, Santa Barbara and Denver, (1994).

Matthew, Herbert, *Half of Spain Died*, Charles Scribner's Sons, New York, (1973).

Mazour, Anatole G., *Rise and Fall of the Romanovs*, D. Van Nostand Company, Inc., Princeton, New Jersey, (1960).

Mazour, Anatole G., *Tsarist and Communist Russia*, D. Van Nostrand Company, Inc., Princeton, New Jersey, (1962).

McCauley, Martin, *Octobrists to Bolsheviks; Imperial Russia, 1905-1917*, Edward Arnold, Publisher, London, (1984).

Medvedev, Roy A., *Let History Judge*, Alfred A. Knopf, New York, (1972).

Modin, Yuri, *My Cambridge Friends*, Farrar, Straus and Geroux, New York, (1994).

Moorehead, Alan, *The Russian Revolution*, Harper and Brothers, Publishers, New York, (1958).

Morris, Boris, *My Ten Years as a Counterspy*, The Viking Press, New York, (1959)

Mosley, Nicholas, *The Assassination of Trotsky*, Michael Joseph, London, (1972).

Nollan, Gunther, *International Communism and World Revolution*, Praeger Publishers, New York, (1961).

Orlov, Alexander, "The Ghastly Secrets of Stalin's Power" *Life*, Nos. I through IV, 6-27 April 1953.

Oswald, J. Gregory and Anthony J. Strover, *The Soviet Union and Latin America*, Prager Publishers, New York, (1970).

Payne, Robert, *The Life and Death of Lenin*, Simon and Schuster, New York, (1964).

Petrov, Vladimir and Evdokia Petrov, *Empire of Fear*, Andre Deutsch, London, (1956).

Pipes, Richard, *The Russian Revolution*, Alfred A. Knopf, New York, (1990).

Poppino, Rollie E., *International Communism in Latin America*, The Free Press, New York, (1964).

Porch, Douglas, *The French Secret Service*, Farrar, Straus and Giroux, New York, (1995).

Powell, S. Steven, *Covert Cadres*, Green Hill Publishers, Inc., Ottawa, Illinois (1987).

Rapoport, Vitaly and Yuri Alexeev, *High Treason*, Duke University Press, Durham, (1985(.

Read, Anthony and David Fisher, *Colonel Z*, Viking Press, New York, (1985).

Reed, John, *Ten Days That Shook The World*, Random House, New York, (1960).

Rexroth, Kenneth, *An Autobiographical Novel*, Doubleday and Company, Garden City, New York, (1966).

Richardson, R. Dan, *Comintern Army*, The University Press of Kentucky, Lexington, (1982).

Riding, Alan, *Distant Neighbors*, Alfred A. Knopf, New York, (1985).

Robins, Harold, *Security at Coyoacan*, Labor Publications, New York, (1977).

Rositzke, Harry, *The KGB; The Eyes of Russia*, Doubleday and Company, Inc., Garden City, New York, (1981).

Rowan, Richard Wilmer, *Secret Service: 33 Centuries of Espionage*, Hawthrown Books, Inc., (1967).

Sanchez Salazar, Leandro A., *Murder in Mexico: The Assassination of Leon Trotsky*, Secker and Warburg, London, (1950).

Sayers, Michael and Albert E. Kahn, *The Great Conspiracy*, Little, Brown and Company, Boston, (1946).

Schapiro, Leonard, *Russian Studies*, Elisabeth Sifton Books, Viking, New York, (1987).

Segal, Ronald, *Leon Trotsky*, Pantheon Books, New York (1979).

Shatz, Marshall S., (Editor), *The Essential Works of Anarchism*, Bantam Books, London, (1971).

Sheridan, Clare, *Mayfair to Moscow*, Boni and Liveright, New York, (1921).

Shklovsky, I.V., *Russia Under the Bolsheviks*, Wilkinson Bros. Ltd., Green Lane, N.16, London, Circa (1920).

Shub, David, *Lenin; A Biography*, Doubleday and Company, Garden City, New York, (1949)

Sisson, Edgar, *One Hundred Red Days*, Yale University Press, New Haven, (1931).

Skousen, W. Cleon, *The Naked Communist*, The Ensign Publishing Company, Salt Lake City, Utah, (1961).

Smith, Page, *America Enters the World*, Volume Seven, McGraw-Hill Book Company, New York, (1985).

Smith, Irving R., (Editor) *Trotsky*, Prentice-Hall, Inc., Englewood Cliffs, New Jersey, (1973).

Spender, Stephen, *Voices Against Tyranny*, Charles Scribner's Sons, New York, (1986).

Stein, Philip, *Siqueiros: His Life and Works*, International Publishers, New York, (1994).

Sterling, Clare, *The Time of the Assassins*, Holt Rinehart and Winston, New York, (1983).

Stevenson, William, *A Man Called Intrepid*, Harcourt Brace Jovanovich, New York, (1976).

Stevenson, William, *Intrepid's Last Case*, Willard Books, New York, (1983).

Stuart, Graham H., *The Department of State*, The Macmillan Company, New York, (1949) .

Sudoplatov, Pavel and Anatoli Sudoplatov, *Special Tasks*, Little Brown and Company, New York, (1994).

Suvorov, Viktor, *Inside Soviet Military Intelligence*, Berkley Books, New York, (1985).

Taylor, Edmund, *The Fall of Dynasties*, Doubleday and Company, Inc., Garden City, New York, (1963).

Thayer, Charles W., *Diplomat*, Harper and Brothers, Publishers, New York, (1959).

Trotsky, Leon, *My Life*, Pathfinders Press, Inc., New York, (1970).

Trotsky, Leon, *Stalin's Gangsters*, New Park Publications, London, (1977).

Turner, William. W., *Hoover's FBI*, Dell Book, New York, (1971).

U.S. Senate, Committee of Judiciary, "Murder International, Inc.," Washington, D.C., (26 March 1965).

U.S. Senate, Committee of the Judiciary, "The Legacy of Alexander Orlov", 93rd Congress, First Session, Washington, D.C. (August 1973)

Unknown CIA Analyst, "Leon Trotsky: Dupe of the NKVD", *CIA Review*, Langley, Virginia, (March 1983).

Valtin, Jan, *Out of the Night*, Alliance Book Corporation, New York, (1941).

Van der Rhoer, *The Shadow Network*, Charles Scribner's Sons, New York, (1983).

Van Heijenoort, Jean, *With Trotsky in Exile; From Prinkipo to Coyoacan*, Harvard University Press, Cambridge, Masschuisetts, (1978).

Volkman, Ernest and Blaine Baggett, *Secret Intelligence*, Doubleday, New York, (1987).

Volkman, Ernest, *Warriors of the Night*, William Morrow and Co., Inc., New York, (1985).

Volkogonov, Dmitri A., "Unlocking Russia's Unknown Histories", *Prologue*, (Winter 1992), pp. 353-359.

Warth, Robert D., *Leon Trotsky*, Twayne Publishers, Boston, (1977).

West, Nigel, *Games of Intelligence*, Crown Publishing, New York, (1989). List of defectors and PNG'ed Soviets.

Williams, Robert Chadwell, *Klaus Fuchs, Atom Spy*, Harvard University Press, Cambridge, Mass., (1987).

Williams, Albert Rhys, *Through the Russian Revolution*, Monthly Review Press, New York, (1921).

Wolfe, Bertram D., *Three Who Made a Revolution*, Dell Publishing Company, New York, (1948).

Wolfe, Bertram D., *The Fabulous Life of Diego Rivera*, Stein and Day Publishers, New York, (1963).

Wolfe, Bernard, *The Great Prince Died*, Charles Scribner's Sons, New York, (1959).

Wriston, Henry Merritt, *Executive Agents in American Foreign Policy*, Johns Hopkins Press, London, (1929).

Wyndham, Francis and Daniel King, *Trotsky; A Documentary*, Praeger Publishers, New York, (1972).

Yakovley, Nikolai, *CIA Target -The USSR*, Progress Publishers, Moscow, (1984).
Zamyatin, Yevgeny, *We*, Avon Books, New York (1972).

INDEX